ORIGIN

ORIGIN

Why Genesis 1–11 Trumps Secular Accounts

Graeme Messer

Geoffrey Messer

RECHRISTIANIZE.COM

Published in 2025 by ReChristianize.com.
First Edition (Errata corrected 2026.)

Paperback: ISBN 978-1-7641508-0-4
Hardcover: ISBN 978-1-7641508-1-1
Kindle: ISBN 978-1-7641508-2-8
EPUB: ISBN 978-1-7641508-3-5

This is what the LORD, your Redeemer
who formed you from the womb, says:
I am the LORD, who made everything;
who stretched out the heavens by myself;
who alone spread out the earth;
who destroys the omens of the false prophets
and makes fools of diviners;
who confounds the wise
and makes their knowledge foolishness …

—Isaiah 44:24–25

Contents

List of Figures

List of Tables

Introduction

After decades of strident anti-religious advocacy, the evolutionist Richard Dawkins recently clarified his views. In a 2024 interview,[1] Dawkins announced that he was a cultural Christian; he "[felt] at home in the Christian ethos". Dawkins expressed satisfaction with the decline of Christianity in the UK, but at the same time, warned that substituting this with any other religion would be "truly dreadful".

What caused this change in positioning? We suspect that Dawkins is one of many people who are waking up to the strong correlation between the decline of Christianity and the emergence of an often bizarre and polarized society. The decline of Western Christianity is a complex topic, but the major contributors include a materialist worldview, capped by Neo-Darwinism, that filters out God; liberal theology that interprets the Bible as nothing more than mythology; and a moral shift to individualist existentialism. Notably, the first two of these dismiss a straightforward reading of the biblical origins account. In other words, the decline of Christianity is associated with loss of belief in a candid reading of the early chapters of Genesis, which contain the biblical origins account and have foundational significance for the Christian faith.

The origins debate has divided the Western church. It has not only pitted Christians against the irreligious, it has also acted as a wedge issue within the church, a situation that has lasted for more than 160 years. Late American Bible teacher John MacArthur coined a memorable phrase—*the battle for the beginning*—to describe this conflict. He used this phrase as the title for a book he authored in 2001 which defended a straightforward reading of Genesis.

For centuries, Christians have recognised that a correct understanding of the Genesis narrative must be broadly compatible with the ever-changing evidence of science. But how is this compatibility to be understood? Three major schools of thought within the church have emerged: theistic evolution, old-earth creation, and young-earth creation, each of which contains a range of competing views and opinions. The non-liberal members of each school bring their understanding of the Bible to the origins debate: they accept the authority of the Bible, while acknowledging that the Bible contains a variety of literary genres (such as historical narrative, song or poetry, and apocalyptic writing). In relation to Genesis, each camp acknowledges that the text contains literary constructs, such as the "chiasm" device of Genesis 1.[2]

---

1 The interview can be viewed at youtube.com/watch?v=COHgEFUFWyg (accessed December 9, 2024).

2 Jordan, James B. (1999). *Creation in Six Days: A Defense of the Traditional Reading of Genesis One*. Moscow, ID: Canon Press, pp. 211–218.

Briefly, theistic evolution is the belief that God used evolutionary processes to create the diversity of life on earth. Some proponents have posited that Genesis 1 communicates theological truth in the genre of poetic language and consider the first chapter of Genesis to be non-historical. The late Pastor Timothy Keller was a supporter of this view; to his way of thinking, theistic evolution resolved the compatibility problem by giving secular origins science priority over a historical understanding of Genesis 1. Keller regarded his understanding of Genesis as tentative and subject to revision, and acknowledged that it led to another issue—how do we decide what parts of the Genesis account are historical narrative?[3]

Proponents of the old-earth camp accept that the universe has existed for the billions of years suggested by secular science. There are a number of frameworks in this camp, but the main objective is to achieve compatibility between the timespans of secular science and the Bible by interpreting Genesis 1 as encompassing "old ages". This was a widely held position among evangelical Christians prior to the 1960s. Well-known Christian apologist and professor of mathematics John Lennox is a contemporary defender of old-earth theology.

The young-earth viewpoint, which we argue for in this book, accepts a "straightforward" reading of Genesis. This emphasizes Genesis 1–11 as historical narrative; the six days of creation and the global Flood happened as written. Young earth creationists interpret Exodus 20:11 and 31:17 as strong evidence for the viewpoint of creation in six days, and phrases in Genesis 1 such as "evening came and then morning" (v. 8) as directly affirming that each day comprised 24 hours. Creationists seek to reinterpret scientific findings relevant to origins from a biblical perspective. The young-earth position is frequently referred to as "creationism" by both Christian and secular scholars.

This diversity of views within the church exists because no clear consensus has yet emerged among Christians about how to respond to secular origins science. This weakness impacts on the church's credibility: if Christians cannot agree how to interpret the very first chapter of the Bible, why should their other beliefs be taken seriously?

Further, from the viewpoint of many *outside the church*, Genesis 1 reads as a historical narrative that is incompatible with evolution. Richard Dawkins holds theistic evolution in contempt, regarding it as a "delusion" of "sophisticated" theologians. Darwinism, Dawkins claims, when correctly understood, is "deeply corrosive to religious faith".[4] On the other hand, Dawkins has expressed grudging respect for young-earth creationism because he believes that of the three schools of thought, it is the most intellectually

3 See biologos.org/articles/creation-evolution-and-christian-laypeople
4 See youtube.com/watch?v=VxGMqKCcN6A, from 4:27 to 4:57.

consistent.[5] For this reason, we regard a robust defence of young-earth creationism as one of the most powerful apologetics for Christianity in contemporary society. The purpose of this book is to equip the church to defend a Genesis-as-history interpretation.

Since the early nineteenth century, and even more so following the publication of Charles Darwin's *Origin of Species* in 1859, the Christian church struggled to offer well-founded answers to issues raised by secular origins science. This left believers in an apparent quandary. How can the Bible be considered a source of truth in an age when science appears to refute its origins account?

The contemporary church is faced with a conundrum. Ignore the origins issue, and the pied piper of the secular worldview will, over time, draw youth away from the faith of their parents. On the other hand, if pastors try to tackle the issue, they will find themselves confronted with a complex multi-disciplinary field that takes considerable time and expertise to master—time that pastors may not have.

Understandably, many pastors adopt a cautious approach to this issue. Protestant seminaries and theological schools—as well as Catholic institutions—have for decades tended to espouse (or tolerate) theistic evolution. Their graduates appreciate how divisive the topic can be, and many prefer to de-emphasize this potential wedge issue. However, while downplaying biblical origins has helped to avoid open division, scientifically minded Christians continue to wrestle with the issue in private. Many are particularly concerned that the youth, exposed to evolutionary thought in compulsory schooling and at university, are leaving the broader church. The era of greying congregations and church closures is now well underway.

In the creationist camp, several non-denominational ministries have been active for several decades. These have made solid inroads into the essential task of harmonizing science with the Bible. Christians are now able to access a wide variety of high-quality material, much of it available freely online, which goes a long way towards defending the young-earth creationist perspective. Furthermore, these materials are suitable for a range of audiences, from school-aged children through to the technically informed. However, as laypeople approaching the topic for ourselves, we felt there was a gap. We could not find a book which served as a one-stop shop—providing a comprehensive, high-level introduction to the topic, including a discussion of the downstream effects of the evolutionary worldview.

This book is our attempt to fill that gap; it covers historical, theological, philosophical and scientific aspects to unravel the complexities of the debate. We have endeavored to write for the church at large, including those who are not necessarily of a scientific bent. Importantly, we hope this book will be useful

5 See youtube.com/watch?v=BAbpfn9QgGA

to the pastors and Christian leaders whose congregations include individuals who read Genesis 1–2 as history.

The material presented in this volume could be helpful to others who are interested in the topic of origins. We encourage atheists, or adherents of non-Christian religions who wish to contrast the Christian origins account with their own beliefs, to read this book. In saying that, we note that Christianity is unique: it is not merely theistic, it is monotheistic and *trinitarian* (meaning that the one true God exists in three distinct Persons). One aim of this book is to draw out trinitarian theology as it relates to creation, an angle that non-Christian belief systems do not provide.

This book shows that the creationist position is stronger and more compatible with science than is commonly acknowledged. Further, we dispute the conventional view that contemporary secular origins science is secure. This has even been acknowledged by the secular camp. The late Sir Fred Hoyle, a non-Christian professor at Cambridge University, stated that the field of biology was used to mount a proxy war against traditional Christianity, and in the process became too ideological. The biologists, or "new believers" as he referred to them, had become "mentally ill", resembling the traditionally religious (as Hoyle saw it) by displaying "intense conviction" and attachment to their own brand of dogma, all the while claiming to be impartial scientists.[6] We will demonstrate that Hoyle's criticism of the ideology besetting some areas of biology is well founded; it has certainly stood the test of time.

The book comprises four parts. The first part sets out essential background that readers need to understand the material that follows. The second part rebuts and exposes the secular origins account, highlighting philosophical and scientific problems that remain unresolved, and in some cases cannot be resolved. In the third part, we analyze the biblical creation account (Genesis 1–11) from a young-earth creationist perspective and show why it makes sense. Our analysis is conducted partly through a biological lens, which provides a useful counterpoint to the secular origins account. Our main task here is to show that a straightforward reading of Genesis as history is both believable and compatible with science.

The fourth and final part reviews the current state and outlook for the Darwinian and creationist worldviews.

Several terms used in the book bear some clarification. We use the term *Christian* to mean a person who holds a Bible-based religious perspective and who seeks to obey Jesus Christ as Lord and Savior. Since this book is aimed at a wide audience, we define essential Bible-based beliefs as those consistent with

6 Hoyle, F. (1999). *Mathematics of Evolution*. Memphis: Acorn Enterprises LLC, pp. 3–4.

the historic creeds and fundamentals, while accepting the importance and precision of various Confessions and Catechisms.[7]

Following from this, we define a *non-Christian* as a person who has a religious perspective which is not Christian. An *anti-Christian* is one whose religious perspective is antagonistic to Jesus Christ, Christian churches, Christian doctrine or the practice of Christian behaviors.

We use the word *secularism* to denote a religious perspective that seeks to banish organized religion from the public sphere. A secularist is one who manifests "indifference to or rejection or exclusion of religion and religious considerations".[8]

The meaning of the term *humanist* has varied across time, but currently means "a doctrine, attitude, or way of life centered on human interests or values".[9]

Combining these latter words gives *secular humanism*, a philosophy which typically incorporates *naturalism*, "the doctrine that scientific laws are adequate to account for all phenomena".[10] Naturalism has an overlapping counterpart called *materialism* which holds that matter and energy are all there is and that all reality can be fully explained by physical laws describing how matter and energy operate.

We understand *secular origins science* to be research and investigation into the origin of the cosmos, life, and the diversity of living organisms, with the presupposition that naturalistic, non-purposeful causes are adequate to account for these processes.

We understand *creationism* to be research or investigation into the compatibility between science and a straightforward reading of Genesis (sometimes called young-earth creationism (YEC) or six-day creationism) underpinned by the presupposition that the Bible is infallible. Creationism covers the origin of the cosmos and life in all its diversity. It intersects many scientific disciplines such as archaeology, paleontology, biology, cosmology and physics, as well as other disciplines such as history, philosophy, linguistics and theology.

---

7 Example of ancient creeds include the Apostles Creed (circa 200 AD), the Nicene Creed (325 AD) and the Creed of Chalcedon (451 AD). A more contemporary statement is the "Five Fundamentals" of 1910 (General Assembly, 2018) which lists the inerrancy of scripture, the virgin birth, the substitutionary atonement of Christ along with His bodily resurrection and the reality of Christ's miracles as non-negotiable. These latter beliefs distinguish a so-called "Christian fundamentalist", or more simply—a Christian. Other statements of belief are more detailed and complex—the Westminster Confession (1647) is of longstanding importance in the Anglosphere, while the Belgic Confession (1561) is important in the Dutch Reformed tradition.

8 "Humanism." *Merrian-Webster.com Dictionary*. Merriam-Webster. Accessed July 5, 2025 from merriam-webster.com/dictionary/humanism

9 "Secularism." *Merrian-Webster.com Dictionary*. Merriam-Webster. Accessed July 5, 2025 from merriam-webster.com/dictionary/secularism

10 "Naturalism." *Merrian-Webster.com Dictionary*. Merriam-Webster. Accessed July 5, 2025 from merriam-webster.com/dictionary/naturalism

Finally, we unapologetically use the biblical terms *man* and *mankind* where appropriate, rather than the gender-neutral terms which are more commonly used today. We understand this choice of language is countercultural. But it's biblical and is eminently suitable for a book which is primarily concerned with the restoration of biblical ideas.

*

Graeme: A brief word on who we are and how this book came to be written. Both Geoffrey and I are Christians who have long maintained an interest in the origins controversy, with our own stance firmly allied with the many teachers and preachers who believe Genesis was intended to be read as history in a straightforward manner. Around the time of the Covid pandemic, I became interested in a fascinating calculation presented by creationist writer Jonathan Sarfati in his book *By Design*.[11] Sarfati showed how it was possible to calculate the total number of atomic interactions that can ever have taken place in the observable universe (assuming the universe's age to be 13.8 billion years), then demonstrated that this astronomical number was manifestly inadequate to account for the emergence of life by random means.[12] Intrigued as to whether this argument could be developed further into a comprehensive rebuttal of the orthodox scientific origins account, I contacted my twin brother Geoff and sought his views. I figured that Geoff's knowledge of biology, gained from his university studies, would come in handy; my own studies in mechanical engineering would only be useful up to a point. As we walked along the beach at Queensland's Gold Coast discussing these things, it was apparent that Geoff shared my interest. A project was born.

With drafting underway, it became clear that Geoff's intellectual interests beyond biology were going to come in handy too. An eclectic range of books lined his shelves: the physics lectures of Richard Feynman, Charles Norris Cochrane on *Christianity and Classical Culture*, and many more besides. In fact, this book has been informed by myriad ideas from a range of authors, spanning both generalists and specialists, and Christian and non-Christian.

It also became apparent that Geoff and I had complementary skillsets; this was a team effort that neither he nor I, working individually, could have completed. Ultimately, writing this book proved to be difficult and demanding—but we believe that God opened the right doors at the right time to allow us to bring the project to completion.

*

One of the most important learnings we took away from this project is that for serious progress to be made in the scientific understanding of Christian

---

11 Sarfati, J. (2008). *By Design*. Powder Springs GA: Creation Book Publishers, p. 159.

12 Our investigation of Sarfati's calculation is presented in Chapter 10.

origins, a community approach is required. The ability to discuss and test ideas in the context of a competent network is crucial, so much so that anyone who weighs into the origins debate single-handedly—whether a pastor, seminary professor or other learned member of the church—runs the considerable risk of presenting material reflective of their scientific, theological or philosophical blind spots. The fact that we teamed up was a step in the right direction. But it wasn't enough, and on that note, we'd like to thank our reviewers, without whose frank advice this book would not have been worthy of publication. Our review team, which possessed, in sum, an almost encyclopedic knowledge of the origins debate, was an essential sounding board.

But as much as our reviewers have assisted us, we must take ultimate responsibility for the content: if there are any errors in this book, they are our own. We would like to thank and acknowledge Scott Thomson, Gary Weston, the late Russell Grigg, Benno Zuiddam, and others who reviewed our book anonymously. Last but certainly not least, we also thank our families for their support throughout this project.

It is our prayer that this book influences the church and its institutions toward a deeper understanding of origins accounts, and a more powerful Christian apologetic to the unchurched. In turn, we hope this convinces the broader society of the compelling case for putting one's faith in the biblical Creator, Jesus Christ (John 1:3).

Graeme Messer
Geoffrey Messer
Brisbane, November 24, 2025

Part 1
Essential Background

1
Why worldviews matter

Everybody has a worldview, a mental framework by which they interpret the world around them. Worldviews are foundational to the way we view human identity and how we think we should live and behave. If a worldview is thought of as a building, then the most basic level—the ground floor—consists of the origins account. An origins account explains how the world around us came to be and also explains the origin of life, in particular human life. These important concepts have far-reaching effects on the higher levels of the worldview structure. American philosopher Paul Kurtz, regarded as the "father of secular humanism",[1] stated that

> ... critics of Darwinism properly perceive that, if the implications of Darwinism are fully accepted, this would indeed mean a basic change in our outlook of who we are, what we are, and also how we ought to live.[2]

Kurtz was saying that a societal shift away from biblical concepts, in favor of Darwinian concepts, would transform society from the bottom up by altering the base level of the majority worldview.

Given the significance of worldviews, we will commence with an investigation into their main features and impact on the origins debate. We'll introduce the key aspects of worldviews by comparing two accounts—one Darwinian and the other biblical.

Two dying bulls

Let us begin by considering the Darwinian belief system which attempts to explain the diversity of life by purely natural means. Suppose a predator in the distant past began pursuing a herd of buffalo. One of the herd is a yearling which happens to have an overshot jaw; this unfortunate condition prevents it from grazing efficiently, and as a result, it has not attained the strength that it might have. As the predator closes in, the yearling is unable to keep up with the rest of the herd. Thus isolated, the exhausted animal is soon overtaken by the predator, killed, and eaten. Remaining members of the herd live to fight another day and pass their genes for well-formed jaws to their offspring. Nature has "selected" them for survival. English naturalist Charles Darwin famously

---

1 Secular humanism is a belief system which denies the supernatural, instead drawing on human reason as a source for ethics and morality.

2 Kurtz, 1998.

postulated that these effects could occur over great time periods, such that not just refined species, but new, better adapted species would arise. Darwin went on to posit that this sort of explanation is sufficient to explain all varieties of life, starting from a simple organism in a primordial past, down to the immense diversity of life we see today.

While at first glance this Darwinian account comes across as entirely reasonable and in line with observed processes, it flies in the face of biblical concepts in several ways. One of these is the view of death as a normal part of life. The Bible teaches an originally good universe. When Adam sinned, God not only cursed mankind, but also (innocent) animal life, the plant world and the earth itself. This cursed world awaited the return of the new Adam who would bring restoration. In the fullness of time, Christ came to reconcile man under a new covenant, and ultimately to reconcile all things. In this biblical scheme, death is a result of sin; death of sentient animals and humans was not originally present in creation.

The Bible also includes accounts of dying bulls. Let's consider one of these.

In the first chapter of 1 Samuel, we read of an ancient Israelite family travelling to Shiloh, the religious center of the day. They have with them three bulls, and quantities of flour and wine. Their intent is to sacrifice these to the Lord when they arrive at Shiloh. We focus on the bulls as they walk placidly along, each step bringing them closer to their destination. Clearly, they have no inkling of the fate that awaits them. Prior to the bull's sacrifice it will be immobilized by ropes around each leg. After a brief prayer with hands laid on the animal's head, the priest will slit its throat. The ensuing blood loss will result in the bull collapsing, then dying shortly after. The priest will sprinkle the blood around the altar, cut the bull's body into pieces, and place these on the altar for burning.

The bull does not die because of any lack in its physical makeup, but for *lack of knowledge*.[3] It has no insight into the human dimension, with its abstract concepts of ownership, law, and sacrifice. Hence the bull took no evasive action prior to its death, although it certainly had the strength to do so. The biblical account is fundamentally different to the Darwinian.

The animal sacrifices of the Old Testament foreshadowed the sacrifice of Christ on our behalf (revealed in the New Testament). In this story, humans are analogous to the bull. Each person's life ticks along, day by day, like the steps of the bull toward Shiloh. Time is gradually running down on each person's life toward certain death and judgement, but people, naturally speaking, are as oblivious to the judgment to come as the bull was to its fate.

How do we know this? Because a message from a higher spiritual dimension—the Bible—tells us so. Knowledge that we could never work out by ourselves has been communicated to us by a sympathetic Friend from the

3 This is analogous to Hosea 4:6a:" My people are destroyed for lack of knowledge."

higher dimension. How can we be saved from that fate? By believing the message and acting on it—accepting a substitution on our behalf. That's the Christian gospel in a nutshell.

Christianity is a revelatory belief system. As a human you cannot by your own reason derive Christian doctrine. You could not logically deduce from first principles that salvation is of the Jews; that the first man was Adam; that Jesus died on the cross to save us from our sins. We know these things only because they have been revealed to us. The accounts of animal sacrifices in the Old Testament symbolize and reinforce New Testament truths.

Conversely, the Darwinian account reinforces key concepts of the secular humanist and atheist. Darwin's account purports to explain the diversity of life without invoking the Divine. In fact, Darwin would often emphasize that his scheme left no room for any notion of plan or purpose behind either animal or human life. His notion that all life forms descended from the earliest life implies that humans and animals (and plants) are relatives. This, of course, contradicts the biblical view that all humans are distinct from the animals, and are the only creatures made in the image of God.

The Darwinian approach is based on human imagination and reasoning about nature. The Christian approach is based on human reasoning about the revealed word of God and nature, subject to illumination by the Holy Spirit. No atheist would ever agree to use the second account of the death of the bull as the basis for an intellectual theory of how life operates, because the biblical account of sacrificial animals leads to biblical concepts. Should Bible-believing Christians use the Darwinian account, which according to Dawkins (quoted in the Introduction), is "deeply corrosive to religious faith"?

Presuppositions

If the origins account of a worldview is the ground floor of the worldview building, the foundations underpinning it are unprovable assumptions known as presuppositions (or axioms). As Christians, we infer many of our presuppositions from the Bible, but we also derive presuppositions from our interactions with the world around us.

Presuppositions consistent with the Bible include the propositions that God exists, that God is the Creator of the universe, that mankind has fallen into sin, that the reasoning of fallen man is not entirely trustworthy, and that God has chosen to reveal something of Himself, and His plans and purposes for mankind, in a written revelation: the Bible. Secular humanists (and atheists), on the other hand, derive their presuppositions solely from human reason and imagination applied to information gained from interactions with the surrounding world.

Presuppositions, as we said, cannot be proven. If a presupposition could be proven, then it is not really a presupposition—it is an interpreted fact—

interpreted in the context of the worldview in question. Instead, presuppositions are both the foundation on which a worldview rests and a lens through which data is interpreted. Many people may not be aware of this. For example, in a debate between the late Cardinal George Pell and well-known atheist Richard Dawkins, the compere asked Pell if he could ever provide Dawkins with scientific proof that God exists.[4] The fact that this question was asked shows a misunderstanding of the presuppositional basis in the origins debate. Neither the atheist nor the theist (nor, for that matter, anyone that holds to a different worldview) can prove their presuppositions or disprove the presuppositions of the other side. They may, however, point to evidence that seems to strongly support their presuppositional stance. For the Christian, the historical and legal/testimonial evidence of Jesus' rising from the dead affirms His authority and divinity, and with it His consistent stance on the total truth of the Old Testament, including its origins account. The atheist may seize upon some seeming evidence in support of the evolutionary narrative. But neither qualifies as any sort of ultimate proof. Such evidences simply act to underpin the interpretive lens of that worldview.

There is one more conclusion that follows from this. Presuppositions are resistant to debate, because any evidence that might imply they are wrong is interpreted by their proponents through that very lens. For example, the Bible when interpreted in a straightforward manner gives a timeline since creation of some thousands of years. But a young-earth creationist, Ken Coulson, who we'll encounter later in the book, points out that scientific evidence exists (uranium radioisotope halos) that seem to require at least a hundred million years of time to generate. Does Coulson abandon his belief in the biblical timeline because of this? No—he interprets the scientific evidence in light of the Bible and provides an explanation which incorporates both lines of evidence.

We see the exact same process in the atheistic camp. Atheists have no demonstrable explanation for the emergence of life by natural means. Do they abandon their presuppositions because of this? No—they reinterpret the evidence, typically arguing along the lines that science has made good progress thus far, so there is no reason to believe this problem will not be solved at some future point.

Let's look at another example. Suppose a social researcher cites statistics showing that most scientists support human evolution.[5] This fact might appear to humanists to support Darwinism; the statistic shows that the most educated and rational people favor Darwinism, hence, Darwinism must be both rational and sensible. Even further, they may view this victory of Darwinism in higher education as a sort of natural selection of stronger ideas over weaker ideas—Darwinism acting in the world of thought. But Christians could interpret this

4 The debate can be viewed at youtube.com/watch?v=QaA5QrZBzWg accessed October 19, 2025.
5 Masci, 2019.

same piece of information differently, pointing out, for instance, that it supports the New Testament principle outlined in 1 Corinthians 1:26–29:

> Brothers and sisters, consider your calling: Not many were wise from a human perspective, not many powerful, not many of noble birth. Instead, God has chosen what is foolish in the world to shame the wise, and God has chosen what is weak in the world to shame the strong. God has chosen what is insignificant and despised in the world—what is viewed as nothing—to bring to nothing what is viewed as something, so that no one may boast in his presence.

In this interpretation, the data corroborates what the Bible says and therefore supports the Christian worldview.

Hence a particular piece of data may appear to strongly support both the humanist's Darwinian worldview and the Christian's Christian worldview. The key takeaway is that presuppositions are not proven, but because they influence how data is interpreted, worldviews are self-reinforcing in a manner that can appear like circular reasoning.

This circularity is evident on all sides. For example, a Christian believes the Bible, in one sense, because it declares itself to be the revealed word of God. (However, Christians typically believe in the Bible for additional reasons, such as the striking abundance of early source documents, its trustworthiness as a historical text, the wisdom of its message, the historical fulfilment of biblical prophecies, and its impact on their personal lives.) The secular humanist and the atheist effectively both maintain that human reason is the only trustworthy authority, which again appears to have some circularity. The conclusion both parties reach very much depends on their starting points. This applies to all worldviews.

Authority

At a basic level, worldviews boil down to what people regard as authoritative sources of information. Is the Bible an important source of truth, or should we rely purely on human reason and observational data? Or something else? Christians believe that a combination of information sources is necessary: while the Bible is essential for accurate presuppositions, empirical evidence and human reason all have a part to play because all truth is God's, and man is a rational being made in the image of God.

Nevertheless, the Bible teaches us to trust God first. It also teaches that the human heart is deceptive because of the effects of sin (Proverbs 3:5–6, Jeremiah 17:9–10). Therefore human reasoning must be assessed very carefully by comparing it with biblical teaching, empirical evidence and logic. This worldview holds that humans are dependent on God and His Word. The Christian upholds the Bible as objective truth, in contrast with human ideas which are subjective and may be revised. The Bible is an important standard against which we assess other sources of information.

Many in the atheist camp would claim that we have no need for the Bible, on the presupposition that humans can devise or discover all required knowledge on their own and then determine their own ways of living. This worldview sees humans as entirely independent, and able to alter their views on correct ways of living in line with the times. The notion of independence from the Bible's authority is something that many people find attractive.[6] In their thinking, they—not God—reign supreme.[7]

The personal aspect to worldviews

Stepping back from these technical considerations, why do some people believe and others don't? Why do some people (but not others) have what philosopher Thomas Nagel terms the *sensus divinitatis*, meaning a sense of the divine? For his part, Nagel admits to lacking this, but he observes others around him who "... see in the world the expression of divine purpose".[8] If we are no more than a collection of atoms, governed ultimately by the laws of chemistry and physics, why don't all people have the same response to faith, the response of automata? Why did the late public intellectual Christopher Hitchens gravitate to atheism, while his brother Peter is a devout Christian?

To illustrate these intriguing questions, we consider two on-the-record examples of people from similar religious backgrounds (Catholic) who, by their mid to late twenties, had lived in a state of lapsed faith for some time. Both were involved with calamitous descents from high places which could easily have killed them. In their hour of extremity, did they reach out to God?

The first example is British mountaineer Joe Simpson, whose 1988 book *Touching the Void* was subsequently made into a docudrama of the same name (2003). Simpson was descending from the summit of the Siula Grande, one of the tallest peaks in Peru's section of the Andean Mountains, after his ascent in 1985. Early in the descent, Simpson fell and broke his right leg, seriously hindering his ability to move. His climbing partner, Simon Yates, used a rope and belay system to lower Simpson in stages. For a while this method worked well, but the situation became untenable when, unknown to Yates, Simpson was lowered over the edge of a cliff and found himself hanging in midair, too far away to alert Yates to his predicament, and unable to climb back to the cliff edge. Yates' own position was beginning to slip; eventually he had no choice but to cut the rope to prevent himself from being pulled over the edge. This sent

6 "Humanists reject the claim that the Bible is the word of God. They are convinced the book was written solely by humans in an ignorant, superstitious, and cruel age" (Sommer, 2012).

7 "While there is much that we do not know, humans are responsible for what we are or will become. No deity will save us; we must save ourselves." (Kurtz & Wilson, *Humanist Manifesto II*, 1973).

8 Nagel, 2012, p. 12.

Simpson into free fall. He landed on a small ledge within a deep crevasse, some 60 meters below.

With his broken leg, Simpson could not climb out of the crevasse. He had no food to sustain him, nor any help from Yates, who was now necessarily focused on his own survival. In this extremely vulnerable position, Simpson used the remaining resources at his disposal—the part of the rope that Yates had cut, and his considerable mountaineering experience—to abseil deeper into the crevasse. From this new location, Simpson found a way out of the crevasse and, over the next 72 hours, somehow managed to make his way back to base camp. He was just in time; Yates, who had reached camp days earlier, was only hours from departing when Simpson arrived.

Brought up in a devout Catholic home, Simpson later stopped believing in God. Nevertheless, he had long wondered if, in a crisis, he would instinctively reach out to God for help. In the docudrama film, Simpson recounts the point at which he realized he might die:

> [Crying out to God] never once occurred to me. It meant that I really don't believe and I really do think that when you die you die, that's it, there's no afterlife; there's nothing.[9]

In his hour of need, Simpson did not turn to God. He relied on his own resources, meagre as they were, and these proved sufficient to survive.

Our second example, Genelle Guzman-McMillan, was the last person to be taken alive from the rubble of the World Trade Centre after the 9/11 terror attacks. Originally from Trinidad, Guzman-McMillan moved to the United States in early 2000 with dreams of making it big in the entertainment industry. But instead of the glamorous life, Guzman-McMillan found herself working a variety of odd jobs to get by. Eventually she took up a role as an administrative assistant at the Port Authority of New York and New Jersey, based in the World Trade Centre complex.

On the morning of September 11, 2001, Guzman-McMillan was seated at her desk on the 64th floor of the 110-storey North Tower, when, at 8:46 a.m., American Airlines Flight 11 crashed into the building some 30 floors above. Amid the confusion that followed, including baffling advice from officials to wait in the office until police arrived, Guzman-McMillan and several of her colleagues did not attempt to leave the building until shortly after 10 a.m. By this stage the only route available to them was the fire escape stairwell. Over the following twenty-something minutes, Guzman-McMillan's group descended as far as the 13th floor. Then, at 10:28 a.m., the North Tower collapsed.

---

9 The relevant part of the docudrama can be viewed at youtube.com/watch?v=nWhmOwGqcMQ (from 3:17–3:50). Accessed September 8, 2025.

When everything came to rest, Guzman-McMillan was alive, trapped in a narrow space between concrete beams. She was almost completely immobilized; other than some movement of her left arm, her head was wedged tightly in place and her legs were "smashed flat".[10] "'How far down was I buried?' Guzman-McMillan asked herself. 'I just stared at the darkness, fading into despair. Nobody was coming to get me, and there was nothing more I could do.'".[11]

Alone and trapped, Guzman faced her mortality, eventually concluding that only a miracle of God could save her. She began to pray, with an increasing sense of urgency as time passed. She mixed in Bible verses that she remembered from her Catholic schooling. Her prayers paused as she napped, probably over the night of September 11/12, then continued even more frantically in her wakeful hours. As she banged with her left fist for emphasis, her hand broke through something above her head. Amidst more impassioned prayer, she felt a hand clasping hers, and listened as a voice spoke

> … the four sweetest words I had ever heard. "I've got you, Genelle," a calm, confident male voice said to me. "… They're going to get you out soon."[12]

By 1:30 p.m. on September 12, rescue workers had cut away the concrete beams and lifted Guzman-McMillan to safety.

Prayer came naturally to Genelle Guzman-McMillan, but it never occurred to Joe Simpson. What causes this difference? Whatever it is, our understanding of the Bible shows that a mix of believers and non-believers is part and parcel of the human condition. These two different types will always be with us and will bring two different approaches to origins science. A Christian origins scientist seeks to harmonize scientific data with his or her interpretation of the biblical account. They take the existence of the biblical Creator-God as a given (a presupposition), and their task is to harmonize the increasing stock of scientifically derived data with the unchanging and infallible biblical origins account.

Secular origins science, on the other hand, seeks to explain the world in purely naturalistic terms. Any invocation of God is generally seen as a cop-out, intellectual laziness of the sort that strangles the investigative process. The notion of a higher power is to be avoided at all costs. In fact, there's a phrase for that: *etsi Deus non daretur*[13]—Latin for "as if God does not exist." That's how secular science, including origins science, is conducted.

---

10 Guzman-McMillan, 2011, p. 49.

11 Ibid, p. 51.

12 Ibid., p. 97.

13 This phrase was used by Dutch jurist Hugo Grotius in 1625.

The two Lazaruses

One of Jesus' parables gives us a clear example of the effect of worldviews on individuals. In Luke 16, Jesus tells the illustrative story of a rich man and a beggar named Lazarus. The rich man leads a luxurious lifestyle, while Lazarus, wracked by ill-health, lies at the rich man's gate, longing to eat the crumbs that fall from the rich man's table. Both men die; Lazarus is taken by angels to a place of comfort with Abraham, whereas the rich man finds himself in Hades. In his wretched state, the rich man begs Abraham to send Lazarus to his five living brothers, to warn them of the torments they will have to endure if they remain unrepentant. Abraham responds: "They have Moses and the prophets; they should listen to them." The rich man is not convinced:

> 'No, father Abraham,' he said. 'But if someone from the dead goes to them, they will repent.' But he told him, 'If they don't listen to Moses and the prophets, neither will they be persuaded if someone rises from the dead.' (Luke 16:29–31).

The rich man thinks his brothers have insufficient information about spiritual matters. Yes, they have Moses and the prophets—what we know as the Old Testament—but this treasure trove of God's word has fallen on deaf ears. The rich man thinks a miracle will convince them: if Lazarus returns from the dead, his appearance and message will persuade them to take spiritual matters seriously. More and better evidence is what the brothers require.

Not so, responds Abraham. Their worldview, which leads them to dismiss God's word, will interpret away all other evidence of spiritual realities—even a miracle. Their problem is not a lack of information; their problem is a worldview which shuts out God and His message.

In John 11, we read of another Lazarus who fell ill and died. But this is the account of a real event. Jesus arrived on the scene four days after Lazarus' death, and, having asked those present to take away the stone that was sealing Lazarus' tomb, famously called the dead man back to life. The amazed onlookers were confronted with a *real* resurrection, a man emerging from a tomb bound hand and foot with grave clothes. "Unwrap him and let him go", commanded Jesus (v. 44b).

It's instructive to see how the people of the time interpreted this. Many who saw the miracle believed in Jesus. Perhaps they had been wavering on the edge of belief, and this miracle was enough to finally convince them. But the leaders took an entirely different view. They figured that if Jesus continued His amazing display of miracles, the entire society would come to believe in Him, whereupon the Roman authorities would "take away both our place and our nation" (John 11:48). With their power base under threat, the leaders were soon conspiring to kill both Jesus and Lazarus (John 12:10–11).

In both examples, it was the wealthy and powerful—in other words, the elite of society—who could not believe. This is noteworthy because the Bible repeatedly warns of the dangers and deceit of wealth. A generalized lesson can be drawn: the wealthy and powerful are at risk of becoming so ensconced in worldly systems that they disregard God entirely. Their worldview becomes resolutely anti-Christian[14]—an effect which is observable to this day.

Applying the Christian worldview

A story is told of a Christian convert who came to faith after learning of scientific evidence related to the gradual shrinkage of the sun. This evidence implied the age of the earth was comparable to what is indicated in the Bible .[15] The same man subsequently gave up his faith after an acquaintance convinced him that the argument for the shrinking sun was unsound. However, creationists subsequently presented further evidence which implied that the sun was indeed shrinking, albeit at a slower rate than previously thought. Based on this, should the man have regained his faith? And if still further evidence emerged that contradicted the preceding evidence, should the man again reject Christianity? Otherwise put, should a person allow themselves to be whipsawed in and out of faith depending on the emergence of new scientific evidence?

The answer is no. The unstable belief of this particular man resulted from a failure to interpret evidence through the lens of the Bible. Instead, the man followed the reverse approach—interpreting the Bible in the light of ever-changing scientific evidence—which is the secular humanist approach. This cautionary tale highlights the importance of "renewing our minds" (Romans 12:2) by including biblical presuppositions in our thinking. Christianity, we reiterate, is a revelatory belief system, with a coherent set of presuppositions revealed in the Bible. A theology based primarily on scientific evidence might all have to be walked back when the scientific evidence is updated. Thus, a Christian ought to be very cautious in modifying their basic worldview simply because a new piece of scientific evidence is published. A secular humanist may not necessarily change their stance either when new scientific evidence appears to contradict their worldview.

This has practical applications for Christians engaging with the origins debate. If a Christian were asked a question along the lines of "Would you give up your faith if ...?" and the questioner cited seemingly impressive evidence such as scientists successfully creating life in the lab from inanimate matter, a

---

14 "The moneyed, respectable, capable world has been steadily anti-Christian and Barabbasque since the crucifixion ..." (Shaw, 1912).
15 Wieland, 1991.

mature Christian should answer in the negative.[16] We are called to evaluate scientific claims in light of biblical truth, and not the other way around.

This is why worldviews matter. As we approach the origins debate, with its various evidences, it is presuppositions, or worldviews, that determine *how that evidence is interpreted.*

Conclusion

Each person has a worldview, a framework for interpreting the world. At some level, worldviews contain circular reasoning, and they rest on presuppositions which are unprovable. Christians take their presuppositions from the Bible, whereas humanist presuppositions are based on human imagination and reasoning. There are no neutral worldviews; each will contradict others in some ways.

The two worldviews which we examined, Christianity and secular humanism, each have an origins account. Christianity posits a Creator God and acceptance of the description of creation in Genesis. Secular humanism posits natural forces and discounts the need for a God, and even the existence of God. Personal factors ultimately determine which worldview individual people will subscribe to. People from similar cultures and almost identical upbringings could, and do, hold to diametrically opposed worldviews. These worldviews are in opposition, and some participants in the origins debate view the other side with deep suspicion. One man's truth is another man's propaganda.

In this book we will uphold the principles of the Christian worldview by taking many of our presuppositions from the Bible and evaluating evidence in light of those presuppositions.

An important question that arises in the origins debate centers around the plethora of religious origins accounts, which secular humanists collectively term "creation myths". Why, they ask, should any one of these creation myths be preferred over the others? In particular, why should the Genesis account be given any special credence? We turn to this important question in the next chapter.

References

Guzman-McMillan, G. (2011). *Angel in the Rubble.* Crows Nest NSW: Inspired Living, Allen & Unwin.

Kurtz, P. (1998, Spring). Darwin Re-Crucified—why are so many afraid of naturalism? *Free Inquiry,* pp. 15–17.

Kurtz, P., & Wilson, E. H. (1973). Humanist Manifesto II. Retrieved November 2, 2024, from americanhumanist.org/what-is-humanism/manifesto2/

Masci, D. (2019, February 11). For Darwin Day, 6 facts about the evolution debate. Retrieved December 7, 2023, from pewresearch.org/fact-tank/2019/02/11/darwin-day/

16 If a scientist were to create life in a test-tube in the future, this would demonstrate that creating life requires processes directed by an intelligence, rather than purely random processes. As such, this feat would strongly support creationism, not secular origins theories. See creation.com/synthetic-life-by-venter for further details.

Nagel, T. (2012). *Mind and Cosmos: Why the Materialist Neo-Darwinian Conception of Nature is Almost Certainly False*. New York: Oxford University Press.

Shaw, B. (1912). *Preface to Androcles and the Lion: On the Prospects of Christianity*. Retrieved September 8, 2025, from gutenberg.org/cache/epub/4004/pg4004-images.html.

Sommer, J. C. (2012, March 9). Some Reasons Why Humanists Reject The Bible. Retrieved November 2, 2024, from americanhumanist.org/what-is-humanism/reasons-humanists-reject-bible/

Wieland, C. (1991, June 4). Swaying in the breeze—Editorial. Retrieved November 2, 2024, from creation.com/swaying-in-the-breeze

2

Sizing up the contenders

How can we assess the many origins accounts that various people groups and intellectual traditions have advanced over millennia? In this chapter we introduce a framework for evaluating a variety of origins accounts. We will consider five criteria for testing origins accounts: the problems of existence, information, coherence, consciousness, and compatibility with a scientifically accepted finite beginning. We could list others, but these are sufficient for our purposes. An origins account which passes all tests is credible. Conversely, an account which fails one or more tests is not credible and should be discarded.

The first criterion: The problem of existence

The problem of existence has been contemplated by philosophers for centuries and, more recently, by physicists. Simply put, what is the ultimate origin of ourselves and the universe? Because something cannot spring into existence out of nothingness by itself, we argue that all credible origins accounts must identify an entity that has no beginning. Absent that entity, explaining how the universe came to exist is impossible—or at least, irrational.

Let's look at some examples. Our first scheme derives from some of the ancient Greek philosophers, including Aristotle, who contended that the universe had always existed.[1] In their view, the entity which had no beginning was the universe itself. This view of a universe without a beginning, an eternal universe, was not unique to the ancient Greeks. English polymath Isaac Newton hypothesized a universe of infinite extent, a system compatible with an eternal universe (even though he accepted the Genesis account). An eternal universe which had always existed became the majority view among Western scientists by the 1800s[2] and prevailed until after World War 2.[3]

Another concept, promoted by the ancient Greek philosopher Plato, held that a divine craftsman (the "demiurge") had fashioned the universe out of pre-existing primeval matter as best he could.[4 5 6] This could imply that both a "god"

1 Dreyer, 1953, p. 109.

2 Edward Clodd wrote in 1897 that "Modern science knows nothing of a beginning, and, moreover, holds it to be unthinkable" (Clodd, 1897, p. 30).

3 Krauss, 2012, pp. 2, 105, 107.

4 Oakley, 1926, p. 71.

5 Doyle, 2016.

6 Zeyl & Sattler, 2023.

and matter had always existed and had no beginning. This scenario differs from the first in that it involves two entities that had no beginning—a god and chaotic primeval matter.

A third concept is the notion of a cyclical universe. In this conception, a universe is born, becomes mature, and then collapses or is otherwise destroyed, after which it is replaced by a new universe. Here the entity which has no beginning is the cyclical process that creates and recreates universes across infinite time or an infinite sequence of times. (Conversely, if the universe is not completely destroyed, but collapses into a 'singularity' before emerging and expanding once more into a fully-fledged universe, one could argue that the universe itself is the eternal entity here, too. It has simply changed form. It never ceases to exist entirely; there is never truly 'nothing'.) This view of a cyclical universe has taken scientific form in recent years, being known as the Big Bang–Big Crunch hypothesis.

Turning to religious texts, the account in Genesis 1–11, shared by Christianity and Judaism, posits a Creator God who brought the world into being at a particular instant in time. This God is described as "inhabiting eternity" (Isaiah 57:15). The biblical view therefore is that God is the entity which had no beginning.

All the accounts listed above contain an entity that has no beginning.

From the contemporary atheistic camp, physicist Lawrence M. Krauss has penned a book on the topic: *A Universe from Nothing: Why there is Something rather than Nothing*.[7] The author shows that matter could have arisen from relativistic quantum fields (exactly what these are doesn't matter for our purposes). Krauss presents evidence that matter and antimatter can spring into existence for vanishingly short timeframes. That might be possible under existing conditions in which matter and its strange ways of behaving are already in place. But Krauss is aware that he must posit an entity without a beginning for his account to be rational—what he references as "First Cause".[8] He is not able to definitively say what that entity is but provides several possibilities. One is "eternal inflation", which refers to a process that continuously creates universes.[9] Another involves a very large number of universes in different dimensions, but Krauss does not say from where these have arisen.[10] Yet another suggestion is that the entity without a beginning is quantum gravity; in quantum gravity fields, Krauss assures us, universes can spontaneously appear from nothing.[11] Krauss also posits that an eternal, pre-existing "fundamental nothingness" may be the entity.[12] We conclude that

7 Krauss, 2012.
8 Ibid., pp. 173, 174.
9 Ibid, p. 176.
10 Ibid., p. 176.
11 Ibid., p. 169.
12 Ibid., p. 174.

Krauss has no clear answer to the existence problem—certainly he cannot identify the entity with no beginning definitively. Perhaps that is why Krauss suggests that the question of where the universe came from may not be particularly useful to ponder,[13] a surprising assessment given the title of his book.

For an origins account to be both credible *and* rational, it must cite an entity which has no beginning. This could be the universe, a cyclical or eternal process, a range of possible entities (as per Krauss), or a Creator God. Entities that do not have a beginning are unbounded or infinite, at least in respect to time. Therefore, the problem of existence requires an entity which is infinite in nature.

Types of infinities

In the Christian worldview, the infinity that created the universe is a personal infinity—a triune God. The Bible describes God in various ways and ascribes various names to Him. One of these is "God Almighty" (Genesis 17:1). The word *almighty* tells us that God's power is unlimited; he is *omnipotent*. Isaiah 57:15, cited above, tells us that God "inhabits eternity"; He is eternal. Job 9:10 teaches that God "does great things past finding out; yes, marvelous things without number." God's actions are both great and infinite in scope. All of these texts point to the infinite attributes of God, which are termed omniscience (all knowing), omnipotence (all powerful) and omnipresence (present everywhere). The creation bears the stamp, or imprimatur, of the Creator. Therefore, we expect the world around us to *incorporate the concept of the infinite*, or to lead us to the concept of the infinite.

In the study of the problem of existence, non-personal infinities can also be considered—infinities with neither mind nor consciousness. These could be the universe itself, or a cyclical or eternal process. The secular humanist (and atheist) camp leans in this direction.

Cause and effect

Part of the existence problem involves the concept of cause and effect. In the Christian worldview, the eternal Creator God is the cause of the universe because He created it. Thus, the universe is an effect of the creative work of God.

In one of Krauss's schemes, quantum fluctuations are the cause of the universe, *ergo*, the universe is an effect of quantum fluctuations. The quantum fluctuations themselves occur in the setting of quantum gravity, which apparently is the cause of the quantum fluctuations. However, neither quantum

13 "Why is there something rather than nothing? Ultimately, this question may be no more significant or profound than asking why some flowers are red and some are blue. … Either way, what is really useful is not pondering this question, but rather participating in the exciting voyage of discovery that may reveal specifically how the universe in which we live evolved and is evolving …" (Krauss, 2012, p. 178).

fluctuations nor quantum gravity are 'nothing', so they themselves must have a cause—unless one or both have eternal existence.

Note that both schemes assume that the idea of cause and effect is valid. Most people accept this. But historically, some notable thinkers have questioned it.

The influential Scottish philosopher David Hume (1711–1776) investigated the idea of cause and effect as a skeptic. He argued that people often see that one event follows another, and that when this occurs repeatedly, they label the process "cause and effect".[14] The cause is simply the event which precedes the effect, and the effect is what follows the cause. Hume suggested that the idea of cause and effect is psychological, not real. In other words, people think in terms of cause and effect, but all we know is that one event happens to follow another. We do not know for certain that one causes the other. Our knowledge is limited to the understanding that one event is associated with the other.

But there is a problem with Hume's thinking. An important part of science is investigating cause and effect relationships in order to describe, explain and model them. Therefore, denying that cause and effect are real undermines the basis of science, and we can even say it is irrational. Why? Because we can see the enormous expansion of knowledge over the last few centuries, which is due in large measure to scientific enquiry. Not only has knowledge increased, but a lot of that knowledge accords very well with reality. Technology works—computers and mobile phones are evidence for that. If science is effective—and it is—then attacking the very basis of science is irrational. But it is not just scientists who use the idea of cause and effect. Everyone demonstrates this notion in their everyday life; it is foundational to rational human behavior.

In fact, cause and effect is a presupposition of science and everyday life, and hence is unprovable. That is why Hume's argument that cause and effect cannot be proven is true, but useless. Nevertheless, this type of skeptical reasoning is corrosive of science if accepted, and Hume's thinking has permeated a lot of humanist discourse. It is good to be skeptical up to a point,[15] and the practice of science requires a certain level of skepticism. But we must be skeptical of extreme skeptics, those who would undermine the very basis of thinking, science and truth.

As we conclude our first criterion, we note that the Aristotelian, Platonic, eternal universe, cyclical universe, and Christian accounts all pass. We could add to this list any creation myths that posit an entity that has no beginning. We also noted that the entity which has no beginning must be infinite in some respect, and that the problem of existence involves the concept of cause and effect. The notion of cause and effect is reasonable and supported by the results of the scientific enterprise which it underpins.

---

14 Hume, 1999, pp. 109–124.

15 The Bible praises the Bereans (in Acts 17) for double checking what they were told.

Krauss's scheme may be considered by some to pass this criterion, but he does not definitively identify the entity with no beginning.

The second criterion: The problem of information

The ancients were aware that the universe contained information, or order. The word *cosmos* has the same Greek root as cosmetic, meaning arrangement or adornment, implying order.[16] But only since the scientific advances of the twentieth century has the true extent of the information content of the cosmos been more fully revealed. Science has revealed that the cosmos, the very matter of the universe, is exquisitely fine-tuned in a way that supports life.[17] On earth, the information contained in the DNA of living organisms is enormously complex and has most of the characteristics of language.

For an origins account to be credible, it must explain how the information content of the cosmos has arisen. And, as for the problem of existence, we will see (in Chapters 9 and 10) that the concept of infinity must also be invoked to account for the large-scale informational basis of life. We offer the following thought experiment as a brief illustration of the concept.

Consider two students, one highly intelligent and another much less so, being set a difficult problem to solve. Who will complete it more quickly? The answer, clearly, is the more intelligent one. But what if the problem were extremely difficult? Suppose if, in art class, the students were asked to complete a painting of equivalent artistic merit to the Mona Lisa. Could they deliver? No—Leonardo da Vinci himself took several years to complete his masterpiece. What if we asked a painter of vastly greater talent than da Vinci (assuming such a painter existed) to undertake the task? Conceivably, he or she might complete an equivalent masterpiece in just a few sittings. The greater the intelligence or capability, the shorter the time required to produce a product with high information content.

On the other hand, with less intelligence, a longer time is required. What if we replaced our moderately talented school student with a machine which applied paint randomly to a surface? How long would this take to produce a Mona Lisa-type masterpiece? It would take forever; we can't assign a specific number. With less intelligence brought to bear, longer timeframes are needed, and for non-intelligent random processes, infinite time is required to explain the origin of information that we see in life forms.[18]

The Christian origins account posits that a Creator of infinite intelligence is the originator of the fine-tuning of the universe and the information content of

---

16 Holt, 2012, p. 20.

17 The fine-tuning is revealed in a number of physical constants which are expressed as extraordinarily precise numeric values. These values represent information about the universe (and implicitly contained within it), but this type of information is in a different category to the digital information found in DNA.

18 Gitt, 2020.

living organisms. A prominent British scientist and atheist, Sir Fred Hoyle, came to an analogous conclusion, stating that "a common sense interpretation of the facts suggests that a super-intellect has monkeyed with physics, as well as with chemistry and biology, and that there are no blind forces worth speaking about in nature".[19] Hoyle and his colleague Chandra Wickramasinghe noted that the "measure of intelligence needed to control the properties of the oxygen and carbon nuclei would be exceedingly high".[20] The Genesis account, which is thousands of years old, addresses this problem perfectly by asserting that God spoke the universe and its constituent parts, including living organisms, into existence. Speaking implies language and information, which dovetails with the informational content of the creation revealed by modern science. In the New Testament, John's gospel emphasizes the informational aspect of creation by declaring that the Divine Word (or *logos* in the Greek) was the creative agent by which the world was created (John 1:1).

In the secular humanist camp, only physical, non-personal, non-purposeful, non-intelligent processes can be invoked. In this scheme, near-infinite time is required to explain the origin of the information content of the universe and living organisms. But secular humanists do not have infinite time. They have a quasi-infinity of 13.8 billion years, the insufficiency of which we'll demonstrate in Chapter 10.[21] Humanists have therefore proposed other impersonal infinities to make up for the inadequacy of finite time. The origin of information has emerged as an enormous problem for secular origins science and severely damaged the foundations of the evolutionary account.

Concluding this criterion, the Aristotelian and Platonic origins accounts, with their infinite time, remain in contention, as does the Christian account. The cyclical universe account is left hanging by a thread. If the universe is destroyed, and hence of finite duration, where is the infinity that explains information? Can information survive to be passed on to the forthcoming instance of the universe? If not, which seems likely to be the case, the theory is out; if so, the theory continues for now, in the absence of hard evidence against it. Creation myths that survived the first criterion and posit some plausible means by which information is imparted to the universe also make the cut.

Our view is that the secular humanist account does not pass this criterion. The random processes that atheists must invoke lack sufficient time to produce the information displayed in even the most rudimentary life form (which will be explained in Part 2).

---

19 Hoyle, 1981, p. 12.

20 Hoyle & Wickramasinghe, 1981, p. 141.

21 Other claims made in this section are discussed more fully in Chapters 9 and 10.

The third criterion: The problem of coherence

The coherence criterion relates to the predictable behavior of the universe. Cause and effect are generally discernible, and various processes of nature are compatible with man's sense of logic, reason and mathematics. The coherence and stability of the universe is essential for science to be possible. Consider the well-known equation, $E = mc^2$. (Here, E stands for energy, m for mass, and c is a constant term equal to the speed of light in a vacuum.) That mass and energy are correlated in a manner proportional to the square of the speed of light is a fascinating concept. It means that these disparate quantities can be measured, investigated and calculated. Of course, this is merely one example; physics and chemistry are replete with such evidence of the coherence of the visible universe.

The biblical origins account explains why the world is coherent, and why the world is accessible to scientific inquiry. The Christian God is rational, a super-intelligence who oversees all things. God does not change; the way He upholds and controls the universe is consistent and orderly, rather than chaotic or random. Because humans are made in the image of God, they can reach a level of understanding of the creation, thinking God's thoughts after Him in an analogical fashion. The fact that humans were given dominion over the earth and the animals implies that man has the capacity to understand earthly processes, including biological processes. Thus, coherence is well explained by Christian monotheism.

However, the mere fact of being a theist does not necessarily mean that one can explain the coherence of the universe. The coherence criterion particularly exposes polytheistic and mythological religions. The Norse myths, for example, name deities such as Odinn, Thor, Baldr, and Loki among a broader array of warring and unprincipled deities. Under this scheme—and countless other mythologies like it—coherence in nature becomes very difficult to explain. Different gods working against each other should result in a bacchanalia of chaos and confusion. Processes would be capricious and unpredictable. If the polytheist were to suggest that one deity who was particularly powerful imposed an unchanging order on the cosmos, then the account is effectively moving in the direction of monotheism.

Monotheism *per se* is not necessarily compatible with a coherent universe either. For example, American writer and political adviser Robert Reilley argues in *The Closing of the Muslim Mind*[22] that historically, some Sunni Muslims developed a theology that Allah's freedom and power were of primary importance, to the point that Allah was presumed to act in a capricious manner. Cause and effect were not considered to be reliable, because all events were directed by Allah in unpredictable, miraculous ways. Reilley concludes that this

---

22 Reilly, 2011.

theology has affected some parts of Muslim science, preventing it from continued progress after its early promising start. If Reilly is correct, beliefs about the nature and behavior of God are just as important as monotheism itself in explaining the coherence of the universe.

Many secular humanists believe that the origin of the world is ultimately random and chaotic. Given this, the accordance of human reason with the behavior of nature strikes them as a conundrum. Nobel prize-winning physicist Eugene Wigner admitted this in his remarkably honest article "The unreasonable effectiveness of mathematics in the natural sciences".[23] Wigner conceded that humanist scientists cannot explain why physics is understandable in mathematical terms. Albert Einstein was of the same opinion, stating that "the eternal mystery of the world is its comprehensibility ... The fact that it is comprehensible is a miracle".[24] Nobel prize winning physicist Richard Feynman said something similar: "Incidentally, the fact that there are rules at all to be checked is a kind of miracle; that it is possible to find a rule, like the inverse square law of gravitation, is some sort of miracle. It is not understood at all, but it leads to the possibility of prediction ... why nature is mathematical is, again, a mystery".[25]

A more recent example of a baffled atheistic scientist is given by K. Scott Oliphint, a professor at Westminster Theological Seminary. After Oliphint had delivered a speech at a university about the Bible and God's revelation through creation, a man approached him at the back of the lecture hall. The man turned out to be a physics professor at the university, and an atheist. He candidly informed Oliphint:

> What struck me about your talk is that as I teach physics as an atheist, I have no way to affirm or argue why the laws of physics are the way they are. All I can teach are the laws.[26]

The physics professor had recognized, perhaps for the first time, that his presuppositions provided no foundation that guaranteed the reliability, regularity, or predictability of natural laws. He could state the laws but not explain why they worked. Oliphint's speech had alerted him to a limitation of the atheistic worldview.

Physics is arguably among the most mathematical and empirical of sciences, and all four of the physicists cited above were keenly aware that their non-theistic worldview provided no basis for physics to work.

Summarizing the third criterion, the Aristotelian and secular humanist accounts both fail here; neither explains coherence. Proponents of polytheistic accounts need to explain how competing deities lead to coherence; in our view,

---

23 Wigner, 1960.

24 Einstein, 1979, p. 292.

25 Feynman, 1998, pp. 22–23.

26 Oliphint, 2017, p. 157.

this type of theology posits processes which intuitively act against coherence. The only remaining accounts in contention from here on are the Platonic and the Christian.

(A criticism commonly levelled against the coherence of the Christian worldview concerns the problem of suffering: if God is all powerful and all loving, why does He allow suffering? Either He is not sufficiently powerful to resolve suffering, or He is too callous, or unloving, to care. We address this question in Appendix A.)

The fourth criterion: The problem of consciousness

Consciousness refers to processes of the mind which result in an awareness of surroundings, awareness of others, and a sense of free will. For the Christian, consciousness is easily explained. God, the Creator, possesses consciousness and has imparted it to various life forms, including humans, who are created in His image. It is not much more difficult than that.

For the secular humanist, accounting for consciousness is "arguably the hardest problem facing science today and the only one that remains almost impenetrable even after two and a half millennia of deliberation".[27] Apparently immaterial processes control our material bodies. The vexing task for humanists is to explain what physical processes result in consciousness, and how consciousness can cause changes in the physical world. Drop a stone, and it reliably falls to the ground under the influence of gravity. But humans, exhibiting the immaterial phenomenon termed consciousness, may hurl the rock into the air, smash it to make a tool, or place one rock upon another to construct a building. Although conscious beings are made of the same stuff as the rest of the universe, their actions do not obey the laws of chemistry and physics in a simple mechanistic fashion.

The materialist view posits that everything forward from the Big Bang, including the emergence of life and the evolution of species, can be explained in terms of the physical and chemical properties of matter at the most fundamental level. However, consciousness shows this explanation to be glaringly incomplete. If matter gradually self-assembled into a primeval life form, which in turn evolved into a variety of higher life forms—we are confronted at the end of this process by "a ghost in the machine", to use philosopher Gilbert Ryle's phrase. The appearance of this "ghost", the immaterial arising from material processes, is not plausible. We know subjectively that our consciousness is not material—we cannot see it or touch it. Consciousness appears as a warning that there is something fundamentally lacking in the materialistic view of life, and indeed, philosopher Thomas Nagel argues at length that a failure to explain consciousness "threatens to unravel

---

27 Davies, 2020, p. 184.

the entire naturalistic world picture".[28] The Christian immediately responds that God is missing from the picture. But the secular humanist, who is committed to an atheistic worldview, does not have that option. Consciousness must be explained purely in terms of natural causes.

The result of this is "a voluminous and intricate literature" that has grown up around the problem of consciousness, one that "serves mainly to confirm how intractable [these problems] are".[29] Biophysicist Werner Loewenstein has expressed an even starker view: "Regarding the nature of the very foundation of mind, consciousness, we know as much as the Romans did: nothing".[30] This raises the question of whether science reaches an inflection point in the study of consciousness—is the human brain able to unpick its own processes? It may well be that a comprehensive understanding of consciousness is beyond the reach of human enquiry.

Nevertheless, we cannot definitively rule out a future scientific understanding of consciousness; a path for fruitful research may be discovered. Nagel, for example, proposes a fundamental shift in humanist origins thinking toward a *teleological* account. This word, derived from the Greek *telos*, meaning goal, refers to an explanation which is goal-directed. But this type of explanation heads in the direction of the Christian explanation, as God created the universe with an end goal in mind. Goals themselves imply foreknowledge, which the Bible explicitly attributes to God (Acts 15:18). Indeed, evolutionary theory is specifically formulated to exclude any overall goal or purpose. Whatever happenstance change in DNA best suits reproductive success, in whatever environment happens to be present, is what is more likely to be passed on to the next generation.

Paul Davies, an English physicist who has written extensively on origins-related topics, eschews teleological explanations for such reasons. He sees the issue finally as this: "Is life built into the laws of physics? Do those laws magically embed the designs of organisms-to-be?"[31] While admitting that the known laws of physics result in a negative answer to both questions, Davies wonders whether "new state-dependent informational laws" may be discovered which somehow facilitate the emergence of complex information-managing systems. Like Nagel, Davies sees a fundamental lack in our current understanding of physics and chemistry but proposes that matter possesses life-friendly properties.

Nagel and Davies, as non-Christian writers, both conclude that known physical laws are insufficient to explain conscious life. Whether science eventually invokes teleological (goal-directed) processes or discovers life-

---

28 Nagel, 2012, p. 35.
29 Ibid., p. 41.
30 This quote, by Werner Loewenstein, is cited by Paul Davies as the epigraph to Chapter 7, *The Ghost in the Machine* (Davies, 2020, p. 184).
31 Davies, 2020, p. 217.

friendly properties of matter that are unknown at present, neither can say for sure. Both were once committed to the belief that a complete scientific understanding of life in terms of the properties of matter is possible in principle and may ultimately be achieved. However, the evidence has moved Davies away from this position.[32]

Rounding out the fourth criterion, the Christian account passes with flying colors. The secular humanist account is unable at present to provide an explanation for consciousness, although we do not rule out the possibility that future advances in understanding will solve this problem. The Platonic scenario holds up—if we are lenient—in that the god without a beginning could credibly have imparted consciousness to living beings.

This brings us to our final criterion. The explosion of scientific understanding has given us a plethora of tools to investigate the world around us. Over the last few centuries, there has been a great expansion of knowledge regarding our world and the living organisms it contains. We propose that any credible origins account must be compatible with scientific knowledge. If the origins account consistently flies in the face of scientific knowledge, then the account cannot be correct and must be discarded.

The fifth criterion: Compatibility with science

Evidence from the observational discipline of cosmology is believed by many to strongly suggest that the universe began 13.8 billion years ago. Otherwise stated, the universe has existed for a finite time and had a beginning. (This overall result matches the Genesis account of the universe beginning at a finite time in the past.) But we can go further: scientific evidence shows that outer space is expanding. Not only is it expanding, but the expansion is accelerating. This strongly suggests that the origin of the universe and subsequent expansion is a one-off process, rather than cyclical. This renders the Big Bang–Big Crunch scenario unlikely.

A further line of evidence pointing towards a finite beginning of the universe is provided by the second law of thermodynamics. This law states that all processes in the universe (such as radiant heat emanating from a star) increase the overall entropy[33] of the universe; total entropy never decreases.[34] If one imagines moving backwards in time, total entropy decreases as one backs up. At some point a minimum entropy state is reached, beyond which one cannot backtrack further. Hence, the world must have had a beginning.

What about the Platonic scheme? It turns out that much of Plato's philosophical methodology was compatible with Christianity, and indeed was Christianized by the early church. Hence, parts of the Platonic scheme that were

---

32 Artigas, 1995.

33 Entropy is a term related to, but not quite the same as, the amount of disorder in a system.

34 Hoyle & Wickramasinghe, 1981, pp. 131–132.

seen as correct were merged into trinitarian Christianity. All orthodox forms of Christianity (including Protestant forms) contain elements of Plato's ideas. But Plato's idea of an eternal universe and demiurge were rejected in favor of the Genesis 1–11 account.[35]

Infinity is outside the scope of empirical science

We saw earlier that all credible origins accounts must incorporate the concept of the infinite. This applies specifically to two key areas—explanations for existence, and explanations for biological information. Before proceeding any further with our investigation of the secular humanist cosmological account, we need to unpack the implications of this important concept.

The concept of infinity (and its converse, the infinitesimal) has proved enormously useful in mathematics, but it is not a direct part of empirical science.

To demonstrate, consider the classic experiment, beloved of junior high school science teachers, that illustrates the conservation of matter. The students are asked to measure out a quantity of salt (say, 20 grams), and a particular mass of water (perhaps 400 grams, or 400 ml). The students then dissolve the salt in the water, causing the salt to visibly disappear. Has matter been annihilated? No—the students weigh the salt solution and find that it checks out at 420 grams. The salt, although no longer visible, nonetheless adds to the mass of the solution. Matter has been conserved.

Let's re-run the same experiment, this time with an infinite quantity of water (ignoring the practical difficulties for the moment). To this infinite mass of water, the students add their 20 grams of salt and dissolve it. What is the resultant mass? It is infinite—because infinity remains infinite when something finite is added. The weight of the water was infinite at the start, and the weight of the salt solution at the end is still infinite. Under these circumstances, the mathematics—and the science—do not give us the answers we expect.

Now consider the practical difficulties. Imagine setting up an apparatus to weigh infinite mass. A moment's reflection is enough to realize that it cannot be done. In fact, whether mass, distance, time, or any other scientific metric, infinity cannot be measured or numbered. This puts it beyond the scope of empirical science.

This raises an important point regarding the basis of origins accounts, because origins accounts, unlike empirical science, must directly incorporate the concept of the infinite:

- All credible origins accounts must directly incorporate the concept of an entity that in one or more characteristics is infinite, for example,

35 In any case, most of the ancient Greeks, including Plato, were polytheistic, and the physicist and science historian Pierre Duhem concluded that "the failure of Greek science was due to the influence of such theological doctrines as the divinity of the heavens and the eternal recurrence of all ..." (Jaki, 1978, p. 13).

the entity with no beginning (which in the Christian worldview is the Creator God).

- The infinite is not a concept directly within the realm of empirical science.
- All credible origins accounts incorporate concepts outside of the realm of empirical science.

This is an interesting result, and one that deserves to be shouted from the rooftops: the origins debate *cannot deal purely in empirical scientific evidence.* Origins accounts must ultimately be based on some non-scientific concepts. We already knew that of course, because presuppositions—essential to science and all origins accounts—are not scientific in themselves. But what we see here is that the key categorical distinction between origin account infinities concerns whether the infinity is personal (i.e. the God of the Bible) or impersonal.

The existence of God

Having come to the end of our five tests and found that the Christian origins account is the only realistic contender, a secular humanist could raise an objection: what evidence is there that the biblical God exists?

Several arguments have been proposed that attempt to address this issue. We will briefly consider one of them, the *Kalam cosmological argument.* This commonly cited philosophical argument for the existence of God is expressed in three propositions:

1. Everything that has a beginning has a cause.
2. The universe had a beginning.
3. Therefore, the universe had a cause.

Our first impression is that this looks unconvincing. Granted, the argument withstands scrutiny from a logical point of view, but we would be astounded if it has ever convinced a single atheist to switch to theism. It is probably useful only insofar as it provides philosophical support for theism which existing believers may find reassuring. However, we find this argument problematic for two reasons. The first is a technicality: the cause referred to in the third proposition is not necessarily the God of the Bible; it could be a non-personal entity. Therefore, the Kalam argument is not necessarily theistic. Moreover, it is not distinctively Christian; the word *Kalam* refers to a branch of medieval Aristotelian Islamic theology from which the argument originated; hence the use of the Arabic word *Kalam.*

The second and more fundamental reason is our general concern with philosophical arguments for the existence of God. There are a range of such arguments, many of which have been in circulation for centuries. Anselm of

Canterbury (1033–1109) devised the ontological argument[36] at around the same time that the Islamic scholar al-Ghazali developed the Kalam cosmological argument. Others are brilliantly written—C.S. Lewis' exposition of the argument from Natural Law,[37] in the book *Mere Christianity*, is regarded as a classic of Christian apologetics. But the issue with all of these is that they rely solely on human reasoning. For every C. S. Lewis putting forth brilliant reasoning for God's existence, there will be an equivalent Bertrand Russell arguing the reverse just as persuasively.[38] As we saw in the previous chapter, the existence (or non-existence) of God cannot be irrefutably proven by abstract human reasoning alone (although its immensely strong likelihood can be assessed by considering real-world evidence as we demonstrate in Parts 2, 3, and 4 of this book). In fact, these arguments concede the territory of the debate to secular humanists: the demand for a well-reasoned argument for God's existence stems from the atheistic presupposition that we live in an impersonal universe. In such a universe, knowledge is attained only by scientific or logical enquiry, and never by Divine revelation.

The Christian should directly attack the foundations of this latter idea. God exists, and God is a person. In personal interactions, information is there for the asking. If you meet a new person and are interested in finding out what kind of music they like, you would not scan their auditory cortex; you would ask them. If you were hosting a dinner party and wanted to find out your guests' dietary requirements, you would not request them to provide results from a skin prick test; you would ask them. Similarly, when we are trying to establish the existence or otherwise of the personal God of the Bible, we do not turn to scientific or logical processes; we ask Him. In Acts 17, Paul declares to the non-Christian Athenians that God desires "that they might seek God, and perhaps they might reach out and find him" (17:27). The Bible does not merely postulate God, it invites us to reach out to Him and promises that He will respond to the faithful seeker.

How can the atheist pray to a God they do not believe in, but who is objectively there? We find an instructive example in Mark's gospel. In the lead up to this example, Jesus had healed large numbers of people (6:54–56). Then, in Mark 9, a father with a desperately unwell son approaches Jesus seeking healing for the boy. "... if you can do anything, have compassion on us and

---

36 The ontological argument for the existence of God states that "when one thinks of God, one is thinking of that than which no greater can be thought. The question is then, is it possible to think of that than which no greater can be thought as not existing? Clearly not, for then an existing being would be greater than it. Therefore, by definition, the idea of that than which no greater can be thought includes its own existence" (González, 2010, p. 370).

37 The argument from Natural Law posits that physical laws, biological laws, and a moral law (the sense of right and wrong) impact on human actions. The moral law is roughly universal for humans but is often disobeyed. Lewis argues that the existence of a widely shared concept of a moral law strongly suggests there is a mind behind the universe – identified as God (Lewis, 2012, pp. 3–32).

38 Bertrand Russell (1872–1970) was a British intellectual, contemporary with C.S. Lewis, whose 1927 essay "Why I am not a Christian" became an influential atheistic text.

help us," he begs (9:22). Jesus replies: "If you can? Everything is possible for the one who believes." (9:23).

In our mind's eye, we imagine all eyes turning to the father. Does he believe? Everyone else was simply being healed; but Jesus challenged this man about belief. Perhaps there was something different about him. The man blurts out, weeping: "I do believe; help my unbelief!" (9:24).

The man had believed enough to give Jesus a shot. We now see him claiming to believe while simultaneously confessing to unbelief. Presumably his faith was very low, almost non-existent, *but it was enough for God to work with.* This passage is the atheist's best friend—if only they knew it—because it shows that God is willing to work with people who completely lack a *sensus divinitatis*—whose faith could almost be characterized as unbelief, other than the willingness to give God a try. Accordingly, we advise atheists to pray along the lines of, "Jesus, if you are out there as the Christians say you are, reveal yourself to me!"

An atheist may respond to this in one of several ways. First, they may refuse point blank to seek God. If so, they should not expect to receive anything. As Jesus said, "Ask, and it will be given to you. Seek, and you will find. Knock, and the door will be opened to you" (Luke 11:9). To forego asking, seeking or knocking is to forego receiving or finding.

Another potential response is to pray. Of course, if this is done in a prideful or mocking way, no answer should be expected. The proud will be resisted by God (1 Peter 5:5).

Still another possible response is a genuine prayer that seeks God in a faithful way. This will be met with God's favor, and God will reveal Himself in some way to these kinds of seekers.

But perhaps a more honest prayer for the atheist might be, "God, I hate what you stand for, I hate the idea that you hold me accountable, but I'm reaching out anyway—what will you do in response?" Our sense is that it's preferable to reach out to God honestly in this life than to pass into eternity without reaching out at all.

But as for textbook-style proofs, no. There are none. Secular humanists, in denying a personal God, have an *a priori* commitment to impersonal origins explanations, and therefore expect an impersonal proof of God's existence—of the same kind that you would find in a mathematics textbook. The Christian should avoid that territory, and instead steer secular humanists to the personal, to the God of the Bible. Seeking God faithfully is their responsibility—as it is for all of us—in our capacity as individuals. And faithful seekers will be provided with all the evidence they need for God's existence.

Conclusion

In this chapter we presented a framework for analyzing origins accounts. Of those we considered, other than the Christian account (and the Platonic approach, elements of which were folded into Christianity early on), all accounts have one or more problems that need to be addressed before they can be considered as viable alternatives. In our view, the onus rests with proponents of origins accounts from polytheistic religions to explain how they do not fail the coherence test. The orthodox scientific account, beginning with a Big Bang, struggles with the existence test because it does not clearly delineate which entity has no beginning. Similarly, the Aristotelian and other accounts all fail one or more tests.

Nevertheless, the "scientific account" is the one held as correct by government, educational, and entertainment sectors in the West. Engaging in the origins debate therefore requires a robust understanding of what science is. "Science", it turns out, is an umbrella term encompassing several distinct lines of enquiry. We will investigate these in detail in the next chapter.

References

Artigas, M. (1995, June 28). How to become a millionaire by talking about God. Paul Davies, award Templeton 1995. Retrieved from en.unav.edu/web/ciencia-razon-y-fe/como-hacerse-millonario-hablando-de-dios-paul-davies-premio-templeton-1995

Clodd, E. (1897). *Pioneers of Evolution from Thales to Huxley*. London: Grant Richards.

Davies, P. (2020). *The Demon in the Machine*. London: Penguin Books.

Doyle, S. (2016, April 30). Plato and Christianity. Retrieved January 26, 2024, from creation.com/plato-and-christianity

Dreyer, J. (1953). *A History of Astronomy from Thales to Kepler* (Second edn). New York: Dover Publications Inc.

Einstein, A. (1979). *Ideas and Opinions*. Calcutta: Rupa Paperback.

Feynman, R. P. (1998). *The Meaning Of It All: Thoughts of a Citizen-Scientist.* New York: Basic Books (Kindle edn).

Gitt, W. (2020). The creation hyperbola. *Creation*, 42(3), 17.

González, J. L. (2010). *The Story of Christianity — The Early Church to the Dawn of the Reformation* (Vol. 1). New York: HarperCollins Publishers.

Holt, J. (2012). *Why does the world exist? An existential detective story*. New York: Liverlight Publishing Corporation.

Hoyle, F., & Wickramasinghe, N.C. (1981). *Evolution from Space: A Theory of Cosmic Creationism*. New York: Simon & Schuster Inc.

Hoyle, F. (1981, November). The Universe: Past and Present Reflections. *Engineering and Science*. Retrieved from calteches.library.caltech.edu/3312/1/Hoyle.pdf

Hume, D. (1999). *An Enquiry concerning Human Understanding*. (T. L. Beauchamp, Ed.) Oxford: Oxford University Press.

Jaki, S. L. (1978). *The Road of Science and the Ways to God*. Chicago: The University of Chicago Press.

Krauss, L. M. (2012). *A Universe from Nothing: Why there is Something rather than Nothing*. New York: Simon & Schuster.

Lewis, C. (2012). *Mere Christianity*. London: Collins.

Nagel, T. (2012). *Mind and Cosmos: Why the Materialist Neo-Darwinian Conception of Nature is Almost Certainly False*. New York: Oxford University Press.

Oakley, H. D. (1926). The Religious Element in Plato's Philosophy. *International Journal of Ethics*, 67–80.

Oliphint, K. S. (2017). *Know Why You Believe*. Grand Rapids, MI: Zondervan.

Reilly, R. R. (2011). *The Closing of the Muslim Mind: How Intellectual Suicide Created the Modern Islamist Crisis*. Wilmington, Delaware: ISI (Intercollegiate Studies Institute) Books.

Wigner, E. (1960). The unreasonable effectiveness of mathematics in the natural sciences. Richard Courant lecture in mathematical sciences delivered at New York University, May 11, 1959. *Communications on Pure and Applied Mathematics*, 13(1), 1–14.

Zeyl, D., & Sattler, B. (2023, September). Plato's Timaeus. (E. N. Zalta, & U. Nodelman, Eds.) Retrieved January 26, 2024, from plato.stanford.edu/archives/fall2023/entries/plato-timaeus/

3
The tool called science

The commonplace belief that "science has disproved God" has become almost an article of faith for the masses today. Picking out an older book (published in 1962) from our shelves at random, we read that the author has "shed beliefs that were turned to myths by Darwin and the science men".[1] The biblical creation account, so the reasoning goes, was considered rock solid in an earlier, more ignorant age, but has crumbled to the dust of myth under the onslaught of modern science. In short, science has disproved God.

Ironically, this belief betrays a misunderstanding of the role and scope of science. Those that wish to understand the origins debate must also understand what science is. The task of this chapter is to set out the essentials, starting with the various categories of science.

Categories of science

Observational science

As its name suggests, observational science refers to collecting data and information via observation, as well as naming and classifying natural phenomena. A well-known example of this is the work of the Swedish naturalist Carl Linnaeus (1707–1778). By carefully observing the structure and function of different types of living organisms (for example, whether the animal had an even number of toes), Linnaeus devised classification systems for the animal and plant worlds. His classification systems were hierarchical, starting with kingdoms and descending through classes, orders and genera to species, the lowest level of the classification. Linnaeus introduced binomial naming (binomial meaning two names) which used both the genus and species names to uniquely identify each species. This naming system has remained in use ever since; humans, for example, are known as *Homo sapiens*.

Observational science has a very long history, stretching back to the dawn of time. The ancient Israelite king, Solomon, described plants and animals (1 Kings 4:33). But even Solomon's work was not without precedent; the earliest chapters of the Bible record Adam naming the animals (Genesis 2:19). The importance of naming becomes clear when we engage with the biblical text. For example, lions are mentioned in the Bible, and modern readers have a clear understanding of what the biblical author meant. On the other hand, creatures

1 Simpson, 1962, p. 5.

with names transliterated as *behemoth* and *leviathan* are mentioned in Job (40:15 and 41:1), but it is not immediately clear what animals are referred to. In this case, the description of the animal is more informative than the specific name in elucidating what kind of animal these names might refer to.

For this reason, accurate description of observations is essential. As science progressed, researchers discovered phenomena for which no words existed. Therefore, an important aspect of observational science is the coining of new words to name these phenomena. Linnaeus's binomial naming system is a good example from the biological field. Chemistry is another field rich in new words. As new elements were discovered, scientists had to coin new names, and this was frequently done by combining two Greek words to form a new English word. For example, when oxygen was discovered, Antoine Lavoisier named it by combining *oxy* (Greek for "sharp") and *gen* (Greek for "producing"). Lavoisier chose these words because he believed they described, to some extent, the properties of this new element.

Observational science may involve logic and mathematics. An example of this is Tycho Brahe's (1546–1601) very accurate naked-eye observational data of the movement of planets, which were recorded numerically.

Descriptive model science

The next step from observational science is descriptive model science—this describes *what* is happening (without explaining *why*). This summarizes, or describes, changes occurring in a system under study—often using mathematics or geometry—and allows predictions to be made. A simple example of this is the calendar. This summarizes the movement of the sun and allows predictions of future solar positions, without providing a causal explanation for the movements. Calendars have been used for these purposes since ancient times. Another example, related to calendars, is the Ptolemaic cosmology of the ancient Greeks. Ptolemy's (100–170 AD) model envisaged the earth at the center of the universe with the distant stars located in a spherical shell surrounding the earth. The planets were thought to be between the earth and the stellar shell. Ptolemy's model can be made arbitrarily accurate in its description and prediction of the planetary motions, although we now know it does not accurately account for *why* the planets move as they do (it is not a *causal model*—the third kind of science we discuss below).

Copernicus (1473–1543) proposed a fundamentally different descriptive model in which the planets orbited the sun in circular orbits. While this was a step towards a causal model, it was not a significant step forwards in accuracy compared with advanced Ptolemaic models, because, as we now know, planetary orbits are not circular. Later, Johannes Kepler (1571–1630) discovered that elliptical orbits around the sun afforded a better match with Tycho Brahe's data on planetary motions. He did not propose causally why this was the case, even though he was able to model the planetary motions with mathematical accuracy and make predictions as to future positions. This was simply a better descriptive model than the Ptolemaic system and an improvement on the Copernican model, but Kepler had not made the leap to our next category, causal model science.

Causal model science

This type of science identifies causal mechanisms which explain *why repeatable phenomena occur the way they do.* (If these explanations are later found to be incorrect or incomplete, that science is demoted to descriptive model science—because it is now seen to no longer accurately explain why the phenomena occur the way they do). The example that follows on from Kepler's elliptical orbits is Newton's (1642–1727) law of gravity and the first two of his laws of mechanics, which allowed the concepts of gravity, inertia, and acceleration to model and explain the planetary motions. These equations gave results as accurate as Kepler's as well as improving scientists' understanding of the processes and widening the areas to which the theory could be applied. This was a major advance in astronomy.

Causal models are very useful because they summarize a lot of experimental and observational data. Predictions based on causal models can in turn be tested against experimental or observational data. Causal models can also stimulate the invention of new theories extrapolated from existing models. Any causal model which does not provide accurate predictions should be discarded.

These three classes of science do not necessarily encompass all of science, but they are sufficient for the purposes of this book.

The lay person typically only interacts with descriptive or causal model science via popular books on science. But the general public benefits tremendously from science through our next category, applied science. The applications of science include technology and forensic science.

Applied Science: Technology

Technology refers to the application of knowledge to provide utility for people. A well-known example of technology is the wheel, use of which predates the scientific revolution. Some technologies are improved upon. In the case of the wheel a late nineteenth century improvement was John Dunlop's invention of the pneumatic tire, which gives wheeled vehicles a smoother ride.

A characteristic of technologies is that they are subject to obsolescence. For example, a person in St. Joseph, Missouri, in 1860 wishing to send an urgent message to someone in California may well have sent a letter by the Pony Express. A couple of years later, a telegraph wire connecting the eastern and western U.S. was in operation, rendering the Pony Express obsolete. It was now possible to get a message into your colleague's hands within a few hours. In the years that followed, many more cities were added to the telegraph network, enabling rapid and cost-effective communications between almost any two points in the continental United States, and eventually, across continents. The telegraph's useful life continued for more than a century. Our grandparents were sent a congratulatory telegram from Queen Elizabeth II in the 1980s on their 60th wedding anniversary. The introduction of facsimile rendered the telegram obsolete by enabling the near-instant exchange of written communications; but the fax machine, in turn, was rendered obsolete by email.

Technologies generally come into widespread use then fade out in a gradual process. As a new technology becomes available, early adopters eagerly make use of it. Later, a larger cohort of adopters makes the technology commonplace. As part of this process the previous technology is phased out of use and may be exhibited in a museum even though it is still functional. In transport technology, steam engines have been obsolete for decades but historical steam trains still chug along today, offering safe and reliable experiences for enthusiasts.

Applied Science: Forensic science

We now turn to an application of science which is of particular interest to our topic: forensics. This uses the other branches of science to reconstruct one-off past events, usually for presentation in a court of law. It allows the scientist to become an "expert witness" in relation to a crime. Archaeologists, paleontologists and biologists also use science forensically when they investigate past historical events. In fact, there is considerable overlap between the skill set of paleontologists and detectives; hominid researcher Louis Leakey used his forensic skills to assist the Kenyan police in solving several cases in the 1940s.[2] Forensic scientists, like detectives, also face similar issues in their gathering of data—events further back in time may be affected by degradation of samples or documents or even outright loss of data. The ability to "triangulate" (establishing multiple independent forensic evidences that point to the same conclusion) helps make conclusions more secure.

Forensic evidence can often be interpreted in more than one way, permitting several different explanations of a past event. To illustrate this crucial point, we will present the disappearance of flight MH370 as a case study.

Forensic case study: Flight MH370

How and why Malaysian Airlines flight MH370 disappeared in 2014 remains an unsolved mystery, as of publication of this book. The late investigative journalist Ean Higgins explored several possibilities in his book *The Hunt for MH370*,[3] including an onboard fire, a terrorist hijacking gone wrong and a rapid

---

2 Berger, 2000, p. 122.

3 Higgins, 2019.

decompression. But in Higgins' view, the most likely explanation—indeed the correct explanation, he thought—is that the captain of the flight, Zaharie Ahmad Shah, hijacked his own plane and deliberately flew it until it reached fuel exhaustion over the remote southern Indian Ocean. In this interpretation of events, Zaharie allegedly murdered the passengers and crew, taking them to a watery grave some 2000 kilometers west of Perth.

Higgins cites a slew of aviation experts that support his view. But he also points out that the notion that an experienced Malaysian pilot could not be trusted with a passenger plane would be a source of considerable embarrassment to the Malaysian government, not to mention the airline. At all events, the government agency responsible for the undersea search, the Australian Transport Safety Bureau (ATSB), based its 2015 search parameters on the theory that the flight crew were incapacitated in some way. This is an extremely important distinction, Higgins notes, because it determines where the aircraft might be expected to have come to rest, allowing a search zone to be defined. The intention of the pilot could be critical in assessing how far the plane might have continued after it ran out of fuel, as Higgins explains:

> If the ATSB is right, the aircraft came pretty much straight down after it ran out of fuel, producing a relatively narrow search zone. [However, if] Zaharie flew the aircraft to the end and ditched it, either under power or gliding it after he ran out of fuel, he could have taken it a much longer distance, perhaps 100 nautical miles or more according to the ATSB, and well outside the search area the bureau defined.[4]

With limited resources and a vast tract of ocean to search, the ATSB needed to make an informed decision about where the doomed plane was most likely to be. If the search parameters were incorrect from the outset, possibly due to flawed assumptions, the hunt could never have succeeded.

As the investigation commenced, officials originally searched in the South China Sea because the plane's final transponder signal was sent from there. Later, new evidence came to light, showing that the plane had turned back over Malaysia, and must have been somewhere along two arcs, one in the southern Indian ocean, the other over south-east and central Asia. Eventually, new data—primary and secondary radar data, and complex calculations based on automatic satellite communications made from the plane—corroborated the theory that MH370 had ended up in the southern Indian Ocean.

As mentioned above, forensic evidence is regarded as circumstantial because it often supports multiple interpretations. This is precisely what happened in the MH370 disappearance. Aviation experts had access to the same basic data, but different experts offered different interpretations. Ean Higgins

4 Higgins, 2019, p. 4.

personally supported the "rogue pilot" interpretation, but, as a seasoned journalist, was careful to preface his outline with this disclaimer:

> [L]ike other theories to be outlined later in the book, [this scenario] is a dramatized reconstruction based on fact, detective work, deduction, and informed speculation. It is not purporting to be what happened, but one version of what *could* have happened consistent with what is known about the flight and the aircraft.[5] (Emphasis in original.)

Another scenario Higgins includes in his book is the "terrorist hijacking gone wrong" theory, which was supported by a former transport accident investigator turned university professor. This expert claimed that not only does this scenario fit the data, but it is also the only scenario which "doesn't require a string of implausible simultaneous unrelated failures and errors".[6]

Yet another expert promulgated the "elopement by parachute" theory, which has the captain of MH370 bailing out of the aircraft near a waiting fishing vessel owned by his mistress's family. As far-fetched as this sounds, its proponent, a former manager of Qantas flight training, who was also a former board member of Australia's air traffic control authority, points out that "there is, in fact, a precedent in aviation history for every key element required to make [the] theory seem plausible".[7]

This brief outline of the MH370 case reveals several similarities with origins research. First, initial evidence supported one theory, which new evidence later ruled out. A similar example from origins research is the theory of the "spontaneous generation" of life, based on observations of flies appearing on carcasses, beetles appearing on animal dung, and the like. Discoveries by Louis Pasteur in the nineteenth century showed that "life comes from life", finally disproving the long-held belief of spontaneous generation.

Second, the forensic investigation is based on data which was generated at a past time; such data may be degraded or incomplete by the time investigators collect it. A parallel to this from origins research is the fossil record, which purportedly supports evolution, but which is widely acknowledged to be only a partial record of past life.

Third, the forensic investigation involves a one-off past event. Repeatable experiments are not possible. One example of this effect from origins research is the beginning of the universe, for which there are a plethora of relevant scientific observations, even though the event itself can never be repeated. In fact, all purported events in the secular history of life fall into this category, including the origin of life from non-living matter (termed "abiogenesis"), and the elements of the Darwinian "evolutionary tree", including the alleged

5 Ibid., p. 17.
6 Ibid., p. 173.
7 Ibid., p. 72.

evolution of man from ape-like ancestors. None of these examples are observable, but merely proposed.

Next, we saw that forensic evidence can admit of a variety of interpretations: all the MH370 scenarios included in Ean Higgins' book fit the data. In secular origins research, the fact that the data could support a variety of interpretations is rarely mentioned. Instead, the invincibility of science is loudly proclaimed. An origins scientist who did admit this was cosmologist George F. R. Ellis, who wrote in *Scientific American* in 1995 that

> People need to be aware that there is a range of models that could explain the observations ... For instance, I can construct you a spherically symmetrical universe with Earth at its center, and you cannot disprove it based on observations ... you can only exclude it on philosophical grounds. In my view there is absolutely nothing wrong in that. What I want to bring into the open is the fact that we are using philosophical criteria in choosing our models. A lot of cosmology tries to hide that.[8]

A final but significant similarity is that forensic investigations can include potential solutions which the establishment finds convenient to ignore.[9] Higgins postulated that the ATSB may have eschewed the "rogue pilot" scenario—despite expert evidence that it was the best fit—to avoid embarrassing the Malaysian government. Some members of the contemporary science establishment may similarly eschew a Christian origins explanation, because—at core—the Christian God is an affront to secular humanists. And, with the Darwinian conception of origins research broadly accepted, and senior Darwinists long since promoted into positions of influence, funding of origins research will invariably flow in the direction that satisfies these authorities.

The importance of eyewitness evidence

We noted above that forensic evidence in a court of law is regarded as circumstantial, as it allows for a variety of interpretations. This is precisely what we see in the MH370 investigation. As stated above, there are several plausible theories. Because of this, locating the aircraft is not an end in and of itself. It will bring closure to the families of those who lost loved ones—a very important outcome—but from the perspective of the investigators, the significance of locating the wreckage is that investigators will be able to recover the black box flight recorders.

The black box recorders, as they are known, comprise the flight recorder, which stores a range of flight parameters and inputs provided by the pilots, and the cockpit voice recorder. It is this second item that provides crucial evidence needed to solve some air crashes. (Indeed, if regulators had their way, a cockpit

8 Gibbs, 1995, p. 55.

9 We are not suggesting that the ATSB erred in its investigative approach.

video recorder would be installed as well.) The importance of the cockpit voice recorder was highlighted in the investigation of the loss of Eastern Airlines Flight 212 in September 1974. In this incident, the plane crashed into terrain several miles short of the runway. Analysis of the flight data showed what the plane was doing but could not establish why. It was the cockpit voice recorder which revealed that the crew had engaged in idle banter during the critical landing approach, and that none of the flight crew made the altitude callouts that are standard procedure during the landing phase. Instead, the cockpit voice recorder revealed that the crew were distracted and lacked awareness of the plane's altitude. The outcome of the investigation contributed to changes in FAA procedures, which mandated the "sterile cockpit rule" in 1981. This meant that on-topic conversation in the cockpit was mandated by law.

In the same way, investigators want to recover the cockpit voice recorder from MH370 in the hope that it will solve the mystery of why the plane went off course. Was it a decompression? A terrorist hijacking gone wrong? Investigators hope that only one theory will be tenable after the voice recorder is analyzed. The "hard data" and the "soft data" will form two consistent streams of evidence which align to solve the mystery.

The need for eyewitness evidence is not limited to aviation investigations. Forensic evidence is often inadequate to solve crimes. For example, Sydney police have been hampered in recent years in their efforts to solve a string of underworld murders. A media report from 2022 explained that organized criminals in Sydney refused to talk to police:

> A "code of silence" continues to frustrate police as the number of unsolved gangland murders in Sydney over the past two years rises to more than a dozen.[10]

We have no doubt that the forensic abilities of the Sydney police are world class; but the constabulary has found that "scientific data" are not sufficient to unravel the events of the past. The lack of eyewitness testimony was a common problem in these underworld investigations, as the media report indicated:

> NSW Police Minister Paul Toole ... acknowledged many of the recent killings linked to organized crime in Sydney remain unsolved months or years after they were committed.

The mystique of "science" in the origins debate relies on people not considering the real-world experience of groups such as aviation regulators, police, and law courts—groups that are responsible for reconstructing past events. In all these domains, eyewitness testimony is key, and forensic evidence scarcely rises above the circumstantial.

---

10 Parkes-Hupton, 2022.

The biblical creation account provides eyewitness testimony that secular humanism cannot. It is comparable to the cockpit voice recorder (and, in a way, a video recorder). It supplies not only a time-stamped readout of what God said, but also a descriptive account which fills out, to some extent, our understanding of how and why the world was created. When secular humanists praise the use of science in the origins debate, their rationale is that "science" is a neutral, rigorously fact-based enterprise, completely free of superstition and dogma. In fact, attempting to solve the mystery of origins with only hard data is comparable to investigating an air crash using the flight recorder—or other available measurements—exclusively, and discounting the cockpit voice recorder.

We have complete confidence that as science progresses, the theories that conflict with the biblical account will prove to be untenable. In other words, as Christians we are strong supporters of the scientific endeavor. Like Ean Higgins' belief that political influences may have affected the choice of search parameters in the MH370 investigation, we believe, and will show in this book, that elements of the science juggernaut are tilted against God, but that the more study that can be done, the better. We do not believe that science will ever get to "prove" the Genesis creation account but that as more comes to be known, the more the "cockpit voice recorder" of Genesis and the scientific account will align. Non-biblical approaches will become less and less tenable as more data emerges.

Conclusion

We conclude this chapter with a summary of the content we have covered in diagrammatic form. Figure 3.1 shows the categories of science and applied science.

The diagram is not intended to be exhaustive, but to at least illustrate the distinction between origins science and the empirical sciences. At the top are the three classes of science; descriptive model and causal model science are the empirical sciences of repeated phenomena. Causal model science seeks to explain "how it works". By contrast, forensic and origins science both involve investigations of one-off past events.

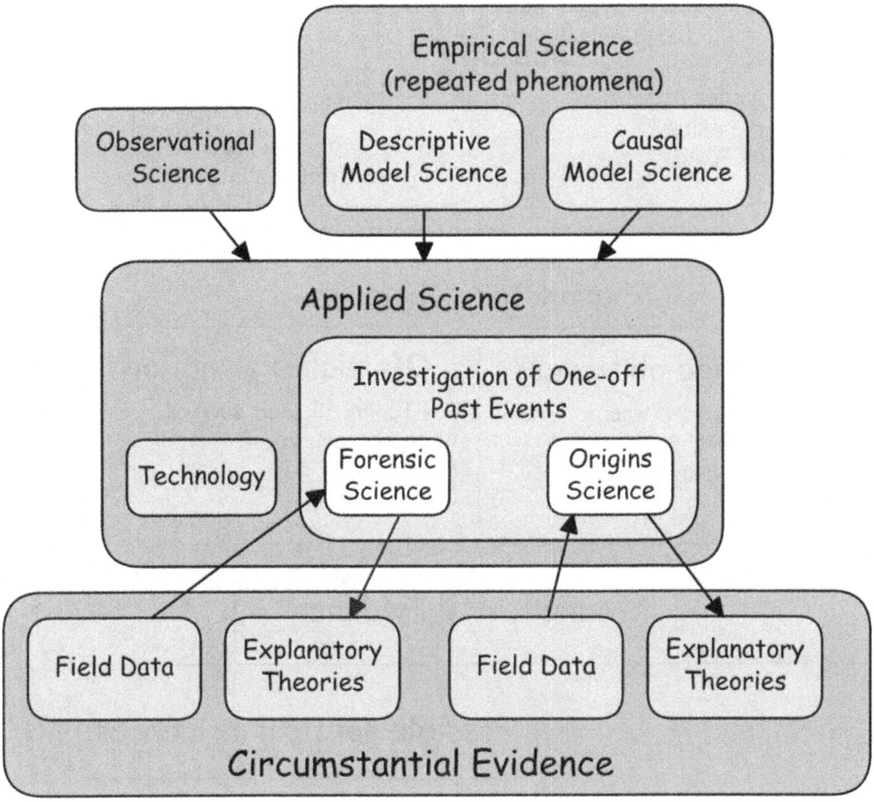

Figure 3.1: Categories of science

Both forensic and origins science utilize field data as inputs to produce explanatory theories as outputs. The field data and the explanatory theories constitute circumstantial evidence, in that conclusions from the field data must be inferred, and any number of explanatory theories which fit the data may be possible. Origins science primarily seeks explanations for "how it came to be," rather than "how it works".

Scientific research into the Big Bang, abiogenesis or holistic evolutionary theories is not empirical because it investigates one-off past events. It is forensic in nature—and thus results in circumstantial evidence.[11]
Figure 3.2 illustrates these concepts as they apply to the MH370 investigation. From the same set of field data, two classes of explanatory theories were developed. The first was what we term the "official theory", or the theory that the ATSB proposed. According to Ean Higgins, the adoption of this theory may have involved political considerations. The second class of theory includes

11 An empirical (experimental) approach can be used to analyze the forensic (or origins) data—for example, putting a carcass in water to test theories of fossilization.

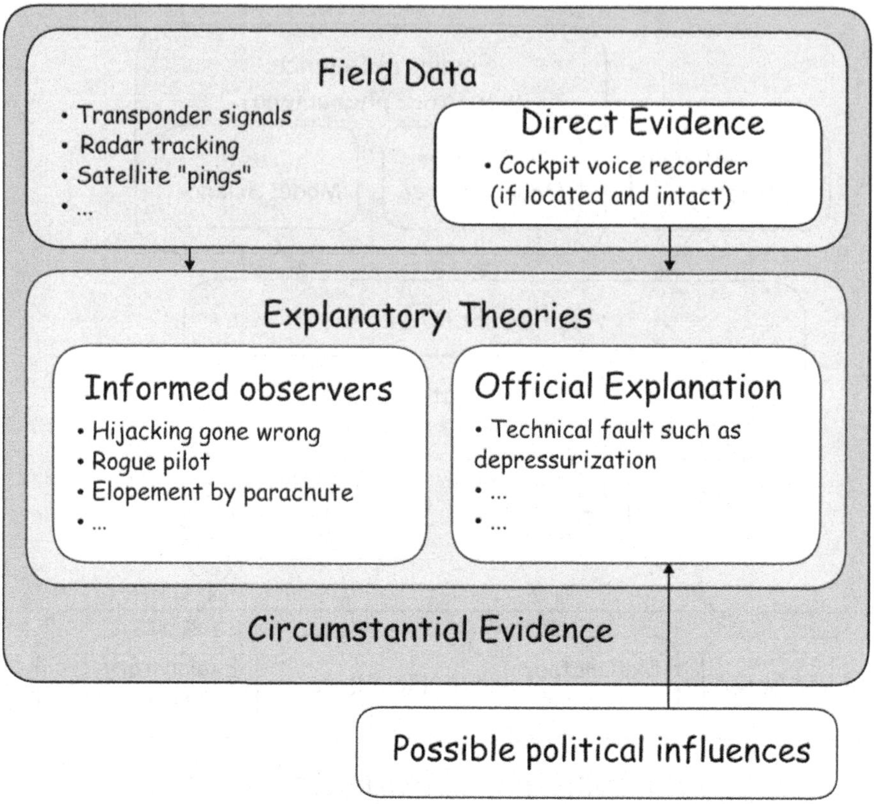

Figure 3.2: Categories of evidence in the MH370 investigation

those promulgated by informed observers. Although these theories were not officially investigated, they all fit the available data. The field data and theories have not solved the mystery of why MH370 disappeared, but investigators hope that the cockpit voice recorder, when recovered, will provide direct evidence which finally solves the mystery.

In this chapter we covered some of the categories of modern science, but further background is needed to properly understand the origins debate. In the next chapter we will investigate what philosophers and scientists have said about science and discuss two important problems that have arisen in the scientific landscape.

References

Berger, L. R. (2000). *In the Footsteps of Eve: The Mystery of Human Origins*. Washington: Adventure Press.

Gibbs, W. W. (1995, October). Profile: George F. R. Ellis: Thinking Globally, Acting Universally. *Scientific American*, pp. 50, 54–55.

Higgins, E. (2019). *The Hunt for MH370*. Sydney: Pan Macmillan Australia Pty. Ltd.

Parkes-Hupton, H. (2022, August 17). Police met with 'code of silence' as Sydney gangland murders remain unsolved. Retrieved February 2, 2023, from abc.net.au/news/2022-08-17/sydneys-unsolved-gangland-murders-continue-to-rise/101338812

Simpson, C. (1962). *Plumes and arrows*. Sydney: Angus and Robertson.

4
The case of the missing philosophy

In a 1997 article, the late paleontologist Stephen Jay Gould discussed the interaction between science and religion. In Gould's view, science and religion concern themselves with separate domains of knowledge and inquiry, and he summed up this idea with the phrase *non-overlapping magisteria*. The Latin word *magisterium* (plural: magisteria) means governance or jurisdiction (although this word is rarely used in any other context than the Roman Catholic Church, which terms the collective teaching of the popes and the bishops "the Magisterium"). In Gould's own words:

> The net of science covers the empirical universe: what is it made of (fact) and why does it work this way (theory). The net of religion extends over questions of moral meaning and value. These two magisteria do not overlap, nor do they encompass all inquiry ... .[1]

Not everyone agrees with Gould; Richard Dawkins is a well-known critic of this view, commenting that it "sounds terrific, right up until you give it a moment's thought".[2] But to properly understand the issue, we need to understand what science is. This is a philosophical question, and a good place to look for answers is in the work of philosophers of science.

We will proceed by investigating four influential philosophies of science and commenting on their relationship to the origins controversy. The first two include thoughts on the demarcation problem—how to distinguish between "science" and "non-science". The second two relate to how science progresses. Of these four philosophies, two originated from scientists, and two from philosophers.

A logical view of science: Popper

Karl Popper (1902–1994) was a philosopher of science who sought to describe how the process of science ought to work in a logical sense. As Popper saw it,[3] science was concerned with developing testable models, works of imagination that attempted to describe aspects of the behavior of the physical world. The key work of the scientist, Popper argued, was to try to falsify a model through experimentation. The more a model was tested, and survived

1 Gould, 1997, pp. 19–20.
2 Dawkins, 2006.
3 Popper, 2002.

those tests, the more scientific credence it accumulated. But even if a model survived a hundred, or a thousand, or ten thousand different tests, it was never assured of passing the next one, and, accordingly, was still at risk of falsification.

This is known as "the problem of induction". Induction refers to the process of observing specific data and then extrapolating from there to a general rule or law. The problem is that repeated observations, no matter how many, can never securely predict what might be observed in the future. The example Popper used to explain induction was the color of swans. Before Australia was discovered it was noted that all swans that had been observed were white. By induction it was argued that all swans are white—as a general rule. Later, black swans were observed in Australia. This invalidated the generalization that all swans are white.

Let's go through that process step by step. A scientist notes that all swans they have seen are white. They go to the library and search for documents about swans. All documents describe white swans. From the available data our scientist concludes, by the process of induction, that all swans are white. Subsequently our scientist becomes aware of data from Australia that some swans are black. At this point the scientist must either modify or discard the original hypothesis, which has been invalidated. Clearly his original induction was based on incomplete data. But this is almost always the case in science; we never have all possible data.

In Popper's view, arguing from specific experimental data to construct general models of cause and effect always involves some doubt, because we may not have yet carried out a specific type of experiment which shows that our general model is wrong. We can never know if some new experiment carried out at some point in the future will show that we were incorrect. For this reason, we can never say from a strictly logical viewpoint that a scientific theory is "true".

Popper was thinking about science in a rigidly logical sense. In a practical sense, however, hypotheses that have stood up to a great deal of testing unscathed are taken to be very firm. In fact, some of these well supported generalizable relationships are recognized as *scientific laws*. It is unquestionable, for example, that technology has made significant progress based on scientific theories. Mobile phones are evidence for that. But Popper was averse to the idea of induction and sought to find another logical basis for science, although arguably he ended up having to smuggle the same concept in by other means.

Popper also sought to answer the demarcation problem—namely how to distinguish science from non-science. He suggested that a theory which is theoretically impossible to falsify is not science. Popper's idea has been very influential, and indeed, the notion that unfalsifiable theories are not science has become a common belief.[4]

For example, a scientist might contend that many different universes exist entirely independent from one another, and that these other universes are entirely outside of and separate from our universe. This proposition—commonly known as the *multiverse*—cannot be tested experimentally and therefore cannot be falsified, so, according to Popper's criterion, such a theory is not scientific.

Popper's falsifiability criterion is very commonly cited in the origins debate, to the extent that laypeople may not realize this was simply the musing of a twentieth century philosopher. The reason for its popularity with secular humanist sympathizers is readily apparent. Popper's falsifiability test is used to subtly support the contention that the biblical account, which includes God's creative acts, is of no value in origins. As an example, consider the National Academies' statement in *Science, Evolution and Creationism*.[5]

> Because ... appeals to the supernatural are not testable [which includes falsifiability] using the rules and processes of scientific inquiry, they cannot be a part of science.

We wholeheartedly agree with this statement *when applied to scientific evidence*. However, if the former statement is extended to imply that because the biblical account is not part of science, it has no value in the origins debate, then we strongly disagree.

This type of implication derives from logical positivism, a now defunct philosophical school that was very influential in the early decades of the twentieth century. Logical positivism held that only ideas that were verifiable through observation or logic had informational value, a proposition that was strongly criticized by Popper for its reliance on induction. However, logical positivism was eventually discarded because it is self-refuting: the central thesis of logical positivism was itself not able to be proved by observation or logic. Nonetheless, logical positivism continues to influence the origins debate in practice, via statements from the secular humanist camp that the non-falsifiable concepts of the Bible cannot be a part of science and *consequently* are of no value in the origins debate.

In contrast with Popper's ideas, the next philosophy of science we examine, from Richard Feynman, is studiously ignored in the origins debate.

---

4 For instance, "Falsifiability in Medicine" is an article which explains the use of Popper's falsifiability criterion in evaluating evidence relating to COVID-19 treatments (Taran, Adhikari, & Fan, 2021).

5 National Academy of Sciences and Institute of Medicine, 2008, p. 39.

A skeptical view of science: Feynman

Richard Feynman (1918–1988) was a Nobel prize-winning physicist whose approach to science was one of skepticism. In a 1966 lecture he defined science as a "belief in the ignorance of experts". Science is the idea "that it is worthwhile rechecking by new direct experience, and not necessarily trusting [man's] experience from the past." Feynman's idea was not new—the motto of the world's most prestigious scientific institution, the Royal Society, founded in 1660, is *Nullius in verba*, a Latin phrase meaning "take nobody's word for it".

Feynman went on to make an important distinction. In his view, science is not synonymous with the forms or procedures used in developing science. Indeed, those forms and procedures can be imitated carefully, along with elaborate, formal scientific reporting: "We write, experiment, and observe, and do this or that", yet the result can be pseudo-science. Feynman was concerned that pseudo-scientific imitation was producing "experts" and "pseudo-scientific advisors" who were "tyrannizing" institutions falling under their influence.

Feynman sought to clarify the use of the term science further:

> When someone says, "Science teaches such and such," he is using the word incorrectly. Science doesn't teach anything; experience teaches it. If they say to you, "Science has shown such and such," you might ask, "How does science show it? How did the scientists find out? How? What? Where?"
>
> It should not be "science has shown" but "this experiment, this effect, has shown." And you have as much right as anyone else, upon hearing about the experiments—but be patient and listen to all the evidence—to judge whether a sensible conclusion has been arrived at.[6] [7]

Feynman suggests that people are free to independently verify purported scientific knowledge. Do not trust every expert, he cautions—verify the information yourself. Clearly, we do not have enough time to check everything, but we need to understand that when we have not checked, we are not dealing with science, but information which may or may not be authoritative. It is another person's interpretation of what the science is. This *is not science*, it is an opinion, and a subjective one at that. Feynman enjoins that "[i]t is necessary to teach both to accept and to reject the past with a kind of balance that takes some considerable skill." In short, Feynman tells us, check the science when you can.

---

6 The principle that Feynman is discussing is exactly analogous to the biblical principle mentioned by the apostle Paul when he praised the Bereans for checking what he said by comparing it with the scriptures. In effect Paul was saying that teaching authority can only be established by careful comparison against the Bible and a weighing of evidence—there is no infallible teaching authority in any given church simply because it proclaims there to be. We must check what ministers of the church tell us by comparing it with the words of scripture ("but be patient and listen to all the evidence," as Feynman said). In a sense, the Bible is our magisterium.

7 Feynman, 2023.

When you cannot, understand that you are on more dangerous territory and need to appraise what you are being told in a balanced way.

Feynman adds that going through the motions of "scientific processes" might look like science, but it is not science. He explained what he meant in another lecture.[8] Any science that is not underpinned by utter honesty to fellow scientists and laymen about all aspects of the scientific question lacks scientific integrity. All relevant information must be provided by the scientist, including about weaknesses, so that others can reliably evaluate the contribution, and repeat the study if necessary. Feynman noted that there are temptations to provide only part of the information, or perform part of the experiment, or to primarily seek "results" rather than truth, for reasons relating to funding, organizational issues and so forth. But he concluded that those who only provide part of the information are more akin to advertisers or public relations experts than scientists.

Feynman's concept is tantamount to a demarcation principle for science which helps to distinguish science from non-science. If science lacks integrity and utter honesty it is not science. Instead, it is pseudoscience.

This observation makes sense in a variety of contexts, because honesty and integrity underpin all sorts of human endeavors. Take as an example a chartered accountant who is employed by a large accounting firm of considerable repute. As this professional taps away at his accounting spreadsheets, any observer would be entitled to assume that the accountant was practicing "accounting." But what if the observer was told that this accountant was corrupt and was engaged in "cooking the books" on behalf of his clients—perhaps deliberately misreporting accounting data to illegally reduce a tax liability? Under these circumstances, is the person engaged in "accounting"? At best, it would amount to "fraudulent accounting"—but arguably, fraudulent accounting is not accounting at all. Feynman points out that the same principle applies to "science". In the absence of complete honesty, or full reporting of data, including the limitations of the data or experimental results, research lacks integrity and therefore is not science. Perhaps it is "fraudulent science", or perhaps it is propaganda.

In approaching the origins debate, laypeople would do well to bear Feynman's critique in mind as they consume science journalism. As we saw in the previous chapter, forensic science can be susceptible to political influence. And in origins science, a branch of forensic science with far-reaching societal implications, the temptation to conceal or misrepresent scientific data is ever present. And what of the typically one-sided portrayal of origins science in the media today? Feynman made an observation which seems directly relevant:

8 Feynman, 1974.

> I think we live in an unscientific age in which almost all the buffeting of communications and television—words, books, and so on—are unscientific. As a result, there is a considerable amount of intellectual tyranny in the name of science.[9]

Feynman was saying that a lot of what passes as science in the mainstream media is not science, and that the prestige of science is misused by some to bludgeon and attack those with whom they disagree. Feynman distrusted experts because expertise is an appeal to authority rather than an appeal to scientific evidence. Feynman would have almost certainly disagreed that science represents a new magisterium and would probably have viewed such a notion as anti-science and tyrannical.

Stephen Jay Gould said something analogous to Feynman in relation to the teaching of science in schools. He admitted that scientists frequently exhibit elitist and dogmatic attitudes:

> It is true that we have often allowed the white-coated, advertising image to represent us...We have not fought it adequately because we derive benefits from appearing as a new priesthood. It is also true that faceless and bureaucratic state power intrudes more and more into our lives and removes choices that should belong to individuals and communities.[10]

Gould further acknowledged that top-down centralized imposition of school curricula was insulting to communities, which deserved to have local input.

Feynman rejected any dogmatic and elitist science (which Gould described) as non-science. Both Feynman's and Gould's statements can be interpreted as suggesting that certain institutions were riding roughshod over democracy and misusing the term "science" to advance their agenda. This propagandistic use of the term "science" only works because the public at large is not clear on what science is.

In the origins debate, Feynman's philosophical approach to science is directly applicable to the statement we introduced in the last chapter: "Science has disproved God." We hear Feynman's response: "Science doesn't teach anything ... How does science disprove God? How? What? Where?" These are fair questions. The answer is that the God of the Bible, an immaterial spirit, is beyond the scope of scientific inquiry. Science cannot disprove the existence of God because it lacks the means to detect Him.

We now turn to two philosophies which analyze the process and progress of science.

---

9 Feynman, 2023.
10 Gould, 1994.

A historical view of science: Kuhn

Thomas Kuhn (1922–1996) was a historian and philosopher of science who sought to describe how science has progressed in history. His theory has been very influential; the phrase "paradigm shift", now a well-known adage, was popularized from Kuhn's work.[11]

According to Kuhn, science progresses relatively slowly and incrementally, expanding and testing the prevailing theories. He called this "normal science." But every now and then, new data come to light that do not fit current theories. Generally, the scientific establishment of the day will attempt to fit this anomalous data into existing theories by incorporating a new aspect to the original. If this does not work, the existing theory is called into question, setting the stage for a paradigm shift. A new theory may be proposed which, by successfully modeling all relevant data, starts to gain ascendancy in the scientific community. Simultaneously, the older theory loses credibility, particularly as the generation schooled in the old paradigm is replaced. It is sometimes said that a paradigm shift progresses one funeral at a time.[12]

There have been several notable paradigm shifts in scientific history. The Copernican revolution, which we considered briefly in Chapter 3, was a historically significant paradigm shift. Prior to Copernicus, astronomers based their work on ancient Greek models, particularly that of Ptolemy. This model envisaged Earth at the center of the universe, and the sun, moon and other stars travelling around the earth in "epicycles". Ptolemy's model was complex but correlated well with the astronomical knowledge of the time.

Ptolemy's model gradually fell out of favor as a paradigm shift, which we now refer to as the Copernican revolution, unfolded over almost two centuries. Copernicus realized that a heliocentric model, that is, one in which the Sun, not the earth, occupies the central position, with the planets orbiting around it, fit the observations better and was simpler. Initially the Copernican-era astronomers believed that the orbits of the planets were circular, but later measurements and theories suggested elliptical orbits. This model soon gained widespread support. Many of the scientists involved are household names today: Copernicus, Galileo, Kepler, and Newton. The paradigm shift was a slow process. It took 183 years from Copernicus' publishing his heliocentric concept in 1543, to the finishing touches added by the English polymath Isaac Newton in the third edition of his *Principia* in 1726.

But from 1905, with the publication of Einstein's theory of relativity, several of Newton's signature theories were themselves on the receiving end of a paradigm shift. Scientists realized to their shock that Newton's theory was not

---

11 As an aside, it was Kuhn's seminal 1962 book, *The Structure of Scientific Revolutions*, (Kuhn, 2012) that provided the final nail in the coffin of logical positivism.

12 Attributed to German physicist Max Planck.

"true". As a model it was—and remains—good enough for astronauts to chart a course to the moon and back, and to make many other accurate predictions, but it is not "true". This was another major paradigm shift; the Einsteinian era of physics had overtaken the Newtonian era. As we shall see later, this paradigm shift has a direct bearing on the origins debate.

Another relevant paradigm shift concerns Darwin's theory of evolution. After Darwin proposed his theory in 1859, the work of Gregor Mendel (1822–1884) won wide acceptance. Over time, scientists realized that Darwin's original theory, which did not incorporate Mendel's ideas, was inadequate. A new evolutionary synthesis, known as Neo-Darwinism, or the Modern Synthesis, was devised to incorporate the latest relevant scientific discoveries. Neo-Darwinism drew from a range of disciplines[13] and emphasized mutations as the origin of the genetic novelty on which natural selection worked. The synthesis utilized statistical concepts concerning changes of gene expression in populations. This expansion is probably an example of the process of normal science, but the Neo-Darwinian synthesis itself is breaking down because of new discoveries in the fields of information science and genetics. We may yet see a paradigm shift away from Neo-Darwinian evolution among the secular humanist scientific community, as the modern synthesis is recognized by some scientists to have crumbled beyond repair.[14]

In the origins debate, we saw earlier that philosopher Thomas Nagel had labelled the Neo-Darwinian cause as "ripe for displacement". What Nagel is saying is that the Neo-Darwinian approach is no longer tenable; a paradigm shift is needed as no amount of add-on tweaks can save it. Of course, Nagel received substantial criticism for this,[15] and this collective defense mechanism of the established scientific community brings us to our fourth philosophical approach to science.

A social view of science: McElreath

Richard McElreath (b. 1973), a professor of anthropology, points out that it is easy to overemphasize the logical aspect of science and under-emphasize the social aspects. According to McElreath, Karl Popper knew that science is not completely described by the falsification criterion: there can be legitimate debate about data—whether it is correct, or whether it has been correctly interpreted.[16]

Instead of being falsified by strict logic, hypotheses are falsified in practice by scientific consensus—a social phenomenon. It is when the scientific

13 For example, paleontology, genetics, ecology, mathematical analysis and cytology (Huxley, 1942, p. 8).
14 For example, Eugene Koonin stated "… the Modern Synthesis is gone. What's next?" (Koonin, 2009).
15 For example, refer article "Do you only have a brain? On Thomas Nagel: a philosopher's broadside against Darwinism and materialism is mostly an instrument of mischief" (Leiter & Weisberg, 2012).
16 McElreath, 2020, p. 4.

community begins to regard a hypothesis or model as wrong that it is considered falsified. In McElreath's view, science is a process of argument, sometimes untidy and chaotic, working toward a consensus on a scientific question over time. These arguments play out in print and at scientific meetings. McElreath states that for this reason scientific knowledge is not definite, and a simplistic Popperian suggestion that science is strictly logical is harmful because it is a misrepresentation of the scientific process.[17]

This is an interesting thought. If science is a social phenomenon and most members of a scientific community are secular humanists, then we should expect that science will have a secular humanistic bent. Conversely, if Christians were the dominant group in the scientific community, we would expect science to have a Christian flavor.

Criticism of Popper's view is not confined to McElreath. For example, the distinguished Hungarian philosopher of science, Imre Lakatos (1922–1974), modified Popper's view—in a nod to the realities of science—by suggesting that the falsification criterion is typically not applied to a smaller "hard core" of theories within a research program which the scientists regard as correct. Instead, falsification applies to a collection of "auxiliary hypotheses" which the scientists are less strongly attached to but are important because they allow empirical predictions to be made and tested.[18]

Combining the insights of Popper, Kuhn and McElreath, we conclude that although scientific models can be useful and provide important predictions, we must always reserve judgement in calling them "true". They can be very firm, but we cannot close the gap to say they are "true" because of the problem of induction and the role of scientific consensus—a social phenomenon. Even some of the most certain hypotheses, such as Newtonian physics, have been replaced with better models.

Darwin's theory of evolution is another case in point. It was formulated in ignorance of genetics, molecular biology, cellular biology and information theory. Will this theory, even in its twentieth century modified form (Neo-Darwinism) stand the test of time? Or will the scientific consensus move on?

We also note Feynman's timely warning against "pseudoscience" that may appear at first glance to go through the motions, but on a closer view turns out to be less than honest. This rings true because it matches our experience of other fields—accountants, engineers and the like may appear to be practicing their discipline, but if they accede to political or other interference, their work must be called into question.

---

17 Ibid, pp. 9–10.

18 Musgrave & Pigden, 2023.

The missing philosophy

All these philosophies make a good deal of sense. They illuminate what science is and how it operates. But then, we find this jarring opinion from evolutionary biologist Richard Lewontin (1929–2021):

> Our willingness to accept scientific claims that are against common sense is the key to an understanding of the real struggle between science and the supernatural. We take the side of science *in spite* of the patent absurdity of some of its constructs, *in spite* of its failure to fulfill many of its extravagant promises of health and life, *in spite* of the tolerance of the scientific community for unsubstantiated just-so stories, because we have a prior commitment, a commitment to materialism. It is not that the methods and institutions of science somehow compel us to accept a material explanation of the phenomenal world, but, on the contrary, that we are forced by our *a priori* adherence to material causes to create an apparatus of investigation and a set of concepts that produce material explanations, no matter how counter-intuitive, no matter how mystifying to the uninitiated. Moreover, that materialism is absolute, for we cannot allow a Divine Foot in the door.[19] (Emphasis in original.)

Here we have a leading scientist—a Harvard professor, no less—describing the scientific enterprise using words like "against common sense", "patent absurdity", "failure to fulfill" and "unsubstantiated just-so stories". The scientific establishment provided tacit support for these views by failing to sanction either the comments, or Lewontin himself. This is the kind of science that Feynman despised, and McElreath sensed—a social phenomenon whose common bond is a rejection of the Divine. This is a profound piece of evidence against Gould's non-overlapping magisteria concept; Lewontin explicitly states that science is in a "struggle" against the supernatural.

This is a position with which we agree. "Science" of the type that Lewontin described is against Christianity, but it is mainly against Christianity in the area of origins, where Christianity makes certain truth claims, which can, at least to some extent, be assessed scientifically. Lewontin's jarring opinion alerts us to something important: it is origins science that tolerates "patent absurdities" and "unsubstantiated just-so stories". Empirical science has no time for either.

That brings us back to the "philosophies of science" cited above. Are these philosophies of empirical science or of origins science? This is a fair question, because the philosophers themselves do not specify that they apply to origins science. And this is where we see a lack in the philosophies of science: as far as we are aware, *no explicit philosophy of origins science has been published.* The absence of a clearly articulated distinction is likely a major contributor to the

19 Lewontin, 1997.

confusion that arises in the origins debate—the kind of confusion that leads to statements like "science has disproved God."

But we can get closer to a philosophy of origins science if we consider what scientists have said about forensics, which is related. In 2009, the American National Academies of Sciences published *Strengthening Forensic Science in the United States: A Path Forward.*

The authors of that publication described the "CSI effect", where the general public's understanding of forensic science has been distorted by the *Crime Scene Investigation* television series. These have led to an inaccurate impression of forensic science having unfailing certainty and making no mistakes.[20] [21] The authors point out that within forensics a great dilemma "concerns the question of whether—and to what extent—there is *science* in any given forensic science discipline".[22] This is because forensics amounts to applied science, somewhat separate from the culture and knowledge within mainstream scientific disciplines.[23] Indeed, some experts have cited " ... the norms of science, which emphasize 'methodological rigor, openness, and cautious interpretation of data,' as norms that often are absent from the forensic science disciplines".[24]

If the philosophies of empirical science do not map to forensics, they will certainly not map to origins science. A separate philosophy is needed. We are not in a position to write such a philosophy of origins science, but, in view of what Lewontin said above, we can at least offer some thoughts on the main features of origins science that set it apart from empirical science:

Presuppositions are key. Imagine a set of experiments aimed at finding the properties of various alloys of nickel—tensile strength, ductility and so forth. Anyone with access to the required apparatus and test specimens can carry out the appropriate experiments and generate results which will match findings from other scientists who perform the same experiments in similarly equipped labs. But throughout this process of repeatable experimentation, it is beside the point whether the people involved are secular humanists, atheists or Christians. Not so with origins science. An atheistic worldview *vis a vis* a Christian worldview makes all the difference, as we see above with Lewontin. Narratives promoted within one worldview will seem highly implausible, or even absurd, to those of another worldview, because of differing presuppositions.

---

20 The equivalent, in origins, would be the "David Attenborough effect". Sir David is, at the time of writing, a highly respected communicator who believes resolutely in evolution, and as such has been very effective in convincing the public of the reliability of evolutionary science.

21 Committee on Identifying the Needs of the Forensic Science Community, 2009, p. 48.

22 Ibid., p. 9.

23 Ibid., p. 38.

24 Ibid., p. 45.

A practical outcome of this is that large-scale paradigm shifts, as Kuhn understood them, do not readily occur in origins science, if at all. There may be shifts to some extent, such as the shift from the Darwinian mechanism of evolution to the Neo-Darwinian one (covered in more detail in Chapter 12), but not a wholesale change. A complete change could only come about if there is a change in presuppositions at a societal or civilizational level—a broad-based *worldview* shift.

Expectations of success. Proponents of the secular humanist (or any other) origins account will be utterly convinced that their narrative will eventually prevail, for the simple, almost tautological reason that their narrative is in line with their presuppositions. As a result, both sides (from the point of view of the opposition) appear to be dogmatic about their beliefs and impervious to evidence that appears to contradict their origins account.

Incorporates "just-so" stories. "Just-so" narratives of origins science attempt to account for unknown or unknowable phenomena and are an essential component in both the Christian and secular humanist (and atheistic) worldviews. The widely held belief that Genesis is no more believable than a child's fairy story only focuses on one side of the debate. On the other side are narratives that atheists have concocted to explain evolutionary outcomes. Perhaps the granddaddy of all the atheistic just-so stories is abiogenesis, the theory that life arose by chance from inanimate matter. That there is no scientific explanation for abiogenesis is water off a duck's back to atheists because their expectation of eventual success (above) inures them to present difficulties.

May include fragmentary evidence of a non-empirical nature. Unlike empirical science, which gathers large volumes of precise data under carefully controlled conditions in the lab, origins evidence is necessarily extremely old, and because of this age, often degraded or incomplete. Interpretations of this evidence are heavily influenced by conjecture and presuppositions.

Harder to prove or disprove. Unlike empirical science, where incorrect results are relatively easy to detect (not always; humans were incorrectly believed to have 48 chromosomes for several decades before a later scientist took the time to check this result and pegged the number at 46), origins science is inherently more difficult to prove or disprove because scientists were not present to witness the events. A classic example of the murky nature of origins was *Piltdown man*, a supposed fossil which was included in textbooks for decades before it was later shown to be a hoax. The secular humanist approach regards plausible causal mechanisms as sufficient evidence for common evolutionary descent from a single primordial organism and downplays the need for actual proof. This is the same approach taken by creationists—we assume the Bible is true, so our interpretation of the Bible or science must reflect compatibility between both.

May incorporate empirical science indirectly to help evaluate origins evidence or hypotheses. The results of empirical (experimental) science can be used to analyze data or test hypotheses for plausibility in forensic and origins research. For instance, the sedimentation experiments performed by French sedimentationist Guy Berthault in a water flume are highly relevant to the origins debate.[25] Similarly, empirical taphonomic studies investigating decay processes, fossilization and post-mortem transport and burial may help to explain certain features of fossils, such as a death pose. This has potential value to all sides of the origins debate.

The distinction between origins and empirical science is of enormous practical importance. The noteworthy *McLean v. Arkansas* court case (1982), which sought to determine whether creation science was science and, as such, should be taught alongside evolution in science classes, hinged on the characteristics of "science", which the judge, William Overton, cited as follows:[26]

- It is guided by natural law;
- It has to be explanatory by reference to natural law;
- It is testable against the empirical world;
- Its conclusions are tentative, i.e. are not necessarily the final word; and
- It is falsifiable.

Here's the rub: Overton applied principles of empirical science to creationism (an origins narrative) and found a disconnect. Our contention is that this approach conflates empirical and forensic/origins contexts, which operate differently. If he had faithfully applied the same principles of empirical science to secular humanist origins science, he would have found a similar disconnect.

Overton's first point, that science is guided by natural law, implicitly privileges an atheistic or materialistic worldview as scientific, and other points of view as unscientific, by ignoring the distinctions between empirical science and origins science. It is true that empirical science is subject to the presupposition of natural law as stated in Overton's first statement. However, in origins science, the narrative must explain *why natural law exists*, which at once brings in reasoning from outside of the domain of empirical science, namely philosophical and/or theological arguments. In Chapter 2 we recounted the opinions of Eugene Wigner and Albert Einstein, who both commented on the difficulty for humanist science in explaining the correlation between the natural world on one hand and mathematics and science on the other. In the

25 Berthault demonstrated empirically that in the presence of a water current, sedimentary layers can form laterally and vertically simultaneously, that current formed layers can have the same appearance as layers laid down over long periods, and that strata are not always a measure of chronology (Berthault, 2000).
26 Overton, 1982.

case of physicist Richard Feynman (also quoted in Chapter 2), he even questioned why there should be any law-like behavior in the universe at all; he considered this a "miracle". In our view these illustrious scientists were questioning why natural law should exist within the humanist scientific paradigm. We note in Chapter 9 cosmologist Stephen Hawking's belief that origins science must account for the fine-tuning of the universe. This question impacts *why natural law operates the way it does in our universe*. Overton's point (1) would be more relevant to origins science if it were restated as:

> Origins science must explain why natural law exists in the universe, and why it operates the way it does.

Judge Overton's second point, that empirical science must be explanatory by reference to natural law, is also correct with respect to empirical science. But we will show that both the secular humanist (Chapters 9 and 10) and creationist (Chapters 15 and 16) origins accounts contain miracles—secular and Divine respectively. A key secular miracle in the origins account of secular humanism is abiogenesis, the supposed emergence of life from inanimate matter by random processes. This secular miracle is described as such by humanists[27] precisely because it is not well explained, if at all, by natural law. Origins accounts also necessarily contain the concept of infinity, as we saw in Chapter 2, which is outside the domain of empirical science. These two examples expose Judge Overton's second statement as incorrect when applied to contemporary origins science.

We accept Overton's last three statements as applicable to origins theories. But in our view, they can be shown to apply to modern biblical creationism.

First, in respect of Overton's third point, that science is testable against the empirical world, the creationist enterprise seeks ways to square empirical science and archaeological and paleontological evidence with a straightforward reading of biblical material, including Genesis. Hence, creationists regard many of their theories and accounts as testable against the empirical world.

However, some components of the creationist worldview are not up for debate. This point relates to the view of science espoused by philosopher Imre Lakatos, whom we briefly met above. In Lakatos' scheme, some presuppositions or theories are so fundamental to a worldview that they may be more in the realm of philosophy than science. As such they are not necessarily directly testable and may not be up for debate within a program of research. An example from the creationist side is God's *ex nihilo* creation of matter on the first day of creation, a non-testable presupposition to which creationists hold. Examples from the secular humanist side are the presupposition that at the level of the cosmos, there is no purpose; and that

27 See quotation of George Wald, Chapter 10.

science potentially can, or will, explain the origin of life, if given enough time and resources. Again, these are not testable propositions.

Lakatos explained that rather than testing these fundamental concepts, a research program often investigates secondary theories, which are related to the fundamental concepts, but at a distance. Therefore, Overton's third point is partly true, but is really an oversimplification of how research programs operate within the origins science arena.

Overton's fourth point, that the conclusions of science are tentative, is true of any serious origins research enterprise. In the case of creationism, the infallibility of the Bible is asserted as axiomatic. However, creationists do not assert the infallibility of their interpretations of the Bible, nor do they assert that all scientific evidence is infallibly settled. Creationist interpretations of the Bible and science are subject to change because much of origins science relies heavily on circumstantial evidence which admits of more than one explanation. Just-so stories are (by definition) tentative because they are made in the absence of compelling evidence; all sides of origins debates utilize such narratives to fill in gaps. (By "just-so stories" we are not referring to the Genesis text, but to narratives that build around the text to flesh out the detail. For example, creationists have suggested that the earth's magnetic field was operational by the fourth day of creation to protect the earth from solar winds. That might be true, but it is not provable scientifically. Similarly, some early creationists such as Luther suggested in relation to Genesis 4:17, that Cain married one of his sisters, another scenario which cannot be scientifically proven. An alternative explanation is that Cain may have married a niece, although there must have been some sibling marriages among the offspring of Adam and Eve). Both secular humanists and creationists generally accept that these "just-so" narratives are tentative and subject to change if new evidence appears.

Overton's final point, that science is falsifiable (derived from the philosophy of Popper), is certainly true for the creationist enterprise, excluding that which is held to be axiomatic: the Bible. For example, the moondust argument for a young moon seemed scientifically sound until falsified by better data and careful analysis. This does not mean that the moon is old, but that the specific argument for youth is *scientifically* falsified.[28] In the nineteenth century, naturalist Philip Henry Gosse outlined a scheme to reconcile the young earth of the Bible with the great age that the geology of the time suggested.[29] Gosse's scheme is considered falsified, not by scientific means, but instead discredited by philosophical/theological arguments.

---

28 CMI, 1993

29 This scenario is expounded in Gosse's book *Omphalos* (1857). Here Gosse suggested God created the fossils in the rocks complete with an apparent, but false, built-in life history. Christians considered this unacceptable because it implied that God had created a deceptive world in the rocks.

Indeed, Creation Ministries International maintains an online listing of creationist arguments considered falsified by scientific or other means, and arguments considered doubtful.[30] Hence, Overton's fifth statement is generally true of creationism, as it is for secular humanist schemes.

Did Judge Overton make a category error in *McLean v. Arkansas*? He judged creationist *origins science* against *empirical science* criteria. The approach he took is unfortunate but understandable, given the lack of a philosophy of origins science. If a United States district judge can view the issue in this way, then it is likely to be rife—if not the norm—among broad swathes of the general public. Lewontin's quote, cited earlier, is evidence that a real distinction exists between these two branches of science.

Scientists behaving badly

We conclude our review of science with an important aside. The day-to-day reality is that science is frequently marred by ethical issues. This should come as no surprise to Christians; the biblical worldview teaches us that every human endeavor, which includes science, will be beset by sin, resulting in problems both moral and intellectual.

The evidence shows that this is the case. Although entire books have been devoted to this topic, for example James Penston's *Stats.con—How We've Been Fooled by Statistics-based Research in Medicine*,[31] we will summarize elements of the final chapter from McElreath's *Statistical Rethinking*. He quotes the editor-in-chief of *The Lancet*, a prestigious medical journal:

> The case against science is straightforward: much of the scientific literature, perhaps half, may simply be untrue. Afflicted by studies with small sample sizes, tiny effects, invalid exploratory analyses, and flagrant conflicts of interest, together with an obsession for pursuing fashionable trends of dubious importance, science has taken a turn towards darkness.[32]

If perhaps half of science may be untrue, the problem is determining which half. McElreath continues:

> … the history of the sciences is equal parts wonder and blunder … Its story is one of error, ego, fraud, and correction … [Some philosophers of science have concluded since] most science has been wrong, most science is wrong.[33]

McElreath lists various concerns with contemporary science:[34]

---

30 CMI, 2024.

31 Penston, 2010.

32 McElreath, 2020, p. 553.

33 Ibid., p. 554.

34 Ibid., pp. 554–555.

- If funding of science projects is determined by those who have been funded in the past, this could lead research in a conservative or even corrupt direction.
- Projects with insufficient sample size or small effects have a high "noise to signal ratio". These generate unreliable findings.
- Leading academic journals may prioritize new or positive findings over true findings. Replication work is regarded as less publishable than novel work. Over time, this can distort scientific literature.
- Incorrect analysis of data is commonly published. Even if criticized and retracted, these studies are often still cited by scientists unaware of the retraction or criticism.
- Poor peer review does not prevent mistakes being published but might prevent publication of brilliant work. Honesty about the weaknesses of a study may be punished while exaggeration may be rewarded with publication.

One interesting peer-review convention is to maintain the anonymity of reviewers involved. The peer review process, which is ostensibly aimed at improving the quality of the publication and reducing the number of errors, will therefore lack transparency. Hidden peer review has been relatively ineffective in stopping scientific fraud.

On a positive note, secular humanists are beginning to acknowledge that "Yes, science is political".[35] A guest blog in *Scientific American* explains that politics decides who can become a scientist, how science is conducted, what topics are funded, and what knowledge is disseminated.[36] Given the level of secular humanist political control, this is an issue that Christians need to understand: origins science has become biased via politicization and other factors. As Feynman concluded, and which the establishment finds convenient to ignore, "we live in an unscientific age in which ... there is a considerable amount of intellectual tyranny in the name of science."

Conclusion

In this chapter we briefly considered several philosophies of science, all of which supported our conclusion that there is no scientific magisterium.[37] We also saw that no philosophy of origins science has yet been promulgated, which has led to widespread confusion regarding the distinction between empirical and origins science (the latter a type of applied science).

---

35 Shearer, Jackson, Ahmad, & Paredes, 2020.

36 Sabbagh, 2017.

37 As Protestant Christians, we might say that the Bible is our magisterium, and the science literature comprises our Deuterocanon (second canon). The latter is helpful, useful, and valuable—but not infallible.

As important as this is, a full background to origins issues must also consider the broad historical context in which the Christian and secular humanist accounts are based. We turn to this in the next chapter.

References

Berthault, G. (2000, October 1). Experiments in Stratification. Retrieved December 14, 2024, from icr.org/article/473/

CMI. (1993). Moon-dust argument no longer useful. Retrieved June 21, 2025 from creation.com/moon-dust-argument-no-longer-useful

CMI. (2024, May 27). Arguments we think creationists should NOT use. Retrieved May 25, 2024, from creation.com/arguments-we-think-creationists-should-not-use

Committee on Identifying the Needs of the Forensic Science Community. (2009). *Strengthening Forensic Science in the United States: A Path Forward*. Washington D.C: The National Academies Press.

Dawkins, R. (2006, October 23). Why there almost certainly is no god. Retrieved from richarddawkins.net/2006/10/why-there-almost-certainly-is-no-god/

Feynman, R. (2023, December 9). What is Science? (R. Leighton, Ed.) Retrieved from feynman.com/science/what-is-science/

Feynman, R. (1974). Cargo Cult Science: Some remarks on science, pseudoscience, and learning how to not fool yourself. Caltech's 1974 commencement address. Retrieved December 9, 2023, from calteches.library.caltech.edu/51/2/CargoCult.htm

Gould, S. J. (1994). Evolution as fact and theory. In S. J. Gould, *Hen's Teeth and Horse's Toes: Further Reflections in Natural History* (pp. 253–262). New York: W. W. Norton and Company.

Gould, S. J. (1997, March). Nonoverlapping Magisteria: Science and religion are not in conflict, for their teachings occupy distinctly different domains. *Natural History*, 106(2), pp. 19–20.

Huxley, J. (1942). *Evolution: The Modern Synthesis*. London: George Allen & Unwin Ltd.

Koonin, E. V. (2009). The Origin at 150: is a new evolutionary synthesis in sight? *Trends in Genetics*, 25(11), 473–475.

Kuhn, T. S. (2012). *The Structure of Scientific Revolutions* (50th Anniversary edn). Chicago: The University of Chicago Press.

Leiter, B., & Weisberg, M. (2012, October 3). Do You Only Have a Brain? On Thomas Nagel: A philosopher's broadside against Darwinism and materialism is mostly an instrument of mischief. *The Nation*. Retrieved December 9, 2023, from thenation.com/article/archive/do-you-only-have-brain-thomas-nagel/

Lewontin, R. C. (1997, January 9). Billions and Billions of Demons. The New York Review. Retrieved January 22, 2024, from nybooks.com/articles/1997/01/09/billions-and-billions-of-demons/

McElreath, R. (2020). *Statistical Rethinking: A Bayesian Course with Examples in R and Stan* (Second edn). Boca Raton: CRC Press.

Musgrave, A., & Pigden, C. (2023). Imre Lakatos. (E. N. Zalta, Ed.) Retrieved December 13, 2023, from plato.stanford.edu/archives/spr2023/entries/lakatos/

National Academy of Sciences and Institute of Medicine. (2008). *Science, Evolution, and Creationism*. Washington: The National Academies Press.

Overton, W. (1982). McLean v. Arkansas Bd. of Ed., 529 F. Supp. 1255 (E.D. Ark. 1982). Retrieved May 25, 2024, from law.justia.com/cases/federal/district-courts/FSupp/529/1255/2354824/

Penston, J. (2010). *stats.con—How We've Been Fooled by Statistics-Based Research in Medicine*. London: The London Press.

Popper, K. (2002). *The Logic of Scientific Discovery*. London: Routledge Classics.

Sabbagh, U. (2017, April 25). Science Has Always Been Inseparable from Politics: Scientific research doesn't take place in a vacuum; it can only happen with society's blessing. Retrieved December 11, 2023, from blogs.scientificamerican.com/guest-blog/science-has-always-been-inseparable-from-politics/

Shearer, A., Jackson, C., Ahmad, T., & Paredes, I. J. (2020, October 8). Yes, Science is Political: Scientists need to acknowledge that fact—and to act on it in these most dire of times. Retrieved December 10, 2023, from www.scientificamerican.com/article/yes-science-is-political/

Taran, S., Adhikari, H. K., & Fan, E. (2021). Falsifiability in medicine: what clinicians can learn from Karl Popper. *Intensive Care Medicine*, 47, 1054–1056. doi.org/10.1007/s00134-021-06432-z

5

The scientific rise and fall of the ancients

An early 1990s engineering professor made the memorable comment that if a person who knew absolutely nothing about cars were asked to build one from scratch, it is unlikely that they could produce a vehicle as good as the Ford Model T.[1] It's an interesting point—absent the learning which results from observing the work of others, a designer or builder will be stuck with a laborious and time-consuming process which may not succeed. Continuing with the professor's line of thinking, we may add that people with only a passing knowledge of cars today would be aware of automotive technologies that probably never occurred to Henry Ford. The common phrase "standing on the shoulders of giants" sums up the acknowledged power of building on others' prior achievements.

Standing on the shoulders of giants not only enables better work, it also accelerates technological or scientific progress. In recent times, for example, China has transformed itself, in a matter of decades, from a manufacturer of cheap consumer items to an industrial powerhouse. The main driver of this transformation was the technology transfer that occurred from more advanced countries. Considering the ease and pace of development made possible by cherry-picking the work of others, another well-worn phrase comes to mind: "Don't reinvent the wheel". This is sage advice—why go to the trouble of developing technologies that are already available?

These issues are not new—the ancients were just as driven by the prospect of free knowledge as any nation today. As a civilization came to an end, scholars from up-and-coming people groups were always on hand to access the knowledge base of the outgoing group. It is widely believed today that Western civilization can trace its lineage back to the ancient Greeks via the Romans. Certainly, the Greek influence on science, mathematics, philosophy, and literature is profound. But what does the Bible say about the lineage of Western civilization, and what lessons can we derive as we trace the lineage of scientific development?

In this chapter we will examine the broad-brush scheme of civilizational history given in the Bible and draw important lessons from the historical process of knowledge transfer. We will compare two case studies—the

---

1 The Ford Motor Company, founded by American industrialist Henry Ford (1863–1947), mass-produced the Model T automobile from 1908 to 1927.

Copernican revolution, which saw the emergence of an empirical causal model of planetary motion, and Darwinism, an example of origins science. This history encompasses a succession of empires over a span of 2500 years, and key scientific developments that occurred over this lengthy period.

A biblical scheme of history

Soon after the ancient Babylonian king Nebuchadnezzar the Great came to power in 605 BC, he had a terrifying dream that left him disturbed and sleepless. But this was no ordinary nightmare; Nebuchadnezzar knew that the dream was of great significance, and wanted its meaning explained to him. He summoned his wise men, ordering that they tell him both what he had dreamed and what it meant. The wise men of Babylon were at a loss, but God revealed the dream and its meaning to the Israelite prophet Daniel in a night vision. Daniel had the captain of the king's guard arrange an audience with Nebuchadnezzar.

The king's dream, Daniel explained, had pictured a human image, awesome in form and splendid in appearance. "Its head was of fine gold, its chest and arms of silver, its belly and thighs of bronze, its legs of iron, its feet partly of iron and partly of clay" (Daniel 2:32–33). Then something dramatic happened. A stone, "cut out without hands… struck the image on its feet of iron and clay and broke them in pieces" (v. 34). The awesome image collapsed, turning into dust which was carried away by the wind. After this, the stone that struck the image "became a great mountain and filled the whole earth." (v. 35).

God used this awe-inspiring vision to lay out for Nebuchadnezzar a broad scheme of civilizational history. Daniel explained to the king that the head of gold was none other than Nebuchadnezzar himself—God had made Nebuchadnezzar a king of kings, with a kingdom, power, strength and glory. The next part of the image was the silver chest and arms. This represented a successor kingdom, inferior to Nebuchadnezzar's, which we now know was the Medo-Persian empire. Further down, the belly and thighs of brass represented the Greek empire. The lower part of the image, the legs of iron, represented the Roman empire.

These lower parts were struck by the stone, which destroyed the entire image. God was revealing that the ancient world, represented by these successive empires, would end, being destroyed by a new kingdom established by God. This new kingdom, which grew to fill the whole earth, represents Christian civilization, whose advent marked the beginning of modernity.

An important feature of history is revealed in the image of the statue. These successive empires all formed part of a larger whole. Although the four empires were distinct from each other, they derived successively from each other in important ways, including their intellectual and scientific development.

Learning was conveyed from one empire to another via manuscripts and libraries. The knowledge conveyed concerned calendars, mathematics and theories of physics and the cosmos.

For readers wondering what relevance ancient Babylonia has in our post-modern era, consider the base-60 (or sexagesimal) numbering system of Babylonian mathematics. This system had several important advantages over other numbering systems (including our own base-10, or decimal system). For example, one half, one third, one quarter, one fifth, and one sixth can all be calculated with ease using the sexagesimal system, whereas the decimal system can only easily represent one half, one fifth and one tenth.[2] The Babylonian system was so useful that successive civilizations continued using it. If you have ever wondered why there are 60 minutes in an hour, 60 seconds in a minute, or 360 degrees in a circle, look no further than the ancient Babylonians and their über-useful sexagesimal numbering system.[3]

To demonstrate how Nebuchadnezzar's dream and its interpretation played out at a civilizational level, consider London's famous landmark, Big Ben. The fact that the clock-faces are marked out at 60 minutes to the hour is derived, as we have just seen, from the Babylonians. Looking elsewhere on the clock, we see the Roman numerals around the clockfaces. Thus, the influence of the Babylonian head of gold, and the Roman legs of iron, are on display for all the world to see, even though these empires have long since crumbled into the dust of history. At the same time, Christians can be found in virtually all countries on earth. Big Ben is a monument to the truth of Nebuchadnezzar's vision—the ancient empires have disappeared, even though their influence continues to this day.

We will consider the scientific rise and (relatively recent) fall of two scientific developments of the ancients: physics and evolution. We will compare these two cases and draw important lessons.

Babylonian developments

Early in his reign, King Nebuchadnezzar assembled a multi-national council of advisers who were "skillful in all wisdom, cunning in knowledge, and understanding science ..." (Daniel 1:4). The Israelite prophet Daniel was among their number. He had been deported to Babylon and, as one of the intellectual elites of Israel, inducted into Nebuchadnezzar's advisory council. The Bible describes these advisers as magicians, astrologers, sorcerers and "the wise men". They were knowledgeable in calendars and mathematics and what we would today call astronomy and astrology. Daniel was a stand-out in this

2 In relation to each decimal digit.

3 Lombardi, 2007.

illustrious group because of his God-given ability to interpret visions and dreams (Daniel 1:17).

The fact that Nebuchadnezzar assembled this cosmopolitan think-tank was emblematic of the Babylonians' focus on knowledge and learning. Charting the sun, seasons and the year was critical to this ancient agricultural society to inform its crop cycles.[4] The Babylonians sought to understand the movement of the planets so that they could make astronomical predictions[5] and developed water clocks which enabled astronomical data to be recorded with higher accuracy.[6] They understood that the solar year was the time taken for the sun to return to the same position in its path relative to other heavenly bodies in the region of the sky they called the zodiac. Armed with this knowledge, and with their reasonably accurate water clocks, they estimated the length of the solar year as 360 days, which fitted neatly with their numbering system.[7] Despite their sophistication, the Babylonian calendars nonetheless underestimated the length of the solar year and became progressively out of step with the movement of the sun. But their achievements were impressive—the Babylonians had invented scientific astronomy. This was an ancient instance of descriptive model science. Subsequent work by ancient Egyptian astronomers showed that the solar year was longer; they added an extra five days to the Babylonian calendar, making 365 days.

The Babylonians made significant advances in mathematics. They could calculate simple areas and volumes, and it is likely they could calculate the surface area of a sphere.[8] Babylonians understood what we now know as Pythagoras' theorem[9] in its full generality, and could calculate the square root of two to a precision of one part in a hundred thousand.[10] Their advanced numbering system also helped them make progress in algebra and number theory, as well as geometry.[11] The Babylonians calculated various values for pi, from an imprecise "3" to more precise estimates.[12]

Like other ancient societies, the Babylonians considered the earth to be a flat disk.[13] This model correlated with simple observations.

With the collapse of the Babylonian empire in the sixth century BC, the intellectual accomplishments of the empire were transferred to the Medo-Persian empire, and later to the Greek empire, under Alexander the Great, in

---

4 Kuhn, 1957, p. 10.

5 "... all the Babylonian astronomical models, as I still feel justified in calling them, were entirely arithmetical in character" (Aaboe, 1974, pp. 34–35).

6 Kuhn, 1957, p. 9.

7 Ibid., p. 11.

8 Aleksandrov, Kolmogorov, & Lavrent'ev, 1999, p. 21 (Vol 1).

9 Struik, 1967, p. 27.

10 Aaboe, 1964, p. 26.

11 Ibid., p. 29.

12 Ibid, p. 30.

13 Blacker & Loewe, 1975, p. 61.

the fourth century BC. The Babylonians, with their merit-based think-tanks, had invented a system of knowledge that could be exported across cultures. Even if progress occurred in fits and starts, their scholarship had kick-started cumulative advances.

The vision of Nebuchadnezzar positioned the ancient Greek civilization as two sections below that of the Babylonians. Instead of gold, this empire was represented by brass. Ancient Greece was a polytheistic, warlike, slave-based society which sometimes resorted to human sacrifice. [14] Nonetheless, it made substantial developments in science, mathematics and philosophy. The influence of the Greeks has continued through to the present. Our broad-brush view of scientific history must therefore trace the Greek influence in several domains of knowledge.

Greek astronomy and cosmology

By the fourth century BC the Greek philosophers and astronomers were largely unanimous that the earth was a small stationary sphere located in the center of a larger rotating spherical shell that contained the stars. This concept has been labelled the "two-sphere universe".[15] The sun, moon and planets were thought to be moving in the expanse between the spheres. The word planet, meaning "wanderer", fitted the observed movement of these bodies across the sky in relation to the relatively fixed position of the stars. The ancient Greeks believed that the sun and moon were planets.[16] These beliefs were incorrect, but some of their other ideas were remarkably accurate. Eratosthenes (276–195 BC), head of the library of Alexandria, famously estimated the circumference of the earth at 252,000 stadia, potentially about 29,000 miles (depending on which stadion length is used), a slight overestimate of the correct value of roughly 25,000 miles.[17]

About AD 150, Ptolemy made a major contribution to astronomy when he published what was later called the *Almagest*. This was an attempt to explain the motion of the seven known planets (including the sun and moon) which the "two-sphere" model could not do. Some progress had been made by Greek mathematicians, who developed systems of extra-circular planetary motions, called epicycles. Ptolemy's was the greatest of these attempts. His system on the motions of the planets and stars formed the foundation for further developments for roughly 1300 years in the Christian West. One of the benefits of the two-sphere model and the subsequent Ptolemaic systems were that they encouraged attempts at accurate measurement of the planetary motions to fine

14 Acton, 1864, pp. 8, 13, 17–19.
15 Kuhn, 1957, p. 27.
16 Ibid., p. 45.
17 American Council of Learned Societies, 1991, pp. 681–684.

tune the various theories.[18] In fact, the Ptolemaic system could be made as accurate as needed by adding extra epicycles.[19] This was useful descriptive model science.

Advancements in mathematics permitted further developments in astronomy. Drawing on the foundation of Babylonian mathematics, the Greeks added formulation and proof of theorems as a central part of the discipline, which has remained to this day.[20] Pythagoras showed how to calculate the length of the hypotenuse of a right-angled triangle and developed a proof for his theorem. Equally well known is Euclid's geometry, dating from the third century BC. Euclid, a mathematician and logician, composed a short treatise on geometry, titled *Elements,* which took a presuppositional, or axiomatic approach based on earlier Babylonian work.[21] From numerous postulates and five basic axioms, Euclid reasoned his way to a substantial theory of geometry, all supported by proofs. Euclid's geometry was considered a work of genius—his text remained a core part of Western education for more than 2000 years. His acclaimed fifth axiom, also known as the parallel postulate, stated that given a line and a point not on it (in the plane), there exists exactly one line through the point parallel to the original line.[22] This implies that two distinct parallel lines never intersect. From his work we have the term "Euclidean geometry" which today indicates a three-dimensional coordinate system with each straight-line axis at right angles to the other two.

The collapse of the Ptolemaic system: the Copernican revolution

The Ptolemaic two-sphere model, with its highly accurate predictive capability, remained state of the art through to the 1500s. It was at this point that Christian scholars discovered flaws with the system. Copernicus began what we now view as a paradigm shift—termed the Copernican revolution—by proposing a heliocentric system, in which the earth and other planets revolved around the Sun in circular orbits. This novel concept was recognized as an improvement upon the Ptolemaic system, even though the predictive capability of Copernicus' system was not superior to that of Ptolemy's. Copernicus' ideas gained a significant boost with the concept of elliptical, rather than circular orbits. This piece of the puzzle fell into place when Johannes Kepler (1571–1630) published *On the Motion of Mars*. This seminal paper demonstrated that Kepler's mathematical model of elliptical orbits matched the

18 Kuhn, 1957, pp. 64–72.
19 McElreath, 2020, p. 71.
20 Aaboe, 1964, p. 33.
21 Blacker & Loewe, 1975, p. 42.
22 Aleksandrov, Kolmogorov, & Lavrent'ev, 1999, p. 98 (Vol 3).

extremely accurate astronomical data measured by the Danish astronomer Tycho Brahe. This major advance gradually swept away the concept of circular orbits that had marred Copernicus' original idea. Not only did elliptical orbits match the planetary position data better, but they also provided a mathematical description of the velocity changes of the planets along their orbits.

The next step forward in the Copernican revolution was provided by Galileo Galilei (1564–1642), who used a telescope of his own design to provide supporting evidence for heliocentrism. Copernicus' original heliocentric model now had empirical evidence to back it up. Galileo also investigated inertia and acceleration. The concept of inertia described the tendency of an object to either remain at rest or to continue at a constant speed travelling in a straight line unless an external force was applied.[23]

The advances of Copernicus, Kepler and Galileo were impressive, but they stopped short of an explanation for *why* the planets orbited the sun. This final piece of the puzzle—the capstone of the Copernican revolution—was to be put in place by Isaac Newton.

The Newtonian Era

The highly influential English polymath Isaac Newton[24] (1642–1727) synthesized the new mathematical conceptions into an even simpler and more general theory—his laws of motion (or mechanics) and gravitation. Newton moved astronomy from descriptive model science to causal model science.

Newton's first law was effectively a restatement of the principle of Galileo's inertia.[25] His second law described how the motion of an object changes when force is applied. He explained that a given force will accelerate a lighter object more rapidly than a heavy object. He also stated that for an object of given mass, a larger applied force will cause greater acceleration. He combined these concepts into a simple equation, $F = ma$, meaning that the force on an object is equal to its mass multiplied by its acceleration. Newton applied this law to the concept of gravitation, a force of attraction between two objects which increases as their mass increases but decreases with the square of the distance between them. He showed that his second law could describe the motion of an apple falling from a tree, and just as successfully explain the motion of planets around

23 Feynman, Leighton, & Sands, 2010, p. 9~1.

24 Despite Newton's distinguished place in scientific history, his private papers have revealed that he was intensely interested in alchemy—his life's main work. Physics and mathematics were merely his sideline pursuits. John Maynard Keynes famously described Newton as "the last of the magicians" (Kean, 2011). Indeed, Newton's gravity is sometimes described as "action at a distance"—a concept in which one object affects another while widely separated by empty space. Prominent physicists found this aspect of Newton's theory unconvincing and bordering on irrational, because it was historically associated more with concepts from "magic" or the occult (French, 2005) than objective science. It is important to note that by our definition, Newton was a non-Christian, because he denied the doctrines of the Trinity (Snobelen, 1999) and the incarnation (Dulles, 2005).

25 Feynman, Leighton, & Sands, 2010, p. 9~1.

the sun, and the moon around the earth. Newton even provided an explanation for tidal motion by invoking the gravitational effects of the sun and the moon on the seas.

Newton's third law is usually summarized as "for every action there is an equal and opposite reaction", from which is derived the well-known principle of the conservation of momentum.

Newton's three laws were the foundation of a simple mathematical system with extraordinary explanatory power. Although Newton never explained what gravity was or why it occurred, his laws of motion and gravitation explained how it behaved. These laws were based on Euclidean geometry. They described motion as occurring in a three-dimensional Euclidean space within a medium called time.[26] At this point in history, it was almost universally held that both Newton's laws of motion and gravity, and Euclidean geometry, were literally true.[27]

Newton also independently discovered the mathematical technique of infinitesimal calculus. This used the concept of infinitesimally small distances along a curved graph line to calculate the slope accurately at any point. Developing this method further, Newton was able to calculate the area of a segment between a graph line and the horizontal coordinate axis. These were powerful analytic techniques that marked the beginning of the use of infinity in modern mathematics.

Analyzing the Copernican revolution

The above discussion summarizes, from our vantage point in history, the paradigm shift called the Copernican revolution. The main features of this paradigm shift include its substantial length—almost two centuries from Copernicus to Newton; the fact that several brilliant scientists contributed; and its progressive nature. But how would a person living in that era have experienced it? Imagine a person born 10 years after Copernicus' original ideas were published, and who lived a scientifically informed life of 80 years duration (i.e. 1553–1633). By the time our person completed their education aged around the mid-twenties, Copernicus' text would have been in wide circulation—as would the awareness that this interesting concept had no better predictive capability than the Ptolemaic system. Kepler published the first of his laws regarding the elliptical nature of planetary orbits only when our person was aged in the mid-50s, and the person would have been around 60 by the time Galileo's proof of heliocentrism was in circulation. But our person would have

26 Ibid., p. 2~3.

27 Even as late as 1906, J.L.E. Dreyer wrote: "From Thales to Kepler philosophers had searched for the true planetary system; Kepler had completed the search; Isaac Newton was to prove that the system found by him not only agreed with observation, but that no other system was possible" (Dreyer, 1953, p. 424).

died in 1633, almost a decade before Newton was born. In other words, the protracted nature of this paradigm shift was such that contemporaries of Copernicus, Kepler and Galileo would not have lived to see it brought to completion by Newton. Paradigm shifts, no matter how clearly discernible from the vantage point of the future, may progress with the main protagonists hardly aware that a shift is underway.

Another important point is the incomplete extent to which the Copernican revolution moved science on from the Greeks. Certainly, the Ptolemaic system had become a relic of scientific history by the time Newton published his laws of motion and gravitation, but crucially, Newton's mathematics were based on Euclidean geometry, with time as an independent quantity. In addition, although Newton himself accepted the Genesis account, others found his infinite static universe compatible with the ancient Greek idea of an eternal universe. Thus, Newtonian physics was still linked directly and indirectly to Greek science at a presuppositional level. The Copernican revolution, therefore, is a sort of half-way house—a more thorough break with the limitations of ancient Greek presuppositions was yet to come.

The final break begins: the emergence of non-Euclidean geometry

Mathematicians had puzzled about Euclid's fifth postulate, also known as the parallel postulate, for hundreds of years. Around 1830, it became apparent to mathematicians that the fifth postulate is not always true.

Consider a straight line in a one-dimensional Euclidean geometry. Travelling along that line takes one progressively further away from the starting point. But what if an extra dimension is added? This was a concept that Euclid had not incorporated in his geometry. In two dimensions, for example, we can consider a circle as an example of a non-Euclidean "straight line" in "one dimension". An observer travelling around the circle will eventually return to their starting point. This is not how the Euclidean one-dimensional geometry behaved.

Now consider a Euclidean plane, with two coordinate axes applied to a flat surface. (In high school math class, this would correspond to a piece of notepaper on a desk.) Any triangle drawn on this surface will feature three internal angles that always add to 180 degrees (Refer Figure 5.1a). If we draw a circle, the ratio of the lengths of the diameter and the circumference is always about 3.14 (to two decimal places), a number known by the Greek character "pi" (π).

However, in non-Euclidean geometries, these axioms break down. For example, consider a non-Euclidean "two-dimensional" example. This incorporates an extra dimension to permit the deformation of the Euclidean geometry. In three dimensions we might consider that the flat plane is wrapped and stretched around a sphere such as a terrestrial globe. Note that adding the third-dimension results in deformation of the Euclidean surface in all dimensions, not just the third. The sphere curves in relation to each of the coordinate axes—the X-, Y- and Z-axes.

With this non-Euclidean geometry in place, we now return to our earlier example of the parallel lines. Imagine two people on the terrestrial globe's equator several kilometers apart. If they both proceed due north, they are both travelling parallel to each other. However, if they continue long enough, they will meet at the North Pole. Therefore, in this scheme, separate parallel lines do meet, which invalidates Euclid's fifth postulate.

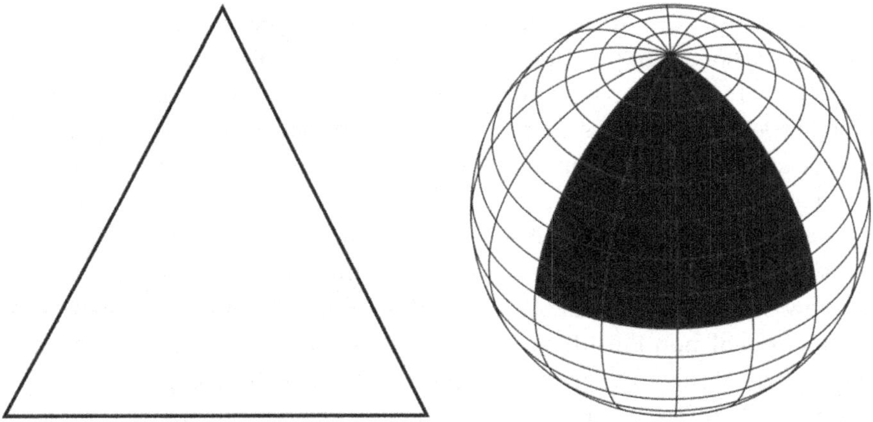

(a) (b)

Figure 5.1: Euclidean (a) and non-Euclidean triangles (b)

Next, consider a triangle drawn on this spherical surface (refer Figure 5.1b). Imagine two people standing at the North Pole. One of them proceeds due south, travelling down the prime meridian of zero degrees longitude, passing through Greenwich and stopping at the equator. The second person travels south by way of the 90th meridian West. Their southward path takes them through North America, until they also halt at the equator. If we consider these two paths as two sides of a non-Euclidean triangle, and the third side as the segment of the equator that joins the two people, we have a triangle whose internal angles add to 270 degrees. Each angle is a right angle (90 degrees), and the sum of the angles, 270 degrees, is different from the 180 degrees found in Euclidean geometry.

An objection might be that the two people are not really travelling in straight lines; the surface of the earth is curved. But we can address that problem. If we use a sphere twice the size of the Earth, then for a given distance, the amount of curvature is reduced. As the size of the sphere approaches infinity, the lines become straighter and straighter until they are in fact straight. Hence, by an appeal to infinity, we can straighten out the curved lines. Our "spherical" geometry describes a geometry with straight lines that conflict with Euclid's fifth postulate.[28]

This non-Euclidean geometry also yields unexpected results for circles. Draw a large circle on the terrestrial globe. The diameter (or radius) of this circle bulges due to the curvature of the earth, and so is slightly longer than it would be in the corresponding flat surface Euclidean geometry (two-dimensional). There is "excess diameter" and "excess radius" compared with Euclidean geometry. If we were to use this geometry to calculate pi, the result would be less than 3.14.

To create a three-dimensional non-Euclidean geometry, we could add a fourth dimension to allow for the deformation of the three-dimensional Euclidean geometry. We will not provide examples, but it is worth pointing out that higher-level dimensions are mathematically possible—even if it is not clear how to visualize them. In fact, systems of dimensions higher than four also work mathematically.

In summary, we have described a non-Euclidean geometry in one and two dimensions and highlighted the possibility of three or more dimensions. We now know there is an infinite number of different non-Euclidean geometries. A two-dimensional non-Euclidean geometry does not need to be based on a convex spherical shape. It can also be concave in shape like a saddle.

Because there are an infinite number of possible geometries, Euclidean geometry is not to be termed "true" as such. Instead, it is simply one logically valid model of geometry among an infinite number of alternatives. However, at Newton's time in history, it was still considered that Euclidean geometry exactly modeled the geometry of the real world.

Enter infinity

We noted that the concept of the infinite was increasingly being used by mathematicians in areas such as the infinitesimal calculus of Newton.

It was a devout Lutheran mathematician, Georg Cantor (1845–1918), who was the first to investigate the concept of infinity systematically. To do so he invented the theory of sets. A set is simply a collection of distinct items; distinct in that there are no exact duplicates within a set. Cantor distinguished finite and infinite sets, with the "cardinality" of a set referring to the number of items it contained. The theory of sets became fundamental to many branches of

28 An alternate approach to "straighten" the lines is to make the observer and their field of view infinitesimally small.

mathematics.[29] German mathematician David Hilbert famously enthused that Cantor "created one of the most fertile and powerful branches of mathematics; a paradise from which no one can drive us out".[30] Cantor classed the natural numbers[31] as countably infinite and was able to prove that the real numbers[32] were uncountably infinite, indicating at least two different classes of the infinite.

Just as Euclidean geometry had been revealed as just one specific type out of the set of all geometries, something similar was occurring with numbers themselves. The natural numbers were understood to be a subset of the rational numbers,[33] because the rational numbers included all natural numbers and all fractions. In turn, the rational numbers were known to be a subset of the real numbers, because the real numbers also contained the irrational numbers—numbers which could not be represented exactly as a ratio or fraction. But finally, mathematicians discovered that all known types of numbers were contained within a super-set called the complex numbers, in which each number is represented as a two-dimensional coordinate on the "complex plane". Numbers were now understood to be anything which behaved like a number; a number did not need to take the "conventional" form. Complex numbers were the "numbers" that were found to behave most like numbers in all conceivable circumstances. They had both a numeric and a geometric representation.

Advances in calculus, probability, set theory and the study of infinity, non-Euclidean geometry and complex numbers indicated that mathematics in the Christian West had progressed far beyond that of the ancient Greeks. But if mathematicians with their non-Euclidean geometries were making a final break with ancient Greece, physicists were not far behind, and it was Scottish physicist James Maxwell who made some discoveries that would never have occurred to the ancient Greeks.

Maxwell's marvel: electromagnetic radiation

James Clerk Maxwell (1831–1879) achieved a synthesis of magnetism, electricity and electromagnetic radiation analogous to what Newton had achieved for motion. He introduced ideas that could not be explained by Newtonian physics, and in that sense, was a steppingstone to Einstein.

---

29 Aleksandrov, Kolmogorov, & Lavrent'ev, 1999, p. 55 (Vol 1).

30 Breuer, 2006, p. 2.

31 The natural numbers are those used for counting, for example, 1, 2, 3, etc.

32 Real numbers constitute a set including all integers, all fractional numbers, and all numbers that are impossible to represent as an exact fraction, such as the square root of two or pi.

33 Rational numbers constitute a set containing all integers, and all fractional numbers, but excluding any number which cannot be represented as an exact fraction.

Maxwell combined the known facts concerning electricity and magnetism into a cohesive theory using a new abstract mathematical representation called *field equations*. From his equations, Maxwell predicted that a form of energy dubbed electromagnetic radiation could propagate at an enormous speed, represented as "c". Realizing that the magnitude of "c" was almost exactly equal to the experimentally calculated speed of light, Maxwell proposed that light was a form of electromagnetic radiation. (Richard Feynman went on the public record stating that Maxwell's discovery was the single most important event in the nineteenth century.[34])

Initially, scientists believed that electromagnetic waves required a substance for their propagation, in much the same way as waves in the sea or sound waves in the air travel through a medium. They dubbed this substance the "ether". The ether was invisible and remarkably resistant to examination. Very precise experiments looking for evidence of the ether drew negative results. This unexpected finding suggested that something strange was going on. Did the ether exist at all? Could electromagnetic radiation move through empty space? In fact, the mathematical concept of fields (from which we derive terms like "gravitational field") removed the need for "ether"; light moves through empty space.

The fall of Grecian physics: Einstein's relativity

In 1905, Swiss government worker Albert Einstein published his paper on *special relativity*, building on the work of the physicist Hendrik Lorentz and the mathematician Henri Poincare. It was called "special" because for simplicity it avoided dealing with the effects of gravity or acceleration. The principle of relativity, which was known prior to Einstein's paper, stated that all physical laws behave the same whether the entire system is at rest or moving in a straight line at fixed speed.

For Maxwell's equations to conform to the principle of relativity, they had to be altered with a mathematical transformation discovered by Lorentz. Under Newton's equations, if you were travelling with a given speed v towards a light source, you would expect the closing speed between you and the emitted light to be $c + v$ because of your motion relative to the light. But Einstein proposed that the speed of light in a vacuum was the same for all observers. Therefore, under this scheme, the closing speed remained c, the speed of light.

When Einstein applied Lorentz's transformation to Newton's equations, he found that Newton's signature formula, $F = ma$, was not correct. Einstein had to replace m with a variable expression which accounts for changes in "rest mass" depending on the velocity of the object. At rest or at the kinds of speeds we typically encounter on earth, the new mass term hardly changes. But as

34 Feynman, Leighton, & Sands, 2010, p. 1~11 (Vol 2).

speeds approach that of light, the mass term increases enormously. In other words, at low speeds, Newton's equations were very accurate. At very high speeds, they weren't. The growing realization that Newton's laws of motion did not hold under all circumstances was staggering.

There were several implications from Einstein's formulation. Suppose you were looking through the window of a rapidly moving spaceship as an external observer at rest. Several non-intuitive effects would become apparent. First, time on the spaceship, as shown by the clock visible within the spaceship, passes more slowly than for you, the stationary observer. Second, the spaceship shrinks in the direction of travel from your perspective. Third, if there are flashes of light emitted from each end of the ship such that a passenger aboard would observe these simultaneously, they will not appear simultaneous to you. If the spaceship is travelling at half the speed of light, and a bullet is fired within the spaceship in the direction of travel, also at half the speed of light, you would expect to see the true speed of the bullet as the speed of light (half the speed of light plus half the speed of light). However, this will not be so. Instead, you will observe the bullet moving at about four fifths of the speed of light. All these effects are predicted by Einstein's theory of special relativity.

Einstein's system differs from the three-dimensional Euclidean coordinate system we saw earlier, because it uses a four-coordinate system where time is the fourth quantity. The four-coordinate system is called "space-time". The four coordinates are taken to be nearly equivalent in nature, except that the time axis is different in two ways from the other axes. Firstly, movement on the time axis can only be in one direction; time never moves "backwards". Einstein also effectively proposed that any movement in any of the dimension axes must always be accompanied by movement in the time axis.

An important result from Einstein's special relativity was the famous equation, $E = mc^2$ (where E represents energy, m represents mass, and c is a constant term equal to the speed of light in a vacuum). This united the law of the conservation of mass and the law of the conservation of energy into a single relationship.[35]

Einstein published his theory of *general relativity* in 1915. This took acceleration and gravity into account, unlike his earlier special relativity. It has often been said that general relativity is effectively a theory of gravitation.

In general relativity, anything with energy responds to gravity. Therefore light, which has energy, but no mass in the conventional sense, can be deflected weakly by gravity.[36]

35 Carmeli, 2002, p. 87.
36 Feynman, Leighton, & Sands, 2010, p. 7~11 (Vol 1).

Another aspect of general relativity is that in a gravitational field, a clock at greater height, or gravitational potential, will run very slightly faster than one at a lesser height, or a position of lower gravitational potential.

But what is gravity? Did Einstein have more to say than Newton?

He did, and again this is related to the geometry he used—a non-Euclidean geometry. As we saw above, Euclidean geometry is characterized by triangles whose angles sum to 180 degrees and circles whose ratio of circumference to diameter is approximately 3.14. Under non-Euclidean geometries neither of these relationships holds. In particular, non-Euclidean circles feature a positive or negative amount of "excess radius", depending on the geometry being considered. Einstein stated that the amount of excess radius in space-time is proportional to the mass of the object. Effectively, space-time is curved by the effect of mass, which we interpret as a *gravitational field*.

Einstein's general relativity ended Newton's occult "action at a distance" idea of gravity and replaced it with the concept of the gravitational field. Einstein had found a higher and more accurate generalization which encompassed Newton's mechanics (or laws of motion). He had essentially demoted Newton's empirical causal model science to useful empirical descriptive-model science.

At a presuppositional level, Euclidean geometry, which had underpinned the study of astronomy for around 2000 years—from the ancient Greeks through to Newton in the eighteenth century—was finally superseded. The ether, too, was out. This marked a major point in science in that not only had the Ptolemaic system been replaced during the Copernican revolution, but the Greek mathematical foundations had also been superseded.

From Ptolemy, to Newton, to Einstein: analyzing the paradigm shifts in cosmology

Starting with the Ptolemaic system as the base case, the Copernican revolution was a revolution *within the overall Greek paradigm*. At a presuppositional level, Newtonian physics relied on Euclidean postulates; it was impossible, given these presuppositions, to progress to Einsteinian relativity. In that sense, Newton represents the perfection of the ancient Greek cosmology, and Einstein's relativity represents the escape of physics from the earlier Grecian paradigm.

It turns out that the exact same effect occurred in origins science. Darwinian evolution, far from being the watershed scientific development that it is made out to be, is arguably the last flutter of ancient Greek biological origins speculations.

Pagan origins science

Various ancient Greek (and Roman) thinkers espoused presuppositions which came to bear heavily on origins science. The first was the eternal universe, which provided a logical basis for understanding existence. Another was spontaneous generation, which held that life regularly emerged from non-living matter. Both these presuppositions survived the end of the Greek empire and were passed down to successor civilizations, including our own.

With the presupposition of spontaneous generation in place, the remaining explanatory task, as the Greeks saw it, was to develop a plausible explanation for the emergence of higher life forms. In other words, if the biological starting point was, for example, small bugs which sprang into life in stagnant water, or maggots on animal carcasses, how did larger, more complex life forms come into being? The Greeks reasoned that the higher life forms must have evolved from the lower. This line of reasoning was very understandable given the presuppositions the Greeks were working with.

The Greeks did not produce scientific papers in the contemporary sense, but musings, sometimes couched in poetry, which revealed their thinking. For example, the Greek philosopher Anaximander spoke of man as "like another animal, namely, a fish, in the beginning".[37] This brief statement, which was provided without evidentiary support, is nonetheless considered to be "well-nigh akin to prevision of the mutability of species, and of what modern biology has proved concerning the marine ancestry of the highest animals." That was the opinion, at least, of British writer Edward Clodd (1840–1930), a committed Darwinist. Clodd provided a brief introduction, then summarized Greek origins musings into 10 main points in his 1897 book *Pioneers of Evolution: from Thales to Huxley*. He also added a brief italicized comment to each based on his understanding of science at the time. We quote Clodd at length:

> Standing well-nigh on the threshold of the Christian era, we may pause to ask what is the sum of the speculation into the causes and nature of things which, begun in Ionia (with impulse more or less slight from the East, in the sixth century before Christ), by Thales, ceased, for many centuries, in the poem of Lucretius, thus covering an active period of about five hundred years. The caution not to see in these speculations more than an approximate approach to modern theories must be kept in mind.[38]

(For our present purposes we skip the first six points of Clodd's summary and present the final four points which are directly relevant to Greek origins

37 Clodd, 1907, p. 7.
38 Ibid., pp. 29–30.

thought.[39] Clodd presents a summary statement of Greek origins, then provides additional thoughts below each statement in italics.)

> 7. Life arose out of non-living matter.
>
> *Although modern biology leaves the origin of life as an insoluble problem, it supports the theory of fundamental continuity between the inorganic and the organic.*
>
> 8. Plants came before animals: the higher organisms are of separate sex, and appeared subsequent to the lower.
>
> *Generally confirmed by modern biology, but with qualification as to the undefined borderland between the lowest plants and the lowest animals. And, of course, it recognises a continuity in the order and succession of life which was not grasped by the Greeks. Aristotle and others before him believed that some of the higher forms sprang from slimy matter direct.*
>
> 9. Adverse conditions cause the extinction of some organisms, thus leaving room for those better fitted.
>
> *Herein lay the crude germ of the modern doctrine of the 'survival of the fittest'.*
>
> 10. Man was the last to appear, and his primitive state was one of savagery. His first tools and weapons were of stone; then, after the discovery of metals, of copper; and, following that, of iron. His body and soul are alike compounded of atoms, and the soul is extinguished at death.
>
> *The science of Prehistoric Archaeology confirms the theory of man's slow passage from barbarism to civilisation; and the science of Comparative Psychology declares that the evidence of his immortality is neither stronger nor weaker than the evidence of the immortality of the lower animals.*

Clodd's seventh point refers to spontaneous generation, and the eighth through tenth points are clearly recognizable as an evolutionary scheme, even if they lack, as they must, the Darwinian mechanism. In fact, this is the thesis of Clodd's book: evolutionary thinking has progressed from Thales (d. 548 BC) to Thomas Huxley (1825–1895), interrupted only, in Clodd's view, by the lamentable rise of Christianity. This halted development for around 1500 years, as Clodd explains in Part 2 of the book, titled *The Arrest of Inquiry— A.D. 50–A.D. 1600*.

The Greeks had apparently provided a template for non-theistic biological origins which was to last for 2400 years. This template comprised spontaneous generation as a starting point, and evolutionary processes to explain the emergence of higher life forms.

But spontaneous generation came under increasing scrutiny as the centuries passed. In 1668, Italian philosopher Francesco Redi experimented with varieties of meat and fish in covered and uncovered jars. Redi's

39 Ibid., pp. 30–32.

observation that maggots did not appear on the covered meat counted against spontaneous generation, and scientists began to believe that, at most, it was limited to "parasitic worms and microorganisms".[40] In 1765, another Italian, Lazzaro Spallanzani, carried out experiments that appeared to refute spontaneous generation and published his results. These were questioned by Irish priest John Needham, whose own experiments appeared to support spontaneous generation.[41] The debate continued.

By the late eighteenth century, Charles Darwin's grandfather Erasmus Darwin made his contribution to evolutionary theory. Perhaps influenced by the ancient Greeks, he expressed his ideas in poetic form.[42] Erasmus believed that organisms had the power to adapt themselves over many generations; that all life descended from a primordial life form, which he termed the "filament", and that evolutionary processes were slow.

Erasmus Darwin's ideas anticipated those of Jean-Baptiste Lamarck (1744–1829). Lamarck espoused the "theory of inheritance of acquired characteristics". For example, imagine two herds of deer, one on an island without any natural predators, and another in a different geographic location inhabited by predators. The deer subject to predation would spend more time running than those on the island, resulting in a leaner build with a more highly developed cardiovascular system. These acquired characteristics, in Lamarck's view, would be passed on to the offspring. Over time, the herd would become considerably more suited to running than the island-based deer. Again, Lamarck followed the template provided by the Greeks—spontaneous generation as a starting point, and an evolutionary mechanism to explain diversity.

The disproving of spontaneous generation

In an irony of scientific history, the curtain closed on spontaneous generation at almost exactly the same time that Charles Darwin was putting the finishing touches on *Origin,* published in late November 1859.

Louis Pasteur, the brilliant French biochemist, began his well-known "swan-necked flask" experiments in 1859. By January 1860, Pasteur believed he had decisively disproven spontaneous generation but wanted to buttress his proof with more robust evidence to convince his most ardent opponents. That month, he wrote to his friend Charles Chappuis:

---

40 Strick, 2000, p. 4.

41 Ibid., p. 6.

42 For example, these lines from Erasmus Darwin's poem "Organic Evolution" describe spontaneous generation:

Hence without parent, by spontaneous birth,
Rise the first specks of animated earth;
From Nature's womb the plant or insect swims,
And buds or breathes, with microscopic limbs. (Darwin, 2024).

> I am pursuing as best I can these studies on fermentation which are of great interest, connected as they are with the impenetrable mystery of Life and Death. I am hoping to mark a decisive step very soon by solving, without the least confusion, the celebrated question of spontaneous generation. Already I could speak, but I want to push my experiments yet further. There is so much obscurity, together with so much passion, on both sides, that I shall require the accuracy of an arithmetical problem to convince my opponents by my conclusions. I intend to attain even that.[43]

Pasteur presented his work on spontaneous generation to the Academie on February 6, 1860. He remarked in a letter to his father that his ideas "seemed to produce a great sensation".[44]

In the summer of 1860, Pasteur continued his experiments in mountainous regions, to investigate his hunch that there are fewer airborne germs at higher altitudes. All his experiments pointed to the same conclusion—that spontaneous generation does not occur. Although some die-hard supporters of spontaneous generation continued to beat the drum for some time afterwards, it is generally accepted that by the early 1860s, Pasteur had disproved this ancient Greek presupposition. Pasteur himself stated that he had disproved the theory in his January 1860 letter cited above.

How does this line up with the timing of Darwin's *Origin*? Darwin is believed to have formed his evolutionary ideas in the late 1830s, and then, for reasons which are hotly debated, delayed publication of his theory. It was only in 1858, when Alfred Russel Wallace sent Darwin the outline of a theory which mirrored Darwin's, that Darwin was prompted to complete his text, and publish without further delay.

Although spontaneous generation was finally disproved by Pasteur in 1859–60, empirical evidence which counted against it was provided by Theodor Schwann in 1837, and by Schroeder and Dusch in 1854. But in an 1837 notebook, right around the time Schwann was publishing, Darwin noted that "[t]he intimate relationship between the vital phenomena with chemistry and its laws makes the idea of spontaneous generation conceivable".[45] Darwin's theory of evolution had roots which began in spontaneous generation, even if, during his long delay in publishing, spontaneous generation had become outmoded.

---

43 Vallery-Radot, 1920, p. 87.

44 Ibid., p. 88.

45 Peretó, Bada, & Lazcano, 2009.

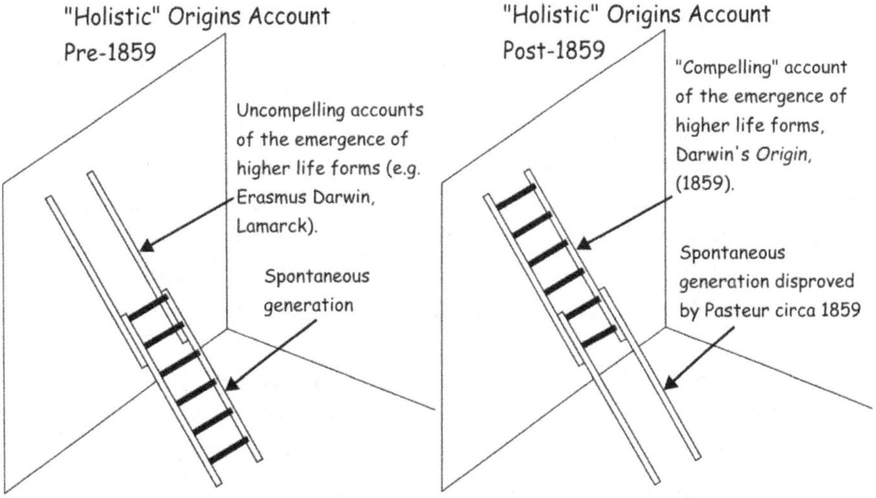

Figure 5.2: Extension ladders representing naturalistic origins accounts, pre- and post-1859

But with Pasteur's work, the ancient idea of spontaneous generation had been decisively overturned. This revolution came just as Darwin was publishing his theory—a theory which was built on the process that Pasteur had just disproved (refer Figure 5.2). The same types of revolution were observed when Newton's laws of mechanics were radically overhauled by Einstein, and also in various technological advancements which made decisive breaks with the ancients.

The reader will recall that technology is an application of science. Therefore, if Darwin represented the capstone of an ancient template, we might expect to see similar examples of ancient technologies being overturned at the same time. And this is precisely what we do see. In the three areas of engineering we discuss below, ancient technologies were being perfected at around the same time Darwin published *Origin*, but then, Darwin-like, they were quickly overturned by technological revolutions.

The first concerns communications. In Esther 8:10, King Ahasuerus, also known as Xerxes (c. 518–465 BC), sent messages across his expansive empire via horse courier. Precisely the same mode of communication was used in America to connect the mid-west and California in 1860–61. The Pony Express, featuring legendary riders such as Buffalo Bill, transported letters to California in 10 days. It is true that the technology behind the telegraph had already been developed, but in the absence of an installed, operable system, the Americans of just 160 years ago relied on an ancient technology to deliver urgent messages.

Turning to the field of construction, we note that the world's tallest building in 1859 was Strasbourg Cathedral, which used the same stone-on-stone construction method as the Pyramids and the Parthenon. Several other cathedrals, which were in various stages of completion, progressively took the title of world's tallest building: the Church of St Nicholas, in Hamburg, in 1874; Rouen Cathedral in 1876, and Cologne Cathedral in 1880. But the Eiffel Tower, completed in 1889, presaged the emergence of tall wrought iron or steel-framed buildings. This construction method, which resulted in buildings known as skyscrapers, completely revolutionized the speed at which buildings were completed and their attainable height. The Empire State Building, with a roof height of 1250 feet, was well over double the height of the tallest cathedrals, and this immense building was completed in 1931 after a construction period of less than two years.

The third technological area is ocean-going transport. The age of sail commenced in antiquity. King Solomon maintained a fleet of ships (1 Kings 9) and the ancient Phoenicians were well-known for their seafaring prowess.

But the crowning achievement of the age of sail—the fastest ships of all—were the tea clippers. An early clipper, the *Rainbow*, was considered a revolution in naval architecture when it was built in 1845. Its performance in the water was in keeping with its sleek lines, easily outpacing all other ships then in service. (In fact, tea clippers may have been the only ships in history that could outpace the highway traffic of their era; the fastest transatlantic crossing by a clipper was completed at an average speed nearly twenty percent faster than a horse and carriage could average on a good road.) The most famous tea clipper, *Cutty Sark*, was launched in 1869 (which, incidentally, was ten years after the publication of *Origin*). Of course, a real revolution was in the offing; by the 1870s, steamships, which, unlike the tea clippers, could utilize the Suez Canal, were voyaging from China to Britain in just over 50 days. This slashed the 90–100 day period of the fastest tea clipper passages.

Both Newton's and Darwin's theories were like tea clippers—apparently revolutionary, they were actually the last and highest forms of ancient templates, all of which were about to be overrun by far larger revolutions. The notion that Darwin followed the template of the Greeks is widely accepted.[46] Edward Clodd, cited earlier, wrote *Pioneers of Evolution From Thales to Huxley* (1897),[47] and Henry Fairfield Osborn, Professor of Zoology at Columbia University, wrote *From the Greeks to Darwin: An Outline of the Development of the Evolution Idea* (1894).

But regarding timing, we can go further and say that Newton's laws of gravitation and motion were like the *Rainbow*, which entered service decades before the Suez Canal opened and the steamships took over. In an analogous way, Newton's laws, published in 1687, seemed assured, as non-Euclidean

46 For an alternate view see: Zuiddam, Benno (2018) "Was evolution invented by Greek philosophers?", *Journal of Creation*, 32(1):68–75.

47 Thales of Miletus was an ancient Greek philosopher.

geometry was only discovered some 140 years later. Darwin's theory of evolution by natural selection, on the other hand, is analogous to the *Cutty Sark*, which launched in the same year that the Suez Canal made available a shortcut to steamships. Darwin published *Origin* at almost exactly the time when spontaneous generation was disproven.

Newton and Darwin: divergent responses

Both Newton and Darwin proposed systems that were partially correct. Newtonian physics is extremely accurate under conditions that we typically encounter on earth. In this "normal" paradigm, Newton's equation $F = ma$ may be used to calculate forces required to accelerate an object, such as a jet aircraft accelerating down a runway. Similarly, Darwin made a useful contribution to biology by showing that natural selection is the mechanism by which some organisms adapt in the wild, without the guiding hand of a human breeder. This was useful empirical causal model science, and Darwin was one of the first to propose this.

However, problems with both Newtonian and Darwinian science become apparent when they are extended beyond their original paradigms. Newtonian physics breaks down when velocities approach the speed of light or when gravitational fields become very strong. At that point, Einsteinian physics must be invoked. It is our contention that, in a similar way, Darwinian evolutionary science breaks down when it is extrapolated to origins issues as a holistic account. Indeed, since the disproving of spontaneous generation, the emergence of higher life forms is no longer the foundational issue that secular origins scientists need to address. The emergence of life is. The situation is, in fact, striking: that Darwin published at the same time that Louis Pasteur disproved spontaneous generation is arguably the greatest scientific wrongfooting in history.

The scientific community has responded to Newton and Darwin in inconsistent ways. Physicists would be cautious invoking Newtonian physics when investigating, for example, Big Bang cosmology. But something different has happened with Darwinism. Contemporary science is yet to acknowledge that the Darwinist scheme breaks down when extended holistically to origins. This leaves dissenters, such as Thomas Nagel[48] and Yale professor David Gelernter,[49] as lonely voices for the time being. For our part, we believe that Darwin's success is in the process of winding down and that Darwinism and its adherents will soon find themselves on the wrong side of an *intellectual divide* which is increasingly manifest in the West. We present details of this forecast in Part 4.

---

48 Nagel, 2012.

49 See "Giving Up Darwin" (Gelernter, 2019).

The recrudescence of pagan origins

It is drummed into modern society that Darwin's *Origin* represented a watershed in the biological sciences. But this understanding is incorrect. The Darwinian link to the ancient Greeks and Romans is highly significant. Darwinism is, in fact, a modern incarnation of pagan origins. Note that one would not describe Newtonian physics as pagan physics, even though Newton's science was linked to the ancient Greeks via Euclidean mathematics. Newton was operating in the field of descriptive and causal model science, which are conducted without reference to religion. Further, Newton accepted the Genesis creation account. Darwin, on the other hand, was operating in the origins science domain, which, as we saw in Chapter 4, is inextricably linked to religious or philosophical presuppositions. Clearly, some of the presuppositions behind Darwinism—that there is no transcendent monotheistic God, that life can emerge spontaneously, that mankind is descended from lower life forms, that mankind is unfallen, that death is not a result of sin—are consistent with the pagan worldview. (Other ideas associated with the Darwinian worldview—that man is made in the image of animals; that the flesh is preeminent; that the fittest shall be first—are notably anti-Christian.) Darwinism therefore represents the recrudescence of ancient pagan origins in our time—sold to the public as neutral, secular science.

*

If Darwinism as a holistic origins explanation ought to be on the way out, we need to account for its popularity. Gertrude Himmelfarb, a historian of the Victorian era, observed that *Origin* was approved by intelligent non-scientists of the time because it appealed to their philosophical prejudices, rather than because of the scientific knowledge or argument it contained.[50] (Of course, Darwinism represented a significant surge in an enormous wave of anti–Christian discourse which had been swelling for several centuries—a wave which has continued down to our own time).

Our view is similar to Himmelfarb's: the success of Darwinism in the origins science realm is largely rhetorical, but this success is sufficient to mask, for now, its theoretical non-success. In the next chapter we review parts of the history of science and changes in its presuppositional basis which eroded societal belief in the Genesis account in favor of naturalism and acceptance of Darwinism.

---

50 Himmelfarb, 1996, p. 296.

References

Aaboe, A. (1964). *Episodes from the early history of mathematics*. Washington D.C.: Mathematical Association of America.

Aaboe, A. (1974). Scientific Astronomy in Antiquity. *Philosophical Transactions of the Royal Society of London, 276*(1257), 21–42. doi.org/10.1098/rsta.1974.0007

Acton, J. (1864). *Human Sacrifice*. London: Robson, Levey, and Franklyn (Privately printed). Retrieved December 14, 2024, from books.google.com.au/books?id=_BZdAAAAcAAJ

Aleksandrov, A., Kolmogorov, A., & Lavrent'ev, M. (1999). *Mathematics: Its Content, Methods and Meaning*. New York: Dover Publications, Inc.

American Council of Learned Societies. (1991). *Biographical Dictionary of Mathematicians* (Vol. 2). New York: Charles Scribner's Sons.

Blacker, C., & Loewe, M. (Eds.). (1975). *Ancient Cosmologies*. London: George Allen & Unwin Ltd.

Breuer, J. (2006). *Introduction to the Theory of Sets*. New York: Dover Publications, Inc.

Carmeli, M. (2002). *Cosmological Special Relativity: The Large-Scale Structure of Space, Time and Velocity* (Second edn). Singapore: World Scientific Publishing Co. Pte. Ltd.

Clodd, E. (1907). *Pioneers of Evolution: From Thales to Huxley*. London: Cassell and Company, Limited.

Darwin, E. (2024, May 29). Extracts from Erasmus Darwin's Epic Poetry. Retrieved from erasmusdarwin.org/wp-content/uploads/2016/11/poem-extracts.pdf

Dreyer, J. (1953). *A History of Astronomy from Thales to Kepler* (Second edn). New York: Dover Publications Inc.

Dulles, A. C. (2005, January). The Deist Minimum. Retrieved September 2, 2024, from firstthings.com/article/2005/01/the-deist-minimum

Feynman, R. P., Leighton, R. B., & Sands, M. (2010). *The Feynman Lectures on Physics: Mainly Mechanics, Radiation and Heat*. New York: Basic Books.

French, S. (2005). Action at a distance. doi:10.4324/9780415249126-N113-1

Gelernter, D. (2019). Giving Up Darwin. Retrieved September 14, 2025, from claremontreviewofbooks.com/giving-up-darwin/

Himmelfarb, G. (1996). *Darwin and the Darwinian Revolution*. Chicago: Elephant Paperbacks.

Kean, S. (2011, January/February). Newton, The Last Magician. *Humanities: The Magazine of the National Endowment for the Humanities*, 32(1). Retrieved December 11, 2023, from neh.gov/humanities/2011/januaryfebruary/feature/newton-the-last-magician

Kuhn, T. S. (1957). *The Copernican Revolution: Planetary Astronomy in the Development of Western Thought*. Cambridge, Massachusetts: Harvard University Press.

Lombardi, M. A. (2007, March 5). Why is a minute divided into 60 seconds, and hour into 60 minutes, yet there are only 24 hours in a day? Retrieved from scientificamerican.com/article/experts-time-division-days-hours-minutes/

McElreath, R. (2020). *Statistical Rethinking: A Bayesian Course with Examples in R and Stan*. Boca Raton: CRC Press.

Nagel, T. (2012). *Mind and Cosmos: Why the Materialist Neo-Darwinian Conception of Nature is Almost Certainly False*. New York: Oxford University Press.

Peretó, J., Bada, J. L., & Lazcano, A. (2009). Charles Darwin and the Origin of Life. *Origins of Life and Evolution of Biospheres*, 39(5), 395–406. doi:10.1007/s11084-009-9172-7

Snobelen, S. D. (1999, December). Isaac Newton, heretic: the strategies of a Nicodemite. *British Journal for the History of Science*, 32(115), 381–419.

Strick, J. E. (2000). *Sparks of Life: Darwinism, and the Victorian Debates over Spontaneous Generation*. Cambridge, Massachusetts: Harvard University Press.

Struik, D. (1967). *A Concise History of Mathematics*. (Third Revised edn). New York: Dover Publications, Inc.

Vallery-Radot, R. (1920). *The Life of Pasteur*. (M. R. Devonshire, Trans.) New York: Doubleday, Page & Company. Retrieved December 11, 2023, from gutenberg.org/cache/epub/60956/pg60956-images.html

6

The great reversal of science: from Christian to anti-Christian foundations

There is no doubt that secular humanist origins accounts have polarized the Western church. The commonly held belief that "science has disproved God" is sufficient to convince a substantial segment of society, including youth in the church, to dismiss Christianity. Consider the personal testimony of Ronald Numbers, a historian of the creationist movement, who shows how loss of belief in a straightforward reading of Genesis was immediately damaging to his faith. Numbers' father was a fundamentalist Seventh-day Adventist minister, and Numbers grew up accepting the Genesis account, including the recent appearance of life on earth. But in the late 1960s, after he attended a lecture on the fossil forest sequences of Yellowstone National Park, his views changed abruptly. Numbers shares his story in the introduction to his book *The Creationists* (1992).

> … [I] stayed up much of the night with a biologist friend of like mind, Joe Willey, first agonising over, then finally accepting, the disturbing likelihood that the earth was at least thirty thousand years old. Having thus decided to follow science rather than Scripture on the subject of origins, I quickly, though not painlessly, slid down the proverbial slippery slope toward unbelief.[1]

Ronald Numbers became an agnostic because he no longer believed Genesis 1–11. No person can relate to Christ in a positive way if His Word cannot be trusted.

Unfortunately, Ronald Numbers had accepted an example of "science" at face value which was almost certainly incorrect.[2] Numbers now describes himself as an agnostic, and it is likely that millions of people in the West could identify with his youthful departure from the beliefs of his parents. Simply put, scientific pronouncements around the age of the earth leave Bible-believing Christians appearing as naive[3]—and worse, wilfully ignorant.

---

1 Numbers, 1992, p. xvi.

2 See "The Yellowstone petrified forests—Evidence of catastrophe" (Sarfati, 1999), available at creation.com/yellowstone.

3 For example, in the 1925 Scopes trial, the old-earth interpretation of William Jennings Bryan was ridiculed.

Within the church and its institutions, secular humanist origins accounts have proven to be an intractable wedge issue, driving members of the church apart in much the same way as a wedge splits wood. Theistic evolution, generally regarded as a deeply problematic doctrine by young-earth creationists, is taught as standard fare in many seminaries.

On the face of it, this situation belies a considerable irony. As we'll show in this chapter, it was the church, and especially the Protestant churches, which gave rise to a sustainable scientific movement. Then, like some invention gone awry, science apparently turned on and trampled the very institution that made it possible.

But is this really what happened? In this chapter we examine the foundations of modern science, and how non-Christians and anti-Christians weaponized *origins* science.

The intellectual entrapment of the ancients

We commence our overview in the early fifth century, the momentous period during which the Western Roman Empire was collapsing. The Goths, a Germanic people from north of the Rhine, were running rampant across large swathes of Europe. They had reached Constantinople in 378, only to turn west and head toward the more vulnerable Western capital. There, under Alaric, descendants of the Goths sacked Rome in 410 AD, a calamity that would have been unthinkable in centuries past. What had happened to the once mighty power of the empire? By what means had their once-invincible legions been sapped of potency? The remaining pagans had a ready answer—the populace had progressively turned away from the Roman gods and embraced Christianity, dooming the empire. The Christian community was under attack.

Augustine, the Bishop of Hippo (in modern-day Algeria), sought to defend the Christians against these charges. Augustine built his ideas upon the trinitarian Christian foundations championed by the Egyptian Christian Athanasius. In his book *The City of God*, Augustine argued that the root cause behind the decline of the classical world was deep-seated intellectual problems with Greek and Roman thought. Rather than the more surface-level issues such as weakened military forces, economic decline, or social instability, the core issue, as Augustine saw it, was the incoherence of classical thought, manifested as intractable incongruities which the great thinkers of the day were unable to resolve.

A key philosophical problem that the Greeks failed to resolve was the problem of the One and the Many. Briefly stated, this problem related to whether there was a unifying concept, or mind, behind the universe, or whether the universe was instead a pluralist entity. For example, there are many people in the world, but they follow a single form. Which is the more fundamental concept—their unity or their manyness?[4]

The Greek philosophers, Plato and Aristotle, proposed different solutions to this issue. For Plato, "ideals" were more important than particular items in the physical world. These ideals were in the theoretical realm, divorced from and (to his mind) more real than hard facts. Aristotle, on the other hand, was more focused on the many. He, and those of his persuasion, placed priority on observable and tangible objects in the physical world.

These starting points, taken to their logical conclusion, yielded divergent approaches to life. The Platonic approach led in the direction of mysticism, whereas the Aristotelian approach led toward empiricism and science.

But this issue also held relevance in the political sphere. Those who veer to "the one" focus on the corporate aspects of humanity, therefore diminishing the importance of the individual. This approach, described hypothetically in Plato's *Republic,* leads politically in the direction of totalitarianism, where individual rights are suppressed so that the state can assume unfettered power. Those who lean towards "the many" favor individual rights. Taken to an extreme, this leads toward anarchism or anarcho-capitalism, where individual rights and freedoms are emphasized over and against the role of the state. So, which is the right approach? The classical philosophers failed to reach a consensus. They were unable to adequately reconcile the two opposing approaches.

Augustine invoked the Christian doctrine of the Trinity to resolve this philosophical dilemma of the One and the Many. Trinitarian doctrine, which envisages God as One, but existing in three Persons, supports the notion that both the one and the many are equally ultimate. The Christian approach incorporates the best of both positions and transcends each one—in the political sphere this implies the individual is important, but so is the state, as the national corporate political vehicle. Both must be respected in their domains, and neither should trample the rights or threaten the existence of the other.

Having pointed out these difficulties with classical thought, Augustine later took care to weed out pagan presuppositions from his own way of thinking. In the area of origins, Augustine would have been aware of Aristotle's concept of the eternal universe. But here was a conundrum: the concept of the eternal universe was entirely rational, and one that addressed the problem of existence, but it had come from the same place—Greek philosophy—that Augustine knew was marred by fundamental inconsistencies. In the *City of God,* Augustine came down on the side of the Bible:

> For as it is not yet six thousand years since the first man, who is called Adam, are not those to be ridiculed rather than refuted who try to

4 "The one and the many" can be expressed in several ways depending on context: unity and plurality; one and several; the forest and the trees. The "many" simply means more than one, not necessarily "a very large number".

> persuade us of anything regarding a space of time so different from, and contrary to, the ascertained truth? For what historian of the past should we credit more than him who has also predicted things to come which we now see fulfilled?[5]

This was a watershed statement. It meant that Augustine, one of the most important early Christian theologians, and a true intellectual heavyweight, had openly come out in favor of what we would describe today as a "young earth".[6] For more than a thousand years this would give an aura of support to all who believed in a plain reading of Genesis.

But another development of historical import was also unfolding. As the Roman empire collapsed, the philosophical works of Aristotle were for a time lost to Western scholarship. The ideas of Plato took hold.[7] This influence was readily observable in the way Christians conducted themselves. Augustine, writing in *Confessions,* laments the time wasted looking at a spider trap a fly in its web. This distraction had taken away from time that he might have spent in contemplation of God. This related to the Platonic approach from ancient Greece in which material things are secondary to the spiritual.

But there was an irony in Augustine's words. He had uncovered, in *City of God,* a different type of web, an intellectual entrapment that had frustrated classical thought. The Christian concept of a triune God led to a resolution of the philosophical problem that had baffled Plato and Aristotle. These brilliant philosophers, who were operating on human reason alone, eventually found themselves trapped in strands of an intellectual web from which progress became difficult.

Christianity did not only resolve the problem of the One and the Many. Pagan gods were thought to act within the world, as a part of it; consequently, many classical scholars looked for ultimate answers within the confines of time and space. In contrast, the God of the Bible is described as the Creator of the world, existing beyond time and space in a spiritual realm. As such, Augustine recognized that it is impossible to reason one's way to presuppositions that underpin the universe from within the universe itself.

History was soon to demonstrate that the trinitarian Christian ideas and presuppositions which Augustine espoused in *City of God* were intellectually sufficient to support the emergence of an extensive Christian civilization. It is said, perhaps apocryphally, that Charlemagne, the Frankish king who was the

---

5 Augustine, 1871, p. 267.

6 Augustine's views are not entirely precursors of modern young-earth creationism, however, his view of the age of the universe is consistent with YEC.

7 Some of the educated early Christians, including Augustine, were led to Christianity via the ideas of Plato and his successors. These ideas served as a steppingstone to the transcendent God of the Bible (Smith, 2018, p. 182).

first Holy Roman Emperor, routinely slept with a copy of that book "beneath his pillow".[8]

Indeed, an intellectual, cultural and artistic renaissance occurred during the reign of Charlemagne and his successors during the eighth and ninth centuries. Learning and scholarship revived after the decline that had followed the fall of the Western Roman Empire. This is known as the Carolingian Renaissance.[9]

Carolingian art combined classical Roman, Byzantine, and Germanic influences, leading to the creation of illuminated manuscripts, sculptures, metalwork, and large-scale church construction.

The Carolingian Renaissance was closely tied to religious reform, as Charlemagne sought to standardize Christian worship practices across his empire. This involved promoting uniformity in church liturgy, music (Gregorian chant), and ecclesiastical law.

Charlemagne gathered scholars from across Europe, including figures like Alcuin of York, Paul the Deacon, and Einhard. These intellectuals were instrumental in advising the emperor and contributing to the intellectual vigor of the court.

Schools were established in monasteries and cathedrals to train clergy and administrators, reviving the study of the liberal arts (grammar, rhetoric, logic, arithmetic, geometry, music, and astronomy). Monks in monasteries were tasked with copying and preserving important classical and Christian texts. The Carolingian Renaissance played a crucial role in laying foundations for further progress during the medieval period and preserved much of the knowledge of classical antiquity for future generations.

Aristotle returns: priming Europe for empiricism

We move forward six hundred years to the high medieval period. Popular historian Justo Gonzalez refers to this period as the "golden age of medieval Christianity",[10] adding that the thirteenth century "marked the apex of papal power ... ".[11] There had been broad unity in many important ways, unity that enabled the Christian world, rightly or wrongly, to engage in large-scale enterprises such as the crusades.

An important development at this time was the reintroduction of Aristotelian philosophy into Western Europe. This was the result of increased contact with the Muslims, who had greater access to Aristotelian philosophy, and whose scholars—particularly Averroes (Ibn Rushd) from Muslim Spain

8 Cochrane, 2003, p. 417.

9 The adjective "Carolingian" derives from the Latin name "Carolus Magnus", who is known in English speaking nations as Charles the Great, or more commonly by the French name Charlemagne. Carl the Great (Carl der Grosse) is a German variation.

10 González, 2010, p. 357.

11 Ibid., p. 369.

(Al-Andalus)—had written extensively on Aristotle's works. With Averroes' commentaries available in Europe, churchmen such as Thomas Aquinas (1225–1274) began incorporating Aristotelian thought into their theology. This fit with one of the theological developments of the era, scholasticism. The Scholastics, or schoolmen as they are also known, sought to apply human reason to theological matters.

It was the Aristotelian revival that cleared the path towards modern science. Aquinas' teacher, Albert the Great (c. 1200–1280), "wrote about animals and plants, about heavenly and earthly bodies".[12] This was a departure from Augustine, who had focused primarily on philosophical and theological concerns rather than the natural world. Instead, observational science came to the fore, and "one could even say that it was Thomas [Aquinas] who opened the way for Western Modernity".[13] But Aristotelianism itself was insufficient to kick-start science because it promoted highly logical but rigid patterns of thought—and armchair-style theorizing, that is, thinking about the natural world and how it works without stepping forth to investigate it.

In 1277, the Bishop of Paris issued a condemnation of 219 philosophical and theological propositions being taught in universities. Many of these propositions were derived from Aristotelian natural philosophy. Effectively, the Bishop had opened the way for Aristotle to be critiqued and for new ideas to be investigated. Pierre Duhem (1861–1916), the French physicist and historian of science, identified this event of 1277 as preparing the soil for the development of modern physics.[14]

Not long after this edict, Jean Buridan (1300–1358) developed his concept of *impetus*, which explained why objects keep moving after they are thrown and why the celestial spheres continued to rotate. He even provided a mathematical formula: $impetus = weight \times velocity$,[15] which is similar to the Newtonian equation for momentum ($mass \times velocity$; mass is directly related to weight). Buridan's ideas were the precursor for Galileo's concept of inertia, and directly challenged Aristotle's ideas about motion.

But along with progress came controversies. The papacy became heavily politicized under pressure from the French monarch, who moved the papal court from Rome to Avignon in 1309. Power and accumulation of wealth were emphasized, and spiritual focus was lost by the church. (This would lead in time to the Western schism which split the Roman Catholic Church.) At a time when church legitimacy was being undermined, a tragedy struck.

---

12 Ibid., p. 379.
13 Ibid., p. 380.
14 Duhem, 1911.
15 Pedersen, 1974, p. 210.

The Black Death

Introduced to Europe via modern-day Western Russia in 1346, the bubonic plague is believed to have spread via the Black Sea to Constantinople (modern day Istanbul), and from there to southern European ports, such as Venice, Genoa and Marseilles, by 1347. From these entry points, the plague spread northwards into Western Europe, as well as south to Northern Africa.

The plague was extremely virulent—highly infectious and associated with high rates of mortality. Most infected people died within a week of onset. Over a period of four years, the plague devastated much of Western Europe, particularly the crowded towns and cities. Somewhere between 30 to 50 percent of the European population is believed to have died, making it one of the most calamitous events since Noah's Flood. Western Europe had been struck by a once in 3000-year disaster.

As the survivors moved about their ravaged towns and countryside, complete with mass graves and empty dwellings, fundamental shifts in the zeitgeist became evident. The elephantine presence in Western Europe, the institution that was everywhere, was the church, and the plague had occurred on the church's watch. While much of the church elite wanted things to continue as they had, there were two other responses, both in opposition to each other, that called for transformative change.

The first of these secondary responses was a reformist impulse. This was embodied in the career of John Wycliffe, an English Catholic priest who increasingly voiced concerns with what he saw as corrupt church practices. But Wycliffe's most significant grievance was the way in which the church prevented lay people from directly accessing the Bible. The scriptural texts were written only in Latin, a language that had been "dead" for centuries among the general population. The Bible was therefore the exclusive preserve of the educated elite. Toward the end of his life, Wycliffe took steps to set this right. Working with several assistants, Wycliffe translated the Bible into the language of the people, what is known today as Middle English. This was revolutionary—the biblical message had now escaped the social stratum of the educated ecclesiastical class and appeared in the hands of laypeople. Wycliffe had started a movement that would reach full flower in the Reformation. By the early 1500s the doctrine of *Sola Scriptura*[16] was championed by the newly formed Protestant denominations.

16 Sola Scriptura—by scripture alone—was one of five solas which briefly summarized important truths concerning the Christian doctrine of salvation. Sola Fide means "by faith alone"; Sola Gratia means "by grace alone"; Solus Christus means "by Christ alone"; Soli Deo Gloria means "Glory to God alone". Sola Scriptura did not mean that church tradition was of no value, rather it meant that scripture trumped tradition where there was a conflict between the two. It also meant that scripture was held to contain all knowledge necessary for salvation.

On the opposite side were those who were antagonistic to the reformist response, and would do away with the church altogether. In the early 1350s, the Italian writer Giovanni Boccaccio (1313–1375) published his *Decameron*, a collection of a hundred satirical short stories that ridiculed the church and its beliefs. This book, which attracted the censure of the church, denoted the emergence of anti-clerical ideas[17] in Europe. Another Italian writer and poet, contemporary with Boccaccio, was Francesco Petrarca (anglicized as Petrarch, 1304–1374). Petrarch coined the term "dark age", a derogatory term referring to the period after the fall of the Roman Empire during which papal Christianity became ascendant. (It was a term that stuck; references to "the dark ages" are still commonly heard today.) These early precursors exhibited several traits that would emerge more fully as an anti-Christian movement by the turn of the sixteenth century. They were both members of the literati—the creative class—and they both went public with an anti-clerical, anti-Christian line. By the time of its heyday around the early 1500s, an anti-Christian form of humanism would be the ideological enemy of organized Christianity. The aim of this brand of humanism was the removal of biblical authority.

This was a significant change. The pre-plague era had seen a broad unity in the church, and an elite which largely supported the tenets of the institution. But in the post-plague era, internal fissures in the church body had appeared, and these would widen until they split the church and gave birth to opposing forces in Western Christendom. This in turn would have a profound impact on origins science. By the time the scientific enterprise was turning up evidence to be evaluated, two opposing presuppositional camps would be seeking to interpret that evidence.

The Renaissance: onset of anti-Christian history

The Renaissance is often presented by secular humanists as a sort of revival of learning and progress. In the fifteenth and sixteenth centuries, the "humanists" were simply those interested in the humanities—and more specifically classical (ancient) Greek and Roman literature, the more ancient, the better.[18] Although Renaissance humanism was diverse, proponents were typically nonplussed by Aristotle and the Roman Catholic scholastics and mocked the tradition of university education.[19] Many humanists rejected science altogether.[20] Among their members were anti-Christians whose purpose

---

17 North, 2015.
18 Hannam, 2010, p. 212.
19 Kuhn, 2003, p. 127.
20 Ibid., p. 127.

was to attack the power of Christianity in society[21] by acting as countercultural dissidents flouting tradition.[22]

These humanists wrote a revisionist history which debased and belittled the contribution to Western civilization made by Christianity and the Bible, instead substituting the myth that the West was fundamentally based on Greek and Roman learning, including classical metaphysics and ethics.[23] The so-called Dark Ages were presented as a period of sterility, ignorance, superstition, obsession with the afterlife and relentless religious oppression. No respect was to be paid to the past.[24] The term "dunce" was invented as a play on the name of Duns Scotus, a scholastic philosopher, whom these humanists regarded as an obscure puzzler.[25]

The humanists did not merely criticize the books of the "medievals", they handed them over to bookbinders as scrap paper.[26] In the 1500s, Oxford University lost most of its medieval collections and sold the bookcases, while in Merton College, three-quarters of the ancient library disappeared.[27] Three hundred years of strenuous effort would have been lost, except that fortunately, copies had been run off in the printing presses before the humanists trashed the originals.[28]

Philosopher of science Thomas Kuhn, whom we met in Chapter 4, notes that the humanists were associated with efforts aimed at "stopping science",[29] although these proved unsuccessful. Irrational ideas such as magic, fate, the occult, belief in witchcraft, and fear of witchcraft flourished in the Renaissance.[30] Progress ground to a halt, and it was only the rise of Christian Protestantism that was able to breathe fresh air into the sails of Western civilization.[31] Even the idea of linear progress in history was foreign to the minds of the Renaissance men.[32]

It was in this milieu that the British Parliament convened the Westminster Assembly of Divines. This distinguished group produced the Westminster Confession (1647), which included, in Chapter 4, a statement of six-day creation:

> It pleased God the Father, Son, and Holy Ghost, for the manifestation of the glory of his eternal power, wisdom, and goodness, in the beginning, to

---

21 Nisbet, 2009, p. 353.
22 Ibid., p. 102.
23 North, 2022, pp. 246–247.
24 Nisbet, 2009, p. 103.
25 Hannam, 2010, p. 217.
26 Ibid., p. 219.
27 Ibid.
28 Ibid., p. 220.
29 Kuhn, 2003, p. 127.
30 Nisbet, 2009, p. 105.
31 Ibid., p. 118.
32 Ibid., p. 102.

> create, or make of nothing, the world, and all things therein, whether visible or invisible, in the space of six days; and all very good.[33]

Why did modern science begin in Europe?

Science has emerged at several times and places in history, but sustained scientific development arose first in sixteenth century Europe. Science communicator David Ellyard noted in his book *Who Discovered What When* that

> … nearly all the discoveries within this book were made in Europe or in regions to which European influence spread, such as North America. Modern science is rooted in the history and culture of Europe; other parts of the world have begun to make their contribution only recently.[34]

Short-lived efforts at scientific progress had previously arisen in various places, notably ancient Greece, which had its share of gifted scientific thinkers. The same could be said of ancient China, India, and the early Islamic civilization. But scientific progress in all these places "petered out", as Ellyard puts it. There were flashes of brilliance, but none of these led to sustainable progress. The common ingredient to all these false starts was a relatively wealthy civilization that afforded some individuals the scope to pursue research and investigation, in contrast to poorer societies which were largely occupied with securing their own survival. Nevertheless, wealth and means were not sufficient to make scientific progress sustainable. A case in point was ancient China, whose technology and exploration went "into eclipse when autocratic central governments shut down the tradition of free enquiry essential for science to flourish".[35]

In Roman Catholic Europe, there was steady progress in philosophy, logic, mathematics and science, with "overwhelming documentation" demonstrating a solid interest in science from the 1100s onward, an interest supported and underpinned by the Christian theism of the medieval period.[36] The humanist philosopher and science writer Loren Eiseley observed that science owed its start "to the Christian conception of the nature of God" leading to the "sheer act of faith that the universe possessed order and could be interpreted by rational minds".[37] Philosopher Alfred North Whitehead (1861–1947) credited the Roman Catholic scholastic emphasis on reason, organization and encyclopedic knowledge as essential preconditions for the emergence of modern science. Further, the scholastic realists viewed God as "utter

33 MacPherson, 1882, p. 52.
34 Ellyard, 2005, p. 11.
35 Ibid., p. 11.
36 Jaki, 1978, p. 13.
37 Eiseley, 1958, p. 62.

rationality" who had planned the cosmos, and since human rationality was in the likeness and image of God, the development of science became possible.[38]

Around 1500, a system of patronage in Italy enabled men of learning to carry out scientific investigations. This was followed shortly after by the Protestant Reformation, sparked in 1517 when Martin Luther nailed his 95 theses to the church door in Wittenberg. The Reformation quickly spread to other parts of Europe. A central idea of the Reformers was *Sola Scriptura,* the notion that the Bible represented the final authority for appeals on spiritual matters, rather than the institutional church and its traditions. Peter Harrison argues that the "Bible—its contents, the controversies it generated, its varying fortunes as an authority, and most importantly, the new way in which it was read by Protestants—played a central role in the emergence of natural science in the seventeenth century".[39] The synergy between the development of Protestant theology and science is intimated by the words of the English philosopher and "father of empiricism" Francis Bacon (1561–1626):

> To conclude, therefore, let no man out of a weak conceit of sobriety, or an ill-applied moderation, think or maintain, that a man can search too far or be too well studied in the book of God's word, or in the book of God's works; divinity or philosophy; but rather let men endeavour an endless progress or proficience in both.[40]

The Puritans deserve special mention here. These Reformed English Christians of the sixteenth and seventeenth centuries had a significant impact on American culture, as well as a significant influence on science. Robert Nisbet noted it is "incontestably clear how close are the roots of Puritan religious faith and the kind of science we associate with Newton and his contemporaries ... [especially the] faith in the permanence and regularity of natural law".[41]

We recall from Chapter 2 that by contrast, atheism gives a person no reason for proceeding on a scientific basis, whether the individuals concerned are the unnamed physics professor that approached K. Scott Oliphint, or Nobel prize-winning scientists like Albert Einstein, Richard Feynman and Eugene Wigner (also quoted in Chapter 2). The modern layperson is taught that science began after the religious ideas and superstitions of the past went into decline, but it can't have—as the statements of the atheistic physicists from the previous chapter, and informed secular humanists quoted above, indicate. Atheistic scientists can join the bandwagon once the scientific enterprise is underway and participate in the scientific processes that Christians trailblazed; but they would

---

38 Russell, 1991, p. 73.
39 Harrison, 1998, pp. 4–5.
40 Darwin, 1871, p. 2.
41 Nisbet, 2009, pp. 125–126.

not have invented science themselves. It is a fact of history that no non-Christian civilization invented a sustainable scientific enterprise.

The Reformers then did something which, as far as we know, had never before been attempted in history: they established universal education. (Other civilizations, like the ancient Israelites, were content to restrict literacy to a select few—refer Isaiah 29:11–12.) The aim was to enable ordinary persons to read the Bible for themselves. As a result, "the literacy rates in the newly reforming populations in Britain, Sweden, and the Netherlands surged past more cosmopolitan places like Italy and France," according to Harvard professor Joseph Henrich.[42] This maps closely to what Ellyard states about the rise of science: "Rich and powerful Catholic Spain made little contribution to science for centuries and was far outdone by smaller powers in the Low Countries and Scandinavia, as well as England and Scotland, where the Reformation took root".[43] There is no doubt that the Protestants cared deeply about literacy; Henrich goes on to add that the Protestant emphasis on education is still visible in divergent literacy rates among Protestant and Catholic missions in different cultures.

> In Africa, regions with early Protestant missions at the beginning of the Twentieth Century (now long gone) are associated with literacy rates that are about 16 percentage points higher, on average, than those associated with Catholic missions. In some analyses, Catholics have no impact on literacy at all unless they faced direct competition for souls from Protestant missions. These impacts can also be found in early twentieth-century China.[44]

The Roman Catholic Church was not caught entirely flat-footed however, as the Jesuits, a Roman Catholic order, had also "involved themselves in education and the establishment of schools".[45] Further, the Jesuits made significant contributions to science, as did Roman Catholic France.[46]

The Protestant tenet of universal education, combined with the educational efforts of the Jesuits in the Roman Catholic world, had opened opportunities for a vastly higher proportion of the population to participate in the intellectual life of their communities. This was undoubtedly an important factor in making the nascent scientific enterprise sustainable. There were simply more educated people around. This effect is somewhat analogous to the sporting prowess of various nations. Norway, a country with a relatively low population, is the top ranked country in the all-time Winter Olympics medal table, beating out much larger countries such as China and the United States. Norwegian children

---

42 Henrich, 2021.
43 Ellyard, 2005, p. 13.
44 Henrich, 2021.
45 Hannam, 2010, p. 228.
46 Ibid., p. 228.

participate in winter sports at a high rate, which flushes out almost all the available talent, and has led to a sustainably high level of performance in these sports. In the same way, the Protestant drive for universal education, a world first, generated a relatively large population of educated people that ultimately helped to make scientific advancement self-sustaining.

Continuing with the winter sports analogy, consider Australia, a country in which these sports are not particularly prominent. Australia produced a winner in the speed skating at the 2002 Winter Olympics, which caused a national sensation at the time, but not enough to spark a sustainable, whole-of-nation interest in winter sports. Predictably, Australia has gone on to achieve only a small fraction of the success of Norway in the Winter Olympics. This is analogous to the occasional genius that has appeared from time to time in various places—such as Al-Khwarizmi in the Muslim world—without sparking sustained scientific progress in the civilization.

The scientific revolution was undeniably the progeny of Christian Europe. The philosophical advances under Roman Catholicism and the Protestant mindset in relation to theology both contributed. As far as we know, European Christian civilization was the first to attempt universal education, a critical component of a sustainable scientific enterprise. Both Roman Catholics and Protestants in Europe went on to make decisive contributions to science.

Presuppositions shift: Deism and Higher Criticism

The anti-Christian impulse resulted in two important intellectual developments unfolding throughout the eighteenth century. The first of these was Deism, "which taught that whatever God there may be is nothing more than the First Cause, a force that made the world the way a clockmaker makes a clock and having set its mechanism to operate according to certain laws, simply winds it up and lets it run".[47] Deists opposed biblical Christianity; their appraisal of God as a hands-off deity precluded any belief in miraculous intervention as the Bible declares. Deism swept through the elite; prominent deists included Lord Shaftesbury (1671–1713) and the Tory leader Lord Bolingbroke (1678–1751) in England, and Thomas Jefferson and Benjamin Franklin in America. Jefferson famously produced his own version of the New Testament, known today as the "Jefferson Bible", with the miracles of Christ edited out.

Deism reached its peak around 1730 with the publication of Matthew Tindal's *Christianity as old as the Creation*.[48] But its appeal was to some extent limited to the educated elite, rather than the masses. Gone were the days of Thomas Aquinas when the entire society held a high degree of respect for

47 Dallimore, 1970, p. 20.
48 Ibid., p. 21.

Christianity. Now, a decided split between the elite and the masses had become discernible. To illustrate this point, consider two accounts from the eighteenth century; the first from the courts of St James's Palace, where, in November 1737, Queen Caroline lay on her deathbed. Among the small group of royalty and courtiers in attendance, the Prime Minister, Robert Walpole, was also present:

> And now we have a painful but very characteristic scene. People wondered that the Queen did not have anyone to pray with her. To stop these remarks, Robert Walpole asked the Princess Emily to suggest to the Queen that Archbishop Potter should be sent for. The Princess hesitated. Then, although about a dozen persons were present, Walpole added: "Pray, Madam, let this farce be played; the Archbishop will act it very well. You may bid him be as short as you will. It will do the Queen no hurt, no more than any good; and it will satisfy all the good and wise fools who will call us atheists if we don't profess to be as great fools as they are".[49]

Our second account, which shows the respect for Christianity maintained by the masses, dates to 1740, when the itinerant Methodist preacher George Whitefield was touring in the New England area of America. He arrived in Middletown, Connecticut, a town located along the Connecticut River around 25 kilometers from Hartford. An "unlettered farmer"[50] from the region, Nathan Cole, recounted his memory of the event, leaving us insight into the reception which the masses were willing to give a Christian minister:

> one morning all of a Suding there came a messanger & said mr whitefield ... is to preach at middletown this morning at 10 o'clock I was in my field at work I dropt my tool that I had in my hand & run home and throu my house and bad my wife to get ready quick to go and hear mr whitefield preach at middletown & run to my pastire for my hors with all my might fearing I should be too late to hear him & took up my wife & went forward as fast as I thought ye hors could bear...
>
> ... when we gat down to ye old meeting hous thare was a great multitude it was said to be 3 or 4000 & when I looked toward ye great river I see ye fery boats running swift forward and backward—when I see mr whitefield come up upon ye scaffold he looked allmost angellical a young slim slender youth before thousands of people and with a bold undainted countenance & my hearing how god was with him everywhere as he came along it solemnized my mind and put me in a trembling fear before he began to preach for he looked as if he was Clothed with authority from ye great god ... .[51]

49 Ibid., p. 23.
50 Ibid., p. 540.
51 Ibid., p. 541.

Such was the difference between the elite and the masses, who remained open to the gospel and responded in large numbers during the Great Awakening of the mid-1700s.

With Deism declining in relevance throughout the latter part of the eighteenth century, a second anti-Christian impulse, Higher Criticism, emerged. This would further erode confidence in the Bible through the nineteenth century and into the middle of the twentieth, starting in universities and seminaries. An important precursor to Higher Criticism was the work of Baruch Spinoza (1632–1677), a Dutch Jew who was excommunicated by his Orthodox community in 1656 for his heresies. To Spinoza, God was not personal. Instead, God and nature were one and the same (pantheism). With respect to scripture, Spinoza rejected supernatural explanations and miracles.

Higher criticism of the Old Testament is generally regarded as starting with the French medical doctor Jean Astruc in 1753. Noting that Genesis contained two different words for God (Elohim and YHWH), Astruc speculated that Moses compiled Genesis from two different source documents.[52] This line of thinking found fertile ground among academics. Astruc's initial ideas spread to German universities where they were expanded and modified.[53] The nascent movement sought to critically appraise the books of the Old Testament using the same analytical methods that were used with classic literature,[54] increasingly bringing to the task a presupposition that the Bible was untrue.[55] The higher critics denied the unity and moral integrity of scripture,[56] proposing instead that scripture was an incoherent patchwork of source documents that contained errors.[57] As might be expected, the higher critics generated theories which agreed with their presuppositions. The early higher critics understood that direct criticism of the Bible would not have been acceptable to "all the good and wise fools" who had similarly spooked Prime Minister Walpole in 1737. Instead, they adopted the subtler approach of appearing to praise the Bible as superb literature, while attacking its origins, authorship and infallibility.[58]

Higher critic Julius Wellhausen published a notable book in 1878 which sought to prove that the Pentateuch (the first five books of the Bible) was "in large measure 'doctored' by the priests under Ezra". According to Wellhausen, the priests aimed to promote a falsehood that Moses wrote the Pentateuch and

52 O'Doherty, 1953, p. 301.
53 Ibid., p. 303.
54 North, 1989, p. 22.
55 C.S. Lewis wrote "I find in these theologians a constant use of the principle that the miraculous does not occur. … I only want to point out that this is a purely philosophical question. Scholars, as scholars, speak on it with no more authority than anyone else. The canon 'If miraculous, unhistorical' is one they bring to the study of the texts, not one they have learned from it. … On this they speak simply as men; men obviously influenced by, and perhaps insufficiently critical of, the spirit of the age they grew up in" (Lewis, 1967, p. 158).
56 North, 1989, p. 33.
57 Ibid., pp. 39–40.
58 North, 1989, p. 3.

that there was centralized worship in the Tabernacle—which Wellhausen denied existed.[59] Wellhausen's technique was to claim that any "passage that fit his theory [was] authentic; all others [were] forgeries." He also proposed that Israel moved gradually from paganism to monotheism. This fit with the general concept of evolution which had become fashionable at the time. It was also consistent with the popular idea of the philosopher Hegel that history progresses from lower to higher stages.[60] Sadly, Wellhausen's work came to dominate Old Testament scholarship in Germany, England and America for a generation,[61] despite the lack of evidence for what he promoted.

Writing in the 1950s, C.S. Lewis, an expert in ancient literature, recorded his dim view of the higher critics and their methods:

> These men ask me to believe they can read between the lines of old texts; the evidence is their obvious inability to read (in any sense worth discussing) the lines themselves. They claim to see fern-seed and can't see an elephant ten yards away in broad daylight.[62]

By the 1970s, the movement was characterized by a growing lack of consensus among its exponents. Conflicting conjecture, a lack of objective external evidence,[63] and contradictory results[64] caused Higher Criticism to decline.[65] However, in discrediting the Old Testament, higher criticism had encouraged resistance to biblical Christianity that was to reach a new level with the emergence of secular humanist origins science.

Neither Deism nor Higher Criticism survived to see the masses fall away from Christianity. In the late 1950s, when Deism had long gone and the Higher Criticism movement was on its last legs, more than 130,000 people flocked to hear evangelist Billy Graham preach at the Melbourne Cricket Ground. That was more than seven percent of the city's population, and reminiscent of Whitefield's immense pulling power in 1740. In 1966, film director John Huston released *The Bible*, a movie covering the first 22 chapters of Genesis. It was the second highest grossing film in North America that year.

But if we move forward another half-century to the present, beyond the reach of Deism and Higher Criticism, Christianity in the West appears to be in rapid, structural decline; the elite have finally persuaded the masses to join them. Rather than a direct attack on the Bible, a major factor in this decline is the far more subtle effect of secular origins science. This emerged, at least in the

---

59 Cardozo, 1995.
60 Ibid.
61 Ibid.
62 North, 1989, p. 36.
63 See Sarfati, 2018, p. 32.
64 Cardozo, 1995.
65 For a summary of arguments refuting key aspects of the Documentary Hypothesis of the Higher Critics see "Documentary Hypothesis Fallacy" (Sarfati, 2018, pp. 22–32).

Anglosphere, in the late 1700s. It would lead to a paradigm shift in anti-Christian discourse, and eventually to the mass de-Christianizing of the West.

The emergence of anti-Christian origins science

Up until the eighteenth century, there was no applied scientific enquiry into the age of the earth. Gottfried Leibniz (1646–1716), the German polymath, had asserted in 1680 that the earth had initially been in a molten state; but it was not until Comte de Buffon's experiments into the cooling time of heated cannonballs, in 1775, that this concept was explored scientifically. This meant that for the entirety of the seventeenth century there was no credible scientific evidence for a long-age earth.[66]

Several early Christians had attributed the fossil record to the biblical deluge, including Tertullian, Chrysostom and Augustine.[67] In the modern European context, the early Danish geologist (and Catholic priest) Niels Steensen (1638–1686), better known by his Latinized name, Nicolas Steno, accepted the biblical evidence regarding the age of the earth and considered fossils to be evidence for the deluge. It was Steno who famously introduced the concept of superposition of sedimentary rock, whereby horizontal deposition occurs in layers, with the position of strata determining their ages relative to each other.[68] As geological field-work proceeded, a few authorities such as Alexander Catcott (*Treatise on the Deluge*, 1768) maintained a view compatible with a straightforward reading of scripture, but increasing numbers of deistic or atheistic investigators began to promote theories which clashed with the Genesis record.[69]

Between 1790 and 1820, the discipline of geology went through several battles. Those who followed the founder of German geology, Abraham Werner, accepted that the formation of rock layers in the earth was largely the result of the chemical and physical action of seawater, a view termed Neptunism.[70] The Scottish naturalist James Hutton, an contemporary of Charles Darwin's grandfather Erasmus Darwin, proposed an alternate theory which was colored by his philosophical beliefs and the local geology of his area. Hutton's geological theory, set out in his 1788 book *Theory of the earth*, posited that the formation of the earth involved repeated cycles in which debris eroded from rock was compacted into sedimentary rock. The sedimentary rock was in turn affected by subterranean heat before being uplifted above sea level. Hutton's view was termed Vulcanism. Hutton was also an early proponent of

66 "Old earth" refers to the much longer timeframes than the approximately 6,000 years of the biblical ("young earth") timeframe.

67 Mortenson, 2004, p. 25.

68 Ibid., p. 26.

69 Ibid., pp. 26–27.

70 Himmelfarb, 1996, p. 83.

"uniformitarianism", the view that changes in the earth have been gradual and constant over a very lengthy period, and that the only processes that have operated over the entire history of the earth are the processes observable today. Neither the Neptunists nor the Vulcanists engaged with the fossil record to any great extent,[71] but both schools accepted an old age for the earth. This was in line with the presumed deistic or atheistic worldview of the founders of these groups.[72]

Hence, by the late 1700s, scientific ideas began to clash with the biblical record as Deism and atheism reached ascendancy among the elite. This marked the first time since the dawn of Christianity that opposition to the faith emanated from a source other than the religious or philosophical.[73] Applied science, which apparently dealt in fact, was beginning to render traditional interpretations of biblical truth as seemingly implausible. This confrontation came via interpretations of geological discoveries which invoked an extremely old earth, well beyond the timeline given in the biblical record.

Hutton rejected the biblical chronology of Genesis, proposing instead that the earth was extremely old. The uniformitarian presupposition that he held required him to reject the creation account, which entails processes which have not occurred since. Hutton's theory met solid resistance because it aroused a suspicion of atheism; at this point Christian belief remained high in the wake of the Methodist revival. Hutton "tried to argue that religion and science should not come into conflict because they held authority in different spheres".[74] Nevertheless, Hutton's closing words in *Theory of the earth* were profoundly significant to those who interpreted Genesis in a straightforward manner:

> The result, therefore, of our present enquiry is, that we find no vestige of a beginning,—no prospect of an end.[75]

Hutton's concepts had cast off the six-day, *ex nihilo* creation espoused in the Westminster Confession like an anachronism. While it is true that initially, Hutton's abstruse writing style limited his notion of an old earth to an intellectually sophisticated audience, a popularizer soon emerged: John Playfair, a Church of Scotland minister. Playfair published *Illustrations of the Huttonian Theory of the Earth* in 1802.

In 1807, the Geological Society of London was formed. This society originally comprised thirteen wealthy London residents who were,

---

71 Mortenson, 2004, p. 29.

72 Ibid., p. 224.

73 The Galileo affair, which is often claimed to represent a clash between Christian faith and science, was in reality a clash between the geocentric Ptolemaic (ancient Greek) and heliocentric systems and had political and personal overtones. "Academic historians are now convinced that this had as much to do with politics and the Pope's self-esteem as it did with science" (Hannam, 2010, p. 3).

74 Baxter, 2003, p. 149.

75 Hutton, 2008, p. 74.

geologically speaking, relatively uninformed.[76] However, there was at least one thing they were sure of—that being organized was essential to their long-term impact. Most contemporary geologists were not members in the early days of the group, but the society gained considerable influence over time, lending credence to old-age theories.

Shortly after this, William Smith (1769–1839) published several volumes based on the fossils uncovered in his work. As a surveyor, he had access to canal and road excavations around Britain; his collection of fossils from such sites led him to define the order of sedimentary layers based on fossils they contained, rather than the nature of the rocks. Smith was ultimately recognized as the "father of British stratigraphy".[77] Stratigraphy, the study of rock strata and stratification, quickly became important in old-age paradigms.

In France, the well-known naturalist Georges Cuvier (1768–1832) explained his geological theory of "catastrophism" in *Essay on the Theory of the Earth* (1813). This suggested that the geological record contained evidence of repeated deluges, reducing the significance of the Genesis Flood. This theory was taken up by William Buckland (1784–1856) and Adam Sedgwick (1785–1873) in Britain and was prominent in the 1820s.[78]

Geology was transforming into a science in which young-earth proponents had almost no influence. It was old-earthers who carried the day. By the 1820s, a descriptive scheme of the geological layers had become broadly accepted:

- The deeper *primary* layer was composed of volcanic and metamorphic rocks. It contained no fossils.
- Next were the *secondary* sedimentary fossiliferous rocks.
- This was followed by the *tertiary* layers which were also fossiliferous, with specimens resembling existing species.
- Finally, the *alluvial* surface deposits of sand, gravel and stones were topped by soil.[79]

In the early 1830s, the lawyer and amateur geologist Charles Lyell published his three-volume *Principles of Geology*, which promoted, in an accessible style, the uniformitarian ideas of Hutton, as well as Lyell's own views implying an extremely old earth. Within a few years, the theory of catastrophism was in steep decline. In many ways, Lyell was the capstone to some 40 years of influential old-age thinking, the result of which was the triumph of philosophical uniformitarianism.

The philosophical uniformitarian presupposition posited by Hutton and Lyell, which ostensibly applies to geology, makes sense for the atheist. Without a God to intervene in natural affairs, the atheist is entitled to assume that the

---

76 Mortenson, 2004, p. 29.
77 Ibid.
78 Ibid., pp. 29–32.
79 Ibid., p. 30.

processes we see today—and only those processes—are the same that have been operating in the past. In his philosophy of uniformitarianism, Lyell further suggested that past geological changes could only be understood by extrapolating, into the past, the rates of geological processes observed today.[80] If we apply the same doctrine more generally, beyond its immediate application in geology, it immediately becomes apparent that Christians cannot accept philosophical uniformitarianism. In the Old Testament we see God intervening miraculously in various ways, a pattern that is repeated in the New Testament with the miracles of Jesus. In John 5, for example, Jesus tells a cripple who had been bedridden for 38 years to pick up his mat and walk, and the man, thus healed, does so.

At this point, the philosophical uniformitarian could interject: *It would take literally months of rehab for that man to regain strength and learn to walk. There is no way he picks up his mat and walks away as described.*

The philosophical uniformitarian rejects this miracle because it is at odds with processes we see today. In fact, the philosophical uniformitarian must reject not only the Genesis deluge but also most New Testament miracles, including—most significantly of all—the resurrection of Jesus Christ. Both the deluge and the resurrection are highly significant to the Christian faith. A consistent philosophical uniformitarian therefore self-selects out of Christianity. As a corollary, it is essential for Christians to apply an interpretive framework that is sourced from and aligns with the Bible.

Lyell viewed his geological pursuits as part of a larger ideological framework. To Lyell, this was "more than just science: he believed he was immersed in nothing less than an intellectual war".[81] And in the battle against Christianity, uniformitarianism afforded a powerful line of attack, precisely because it roped in an independent science—geology—against the Bible. Lyell's friend Charles Darwin fully apprehended and supported this approach, writing that

> ... Lyell is most firmly convinced that he has shaken the faith in the Deluge &c [Noah's Flood] far more efficiently by never having said a word against the Bible, than if he had acted otherwise.[82]

Darwin relished this breakthrough in anti-Christian argument against which the attacks of the past seemed crudely ineffective. Darwin wrote:

> I have lately read Morley's Life of Voltaire & he insists strongly that direct attacks on Christianity (even when written with the wonderful force &

80 Ibid., p. 33.
81 Baxter, 2003, p. 204.
82 Darwin, 1873.

> vigour of Voltaire) produce little permanent effect: real good seems only to follow from slow & silent side attacks.[83]

Christian pushback: the scriptural geologists

By the 1820s, resistance to the old-earth ideas—which were gaining dominance—began to emerge. A smattering of *scriptural geologists,*[84] varying from amateurs to experts, started to express opposition to the old-age view, but crucially, they did not organize themselves as the Geological Society of London had done some two decades prior. Nevertheless, they objected to the precedence that scientific theory had been given over the revealed truth of the Bible and raised a number of concerns with old-earth geological theories.

First and foremost, the scriptural geologists contended that old-earth theories were either incompatible with or ignored scripture outright. To their mind, the fourth commandment (Exodus 20), which instructed the Israelites to keep the Sabbath, and included the statement that God made the earth in six days, severely undermined any day-age or gap theory of Genesis which enabled such old-earth views.[85] In addition, old-earth interpretations of Genesis glossed over the implication that sentient animals had suffered and died before Adam's fall, thereby impugning the character of God.[86]

The scriptural geologists also produced evidence which undercut old-earth interpretations. Several argued that the transitions between geological layers were generally vague and smooth, without interposed erosion or soil which would be expected if great ages had occurred between layering.[87] Another evidence was the existence of polystrate fossil trees, fossilized trees which traverse many layers in the geological record. Such fossils are widespread, and particularly concentrated within coal deposits. The scriptural geologists regarded this as strong evidence for the Genesis deluge, which would have laid down thick sedimentary layers quickly, and over large areas. In any case it counted as evidence against long-age theories because rapid burial was a more likely explanation for this phenomenon.[88]

The strong reliance on shells for dating in stratigraphy caused disquiet in both the scriptural geologist and catastrophist camps. The classification of shells was in its infancy, and differentiating hereditary from environmental causes of shell appearance was difficult enough, let alone distinguishing marine

---

83 Ibid.

84 Mortenson, 2004, p. 199, lists several competent scriptural geologists including Young, Rhind, Fairholme and Murray.

85 Ibid., pp. 200–201.

86 Ibid., p. 202.

87 Ibid., p. 204.

88 Ibid., p. 205.

from freshwater shells.[89] Despite this, atheistic geology placed enormous weight on these questionable distinctives as evidence for old-earth theories.

The old-earthers had claimed that no human fossils had been found among the older layers in the geological record. Several scriptural geologists noted that to the contrary, there were multiple examples of fossilized human bones found in layers with extinct animals. Later, when the old-earthers had seemingly won the debate, they agreed with the contention of the scriptural geologists, but then used that evidence to argue for the great antiquity of man.[90]

Finally, scriptural geologists regarded geology as an emerging discipline and contended that it was too early to propound a theory of the earth based solely on very limited data.[91] This argument was based on the cautious empirical strategy that Francis Bacon (1561–1626) had espoused.

But as the discipline of geology matured within universities, with departmental appointments, growing professionalism and specialization, the scriptural geologists found themselves either ignored[92] or harshly criticized[93] by the geologists now ensconced in Christian universities. By the 1850s, scriptural geology had largely petered out. The whole episode stands as an object lesson in McElreath's social construct view of science. The often solitary, disorganized voices of the scriptural geologists were no match for the well-resourced juggernaut of old-earth geology whose presuppositions were supported by the non-Christian elite.

This was a watershed moment for the church. The concept of long ages, underpinned by philosophical uniformitarianism, had not just won over the scientific community. Hutton and Lyell's long-age ideas also prevailed in the very heart of Protestantism: the seminaries and universities where future ministers would be educated. In the late nineteenth century, the entire faculty of the Presbyterian Princeton Theological Seminary accepted old-age geology. Charles Hodge, the highly regarded theologian and longtime principal of the seminary, included old-age geology uncritically in his *Systematic Theology*.[94] Protestantism had failed to detect the ripple which would become a tidal wave. The leaders had become Huttonian in outlook, and the presuppositions which they had accepted unquestioningly would later undermine their worldview.

As long ages became accepted more broadly, the church and its hierarchy began to be seen as yesterday's men. This was exemplified by Auguste Comte's

---

89 Ibid., pp. 206–209.

90 Ibid., p. 209.

91 Ibid., pp. 210–211.

92 "Lyell … felt that the best way to oppose [the scriptural geologists] was to ignore them" (Mortenson, 2004, pp. 218–219).

93 Sedgwick generalized that the scriptural geologists manifested "bigotry and ignorance" while Buckland complained that they "look for a detailed account of geological phenomena in the Bible" and "regard with jealousy and suspicion the study of any natural phenomena" (Mortenson, 2004, p. 216).

94 North, 2021.

theory of social change, published in a series of papers between 1830 and 1842. Comte argued that society developed in three broad stages: theological, then metaphysical (the transitional stage), and finally scientific. Comte's rationale is plain enough: in the face of scientific pronouncements, Bible-believing Christians, particularly those who had a straightforward understanding of Genesis, appeared as anachronistic and naive.

The notion that Christianity was at war with science became standard thinking after the publication of *The History of the Conflict Between Religion and Science* in 1873.[95] Elsewhere, Andrew Dickson White[96] argued that religion and science were irreconcilable[97] in his two-volume *History of the Warfare of Science with Theology in Christendom* (1896). Another erroneous idea which became conventional wisdom is today known as the "flat earth error", which is the belief that most Christians up until the time of Columbus believed the earth was flat.[98] An early version of this error is found in Washington Irving's 1828 *A History of the Life and Voyages of Christopher Columbus*.

The Christians certainly were confounded. Edward Clodd, the atheistic author of *Pioneers of Evolution*, gives us a contemporaneous view of the Genesis-as-history interpretation. Clodd cites Samuel Rolles Driver (1846–1914), Professor of Hebrew at Oxford, circa 1897:

> With candour as commendable as it is rare, the Reverend Professor Driver ... admits that, "read without prejudice or bias, the narrative of Genesis i. creates an impression at variance with the facts revealed by science"; all efforts at reconciliation being only "different modes of obliterating the characteristic features of Genesis, and of reading into it a view which it does not express".[99] [100]

Origins science had been placed on a foundation of anti-Christian presuppositions. In this interpretation, the Genesis account was falsified, and hence, the entire Bible was untrustworthy. Christians who held to a straightforward reading of Genesis were unable to reconcile anti-Christian origins science with the Bible. What many failed to realize was that a presuppositional shift had taken place, a shift that was in the realm of the philosophical or religious rather than "science".

The West was shifting to an "out of season" era for the Bible (2 Timothy 4:2). The anti-Christians of the West were in the ascendancy, gradually capturing the seats of influence within society.

---

95 Russell, 1991, pp. 36–38.
96 Ibid., p. 28.
97 Ibid., p. 42.
98 Ibid., p. x.
99 Clodd, 1904, p. 50.
100 Samuel Driver was not personally confounded by old-age pronouncements because he was a proponent of Higher Criticism.

But worse was to come for Christianity. The long ages of the uniformitarians had opened up the possibility of an even more devastating line of attack, one that pushed Lyell into the shadows. As Thomas Huxley, a contemporary of Lyell, said, "Lyell was, for others as for myself, the chief agent in smoothing the way for Darwin".[101]

References

Augustine, A. (1871). *The City of God* (Gutenberg edn, Vol. 2). (M. Dods, Ed.) Edinburgh: T. & T. Clark.

Baxter, S. (2003). *Ages in Chaos: James Hutton and the Discovery of Deep Time*. New York: Tom Doherty Associates, LLC.

Cardozo, N. L. (1995). On Biblical Criticism and Its Counterarguments—a short history. Retrieved December 30, 2023, from aishdas.org/toratemet/en_cardozo.html

Clodd, E. (1897). *Pioneers of Evolution: From Thales to Huxley—With an intermediate chapter on the causes of arrest of the movement*. London: Watts & Co.

Cochrane, C. N. (2003). *Christianity and Classical Culture: A study of thought and action from Augustus to Augustine*. Liberty Fund Inc.

Dallimore, A. A. (1970). *George Whitefield* (Volume 1). Edinburgh: The Banner of Truth Trust.

Darwin, C. (1871). *Origin of Species*. New York: D. Appleton and Company.

Darwin, C. R. (1873). To G.H. Darwin 21 October [1873]. Retrieved December 31, 2023, from darwinproject.ac.uk/letter/?docId=letters/DCP-LETT-9105.xml

Duhem, P. (1911). History of physics. In Catholic encyclopedia. New York: Robert Appleton Company. Retrieved from newadvent.org/cathen/12047a.htm

Eiseley, L. (1958). *Darwin's Century: Evolution and the Men Who Discovered It*. New York: Doubleday

Ellyard, D. (2005). *Who Discovered What When*. Sydney: Reed New Holland.

González, J. L. (2010). *The Story of Christianity: The Early Church to the Dawn of the Reformation*. (Vol. 1). New York: HarperCollins Publishers.

Hannam, J. (2010). *God's Philosophers: How the Medieval World Laid the Foundations of Modern Science*. London: Icon Books Ltd.

Harrison, P. (1998). *The Bible, Protestantism, and the Rise of Natural Science*. Cambridge: Cambridge University Press.

Henrich, J. (2021). Evolution of literacy: How Protestantism and the Bible rewired Medieval brains. Retrieved July 5, 2025, from geneticliteracyproject.org/2021/03/03/evolution-of-literacy-how-protestantism-and-the-bible-rewired-medieval-brains/

Himmelfarb, G. (1996). *Darwin and the Darwinian Revolution*. Chicago: Elephant Paperbacks, Ivan R. Dee, Publisher.

Hutton, J. (2008). *Theory of the Earth* (Kessinger Publishing's Rare Reprints). Whitefish: Kessinger Publishing LLC.

Jaki, S. L. (1978). *The Road of Science and the Ways to God*. Chicago: The University of Chicago Press.

Kuhn, T. S. (2003). *The Copernican Revolution: Planetary Astronomy in the Development of Western Thought*. Cambridge Massachusetts: Harvard University Press.

Lewis, C. S. (1967). *Christian Reflections*. Grand Rapids: William B. Eerdmans Publishing Company.

MacPherson, J. (1882). *Westminster Confession of Faith* (Second edn). Edinburgh: T. & T. Clark.

Mortenson, T. (2004). *The Great Turning Point: The Church's Catastrophic Mistake on Geology— Before Darwin*. Green Forest: Master Books.

Nisbet, R. (2009). *History of the Idea of Progress*. New Brunswick: Transaction Publishers.

North, G. (1989). *The Hoax of Higher Criticism*. Tyler, Texas: Institute for Christian Economics.

North, G. (2015, March 23). Moore's Law vs. Apocalypticism. Retrieved December 30, 2023, from garynorth.com/members/13585.cfm

101 Himmelfarb, 1996, p. 97.

North, G. (2021, May 15). Why Christian Homeschool Families Need a Free Online Christian Curriculum. Retrieved December 31, 2023, from garynorth.com/public/22233.cfm

North, G. (2022). *The Biblical Structure of History*. Dallas: Five Point Press.

Numbers, R. L. (1992). *The Creationists*. New York: Alfred Knopf.

O'Doherty, E. (1953, July). The "Conjectures" of Jean Astruc, 1753. *The Catholic Biblical Quarterly*, 15(3), pp. 300–304.

Pedersen, O. (1974). *Early Physics and Astronomy—A Historical Introduction*. Cambridge: Cambridge University Press.

Russell, J. B. (1991). *Inventing the Flat Earth: Columbus and Modern Historians*. Westport: Praeger.

Sarfati, J. (1999, March). The Yellowstone petrified forests—Evidence of catastrophe. *Creation*, 21(2), pp. 18–21. Retrieved from creation.com/the-yellowstone-petrified-forests

Sarfati, J. D. (2018). *The Genesis Account: A Theological, Historical, and Scientific Commentary on Genesis 1–11* (Third edn). Powder Springs, Georgia: Creation Book Publishers.

Smith, S. D. (2018). *Pagans & Christians in the City: Culture Wars from the Tiber to the Potomac*. Grand Rapids: William B. Eerdmans Publishing Co.

7

Darwin's *Origin of Species* and the emergence of the creationists

Charles Darwin (1809–1882) is such a celebrated member of the scientific A-list that he barely needs introduction. An amateur naturalist, Darwin formulated his theory of evolution in the late 1830s.[1] The mechanism Darwin proposed, natural selection, involved only natural causes, and after a lengthy delay, Darwin finally published his theory in *On the Origin of Species* in 1859.

Darwin's importance to the secular worldview is readily apparent. Where Hutton and Lyell had concerned themselves with geology, Darwin's ideas extended to life, including human life. As such, Darwin's work represented the capstone of an apparently comprehensive origins account—one that would underpin a completely non-theistic worldview.

Darwin's theory

Darwinism as an origins theory—as distinct from the empirical causal model theory of variation by natural selection (refer to Chapter 8 for an in-depth explanation of this crucial distinction)—was a speculation that all living organisms are derived from a common ancestor. It relied on the eons of time supplied by Hutton's and Lyell's uniformitarian interpretation of geology. This immense time-span—potentially infinite—rendered credible the notion that a succession of small changes could gradually accumulate into very large effects. Another basic idea was that there are always variations in the inherited characteristics of offspring. Those traits which favored survival would be passed on to future generations more effectively than those less favorable for survival. This mechanism, "natural selection", was a reworking of Thomas Malthus' idea that competition for food sets a limit to population growth.

Darwin supplanted earlier evolutionary ideas such as those proposed by his grandfather Erasmus Darwin, and French naturalist Jean-Baptiste Lamarck. Darwin's linkage of the mechanism of natural selection with naturally occurring variation was an ingenious development, but the speculation that he continued with—that all life was descended from a common ancestor—was far

1 Darwin's famous "I think" note, which outlined the evolutionary tree of life, dates to 1837.

more difficult to establish. Nevertheless, in popular imagination, Darwinism was the final piece of the puzzle which enabled scientists to explain the diversity of life without the need for a Divine creator. The irony is that even if Darwinism were true, Louis Pasteur's circa 1859–60 disproving of spontaneous generation meant that secular origins science has *never offered a holistic explanation for the diversity of life.*[2] This is shown in Figure 5.2 (Chapter 5), where the two parts of the atheistic ancient Greek origins account are represented by a two-part extension ladder. Darwin supplied the top half—an apparently plausible evolutionary mechanism—just as the lower half—spontaneous generation—fell away.

But there was much to suggest, right from the start, that Darwin's speculation was highly problematic. In Darwin's scheme, if an anatomical component was useful to survival—such as camouflaged feathering on a bird—the mechanism responsible for ensuring its passing to posterity was natural selection. If it was potentially harmful for survival, such as brightly colored feathers, the mechanism was sexual selection. (That is, brightly coloured feathers were thought to confer advantages in the contest for mates.) There was always a hypothetical explanation; the theory was remarkably "plastic". Darwin's modus operandi was to propose mechanisms for Darwinian evolution which seemed plausible, rather than providing any evidence for the processes. In particular, there was a notable lack of evidence from the fossil record for the gradual processes which Darwinism required, a lack which Darwin felt compelled to address:

> Geology assuredly does not reveal any such finely graduated organic chain; and this, perhaps, is the most obvious and gravest objection which can be urged against the theory. The explanation lies, as I believe, in the extreme imperfection of the geological record.[3]

Darwin proposed reasons for why the evidence did not support—or even contradicted—his theory. This process has been termed a "logic of possibility", an inversion of normal logic, in which a number of remote possibilities are assumed to add up to probability. Darwin summarized the various difficulties with his theory:

> The several difficulties here discussed, namely our not finding in the successive formations infinitely numerous transitional links between the many species which now exist or have existed; the sudden manner in which whole groups appear in our European formations; the almost entire absence, as at present known, of fossiliferous formations beneath the Silurian strata, are all undoubtedly of the gravest nature. We see this in the

2 In the 160-odd years since Pasteur disproved spontaneous generation, secular science has not provided any experimental or theoretical basis for a replacement 'primordial abiogenesis'.

3 Darwin, 1909, p. 334.

> plainest manner by the fact that all the most eminent palaeontologists, namely Cuvier, Owen, Agassiz, Barrande, Falconer, E. Forbes &c., and all our greatest geologists, as Lyell, Murchison, Sedgwick, &c., have unanimously, often vehemently, maintained the immutability of species. But I have reason to believe that one great authority, Sir Charles Lyell, from further reflection entertains grave doubts on this subject.[4]

Darwin admitted that many eminent palaeontologists and geologists, with the exception of his friend Charles Lyell, had already concluded from fossil evidence that evolution of higher species such as mammals has not occurred. This was enormously problematic.

But it wasn't the only problem. In the mid-1830s, the respected British zoologist Edward Blyth had raised a different kind of query for the evolutionist camp, one for which they had no answer. Blyth observed that species tend to be well adapted in the geographical areas in which they lived, but, at the edges of their habitats, where climate and ecologies transition, they were less well adapted. Blyth's question was noted by Sir Fred Hoyle: "Why, Blyth asked, if species can evolve to the great extent that would be needed to explain the differences between genera, families, orders, and classes, can they not evolve to the lesser extent that would maintain adaptation to and beyond the boundaries of their respective areas?" This was the rule, not the exception, and according to Blyth it rendered evolution untenable.[5]

Undeterred, Darwin took a strategic approach to the marketing of his theory, choosing not to convince the scientific community directly, but have scientists (whom he referred to as "naturalists") dragged along by the weight of consensus:

> ... I think it of importance that my notions should be read by intelligent men, accustomed to scientific argument, though not naturalists. It may seem absurd, but I think such men will drag after them those naturalists who have too firmly fixed in their heads that a species is an entity.[6]

Readers will recall McElreath's concept of scientific consensus. Darwin was looking for something different here; this marketing effort shows Darwin not in the guise of a scientist (recall Feynman's demarcation-like criterion) but as an ideologue whose aim was to *shift the worldview* of the scientific community. This was to be achieved by putting in place a softly coercive ideological consensus that would relegate dissenters to the status of lonely voices in the wilderness.

---

4 Darwin, 1871, p. 294.

5 Hoyle suggests that Darwin did not want to risk open confrontation with Blyth's argument, resulting in the famous 20-year delay in the publication of his theory (Hoyle, 1999, p. 105).

6 Himmelfarb, 1996, p. 296.

The failed pushback of Philip Henry Gosse

Darwin's theory did not fit what was known of the fossil record at the time, nor did it answer Blyth's query, so Darwin feared that the opprobrium of a few scientific heavyweights might have snuffed out his origins speculation from the outset. What Darwin needed, prior to the publication of *Origin*, was a core group of sympathizers who could provide crucial ideological support for his embryonic theory. This group was to consist of the Victorian era's equivalent of Sir David Attenborough: well-known and held in great regard by the public for their accessible and interesting presentation of nature to general audiences. One such person targeted by Darwin and his associates was Philip Henry Gosse, a well-known naturalist in nineteenth century Britain. His cachet with the public gave him broad influence, exactly the kind of influence that Darwin needed.

Gosse's credentials were impeccable. He was a widely published naturalist, with titles such as *Popular British Ornithology* (1850), *A Naturalist's Sojourn in Jamaica* (1851), and *The Aquarium: An unveiling of the wonders of the deep sea* (1854) to his credit. This latter book was a bestseller, and Gosse's practical experience with aquariums—he stocked the first aquarium at the London Zoological Gardens[7]—might have been a factor in his election to the Royal Society in 1856.[8] Gosse's son, Edmund, recorded that it was in the summer of 1857, when Darwin, Lyell and their acolytes were covertly assembling a "bodyguard" of influential science communicators, that his father Philip was approached:

> In this period of intellectual ferment, as when a great political revolution is being planned, many possible adherents were confidentially tested with hints and encouraged to reveal their bias in a whisper. It was the notion of Lyell, himself a great mover of men, that, before the doctrine of natural selection was given to a world which would be sure to lift up at it a howl of execration, a certain bodyguard of sound and experienced naturalists, expert in the description of species, should be privately made aware of its tenor. Among those who were thus initiated, or approached with a view towards possible illumination, was my Father. He was spoken to by Hooker,[9] and later on by Darwin, after meetings of the Royal Society in the summer of 1857.[10]

Gosse, however, was a Christian who supported a Genesis-as-history interpretation of Christian origins. He could not accept Darwinian evolution, and he recognized, correctly, that it was Lyell's long ages that rendered evolution plausible. In Gosse's mind, it was long ages that had to be addressed.

---

7 Gosse, 1890, p. 244.

8 Ibid., p. 263.

9 Joseph Dalton Hooker (1817–1911), proved to be one of Darwin's most loyal supporters. Like his father, Sir William Hooker, Joseph Hooker was the director of the Royal Botanical Gardens, Kew.

10 Gosse, 2004, p. Chapter V.

He decided to do this by reconciling the young earth implied by the Bible with the long ages implied by Lyell's geology. Working alone, Gosse developed a solution—or at least, a solution that made sense to him. In his conception, God created the earth with the appearance of age, by crystalizing a "virtual history" he termed *prochronic* at the instant of creation.[11] Gosse's son Edmund summed up this concept with these chiding words: "God hid the fossils in the rocks in order to tempt geologists into infidelity".[12] Gosse (Snr) expounded this thesis in a book titled *Omphalos* (1857). Omphalos, the Greek word for navel, was meant as an illustration of Gosse's thinking: God surely made Adam with a navel, even though as the first man he never had an umbilical cord; *ergo*, God included certain elements in His creation that were not functionally necessary.

The public reaction to *Omphalos* was instant and unanimous. Atheists and Christians alike found Gosse's explanation absurd. Christians were unhappy with the suggestion that God had acted in the guise of a deceiver; scientists were nonplussed because Gosse's thesis had no practical implications.[13]

In fact, *Omphalos* was the disastrous mistake that Gosse is today (mostly) remembered for. In the first place, he had advanced a philosophical/theological scheme which undercut the investigative process. Worse, it appeared that Gosse was intentionally proposing a "science stopper"—any troublesome evidence from the fossil record could simply be waved away. *Omphalos* appeared as the musings of a man clutching at straws—which was unacceptable to both sides of the debate. This book was an object lesson for creationists in what not to do, but it was a lesson that was not always remembered.

A shift in worldview

At all events, Darwin was able to drum up a critical mass of support prior to publishing *Origin*, and his intuition that these supporters could "drag after them" the men of science proved correct. This largely unknown episode reveals Darwin in the guise of ideologue; the Darwinian revolution was not the scientific revolution of popular imagination, but an organized effort to shift the worldview of Western society away from Christianity.

With secularization as a goal, the top scientists of the day were of limited relevance to the Darwinists. "Cuvier, Owen, Agassiz, Barrande, Falconer, E. Forbes &c.", to quote Darwin again, were sidelined in favor of a carefully vetted "bodyguard" that would provide support to Darwin's theory. Writing in 1888, evolutionist Grant Allen gives us a hint as to the rhetoric deployed against those who did not fall in line with Darwinism:

---

11 Sarfati, 2015.
12 Gosse, 2004, p. Chapter V.
13 Sarfati, 2015.

> A few of the elder generation of naturalists held out ... for various shorter or longer periods ... left behind, like stragglers from the onward march ... The elder men, in fact, had many of them lost that elasticity and modifiability of intellect which is necessary for the reception of new and revolutionary fundamental concepts .... Moreover, some of the elder thinkers were committed beforehand to opposing views, with which they lacked either the courage or the intellectual power to break; while others were entangled by religious restrictions, and unable to free themselves from the cramping fetters of a narrow orthodoxy.[14]

These elder statesmen of the scientific establishment were no doubt used to having the quality of their ideas criticized, but from 1859, a different kind of critique had appeared. A key yardstick, henceforth, was whether a scientist acknowledged Darwin's theory. To deny Darwin was to fall behind, to become a lonely "straggler". To deny Darwin was to show to all the world that one had lost the "elasticity" of one's mind, or perhaps lacked courage, or intellectual power. To disagree on religious grounds was to demonstrate how "fettered" one's mind had become by "narrow orthodoxy". Scientific critique had been replaced by personal attack. And it worked. Scientists either kowtowed or disappeared into professional oblivion. Darwin's bodyguard had achieved what it was intended to do: "drag after them those naturalists" who disagreed with Darwin on scientific grounds. Evolutionary theory soon captured the scientific establishment and became, like Lyell's long ages, an unstoppable juggernaut.

The Darwinists organize: money, connections and power

If Darwin's ideas had required a "bodyguard" to defend them upon publication, they also received ongoing support from a well-connected clique in the 1860s and beyond. Historian Frank M. Turner identified a core group of Darwinists, whom he collectively termed the "naturalistic publicists", as the key drivers of this change after 1859. The group included Thomas Henry Huxley, professor of biology at the Royal School of Mines and strident apologist for Charles Darwin; Sir Francis Galton, half-cousin of Charles Darwin, eugenicist, statistician, and advocate of professionalism in science; John Tyndall, a physicist, superintendent of the Royal Institution (successor to Faraday); and W. K. Clifford, mathematician at University College, London.[15]

This multidisciplinary group of highly placed intellectuals ostensibly called on society to embrace the cause of science. But their actual aim, as Turner points

14 Allen, 1888, p. 116.
15 Turner, 1974, p. 9.

out, was to secularize society—which effectively meant the de-Christianization of the Anglosphere, or perhaps the whole West. Turner again:

> The naturalistic publicists sought to expand the influence of scientific ideas for the purpose of secularizing society rather than for the goal of advancing science internally. Secularization was their goal; science, their weapon.[16]

Some of the key players in the early Darwinian movement were radically opposed to Christianity. Darwin, for his part, revealed why he hated Christianity in a letter to his son:

> I can indeed hardly see how anyone ought to wish Christianity to be true; for if so, the plain language of the text seems to show that men who do not believe, and this would include my Father, Brother, and almost all my best friends, will be everlastingly punished. And this is a damnable doctrine.[17]

Darwin viewed Christianity as a "damnable doctrine" because he believed it promised judgement upon members of his family. As for Huxley, he regarded the strongly anti-Christian credentials of "evolution" as one of its greatest merits:

> In addition to the truth of the doctrine of evolution, indeed, one of its greatest merits in my eyes, is the fact that it occupies a position of complete and irreconcilable antagonism to that vigorous and consistent enemy of the highest intellectual, moral, and social life of mankind—the Catholic Church.[18]

Although Huxley mentioned the Catholic Church by name because it was convenient in a Protestant nation, he almost certainly meant Christianity in general; he later specified that he was referring to any theology based upon the book of Genesis.[19] Darwin referred to Huxley in a letter as

> My good & kind agent for the propagation of the Gospel ie the Devil's gospel.—Ever Yours, C. Darwin[20]

This characterization of Darwinism as "the Devil's gospel", by Charles Darwin himself, should leave readers in no doubt as to the anti-Christian nature of Darwinism, or that Darwinism and Christianity are mutually inimical. Although atheists and many Bible-believing Christians were both agreed on these points, the liberal Christians of Darwin's day claimed that evolution was compatible with Christianity.[21]

---

16 Ibid., p. 16.
17 Himmelfarb, 1996, p. 385.
18 Huxley, 1893, p. 147.
19 Ibid., p. 147.
20 Darwin, 1860.
21 Himmelfarb, 1996, p. 397.

Of the publicists identified by Turner, the standout was perhaps Thomas Henry Huxley, who acquired the moniker "Darwin's bulldog" for his combative approach to the defense of Darwin and his theory. Huxley was a master political operator who did not shy away from confrontation with those that disagreed with him. His quick wits and public speaking abilities made him a very effective ambassador for "the Devil's gospel".

In 1864, Huxley's friend Joseph Hooker, who was assistant director of the Royal Botanic Gardens at Kew under his father Sir William Hooker,[22] had the idea of starting a dining club of like-minded friends. Eventually known as the X Club, partly to conceal its reason for being, this elite group of nine men[23] was not merely to be a social club; the group had an agenda. A key goal was "[p]romoting Darwin's reputation and developing his theory in new directions".[24] The X Club was able to achieve these goals because of its outstanding reach. It had "direct access to the City, Parliament, medicine, industry and the liberal church—to the cultural heart of the country".[25] Like the uniformitarian geologists before them, the Darwinists were getting organized. The X Club met for almost 30 years (1864 through 1892) and wielded considerable influence in favor of Darwin and his theory. It has been described as "a sort of masonic Darwinian lodge".[26]

But even before the X-Club met for its first dinner, "subgroups of the friends had engaged in coordinated action".[27] Their first order of business was to have Charles Darwin awarded a Copley medal.[28] After two unsuccessful nominations in 1862 and 1863, they finally succeeded in having Darwin awarded the Copley—broadly equivalent to a Nobel prize today—in 1864. This helped to cement Darwin in the popular imagination of Victorian Britain as an outstanding scientist.

The X Club also endeavoured to help establish a periodical to promote naturalistic explanations, achieving success with the launch of *Nature* in 1869. One of the primary objectives of this magazine was to present Darwinian thought in the most favourable light possible—by "emphasiz[ing] the pros and hiding the cons".[29] *Nature* went on to become an influential staple of materialist

22 Barton, 2018, p. 3.

23 George Busk (1807–1886), Joseph Dalton Hooker (1817–1911), Herbert Spencer (1820–1903), John Tyndall (ca. 1822–1893), Thomas Henry Huxley (1825–1895), William Spottiswoode (1825–1883), Edward Frankland (1825–1899), Thomas Archer Hirst (1830–1892) and John Lubbock (1834–1913) (Barton, 2018, p. 1).

24 Barton, 2018, p. 16.

25 Desmond, 1994, p. 328.

26 Desmond & Moore, 1991, p. 526.

27 Barton, 2018, p. 6.

28 The Copley Medal is awarded by the Royal Society for achievement in any field of science. First awarded in 1731, the medal was probably the most prestigious award in science until the Nobel Prize, which is awarded in multiple categories (including physics and chemistry) debuted in 1901.

29 Hoyle, 1999, p.138.

scientific literature, and remains so today, more than 150 years after its first issue.

The X Club had extensive connections in science, government and the church. "The most remarkable example of mutual power is the five years in the mid-1870s when Hooker was president, Spottiswoode treasurer, and Huxley the biological secretary of the Royal Society, and the subsequent succession of Spottiswoode and Huxley to the presidency, making a continuous twelve-year reign by the X Club.".[30] John Lubbock, son of wealthy banker Sir John Lubbock, 3rd Baronet, entered Parliament in 1870 and remained there until 1900. In relation to the church, Huxley had helped a liberal Anglican, William Henry Flower, into the post of Conservator at the museum of the Royal College of Surgeons of England in 1862,[31] and further assisted a vicar's son, George Rolleston, into the Linacre Chair of Anatomy at Oxford. Flower and Rolleston had links to the Dean of Westminster, Arthur Stanley; Stanley was godfather to Flower's children. The X Club, therefore, was able to exercise influence with some of the most senior clergy in Britain. It was a connection that they would use repeatedly.[32]

The X Club schemed behind the scenes to ensure that their men were posted to key positions in the Royal Society and other influential scientific institutions. Prior to the annual Royal Society meetings that elected new officers, the X Club would meet covertly to discuss and plan for the election of their preferred candidates. "To all outward appearances, nominations and elections to scientific office and recognition within the Royal Society and other organizations proceeded through established democratic procedures. Rules and procedures were adhered to, yet the X Club agenda was attained".[33] Unbeknown to the science community and the populace at large, Victorian science was being manipulated behind the scenes for private ends.

However, the influence of the X Club had its limits. While some members of the liberal church were receptive to their ideas, more conservative Christians remained largely unmoved. This divide is exemplified by Gosse, mentioned earlier, whose association with the Plymouth Brethren reflected a more fundamentalist outlook. Although his *Omphalos* argument failed to persuade, its publication nonetheless revealed that not all Christians accepted Darwinian evolution. For the Darwinists, dialogue with the liberal church proved relatively accessible, but engagement with fundamentalist groups was far more challenging. This tension has persisted in various forms over the decades.

---

30 Barton, 2018, p. 7.

31 Desmond, 1994, p. 305.

32 The X Club advocated successfully to have Charles Lyell and Charles Darwin buried in Westminster Abbey following their deaths in 1875 and 1882 respectively.

33 Thyer, 1995.

Darwin's timing

The X Club and its networks were able to bring important British institutions under their influence and use that influence to lobby for the Darwinian cause for almost 40 years.[34] "Without the Xs,", writes Darwin biographer Janet Browne, "Darwin's ideas would never have percolated British culture quite as far as they did".[35] With this in mind, we need to revisit the discussion on the timing of the publication of *Origin*.

In Chapter 5, we saw that Darwin formulated his theory of evolutionary descent in 1837, but delayed publication for two decades, until 1859. We noted that this timing almost exactly matched Louis Pasteur's disproving of spontaneous generation, which deprived the Darwinists of a holistic origins account. From a scientific point of view, this was awkward—the ladder shown in Figure 5.2 looks like a prop from a Charlie Chaplin skit. But from a public relations point of view, the 1859 publication of *Origin* was extremely fortuitous.

In the first place, only two of the nine X Club men were aged beyond their teens in 1837. John Lubbock, the youngest, was a mere toddler. But by 1859, all of these were formidable intellects in their primes. If Darwin had published earlier, would a group equivalent to the X Club have emerged in the 1840s? Perhaps not—in 1844, a bestselling book, *Vestiges of Creation*, had put forward a progressive evolutionary scheme, but no influential group rallied around the concept. However, *Vestiges* had primed the public with proto-evolutionary concepts. By the late 1850s, the intellectual milieu had reached a tipping point. Darwin's theory was, from a sociological and public relations point of view, the right idea at the right time—for atheists and agnostics. For fundamentalist Christians, like Gosse, it was a disaster.

Given Darwin's revered place in the pantheon of science, it is an astounding fact of history that the eminent scientists of the day objected to Darwin's theory on both scientific and religious grounds.[36] Even more astounding, from the point of view of Christians who held to a straightforward reading of Genesis, was the fact that the clergy and religious journalists by and large were more accepting of Darwin's theory than the top scientists, with some theologians exhorting the public to respect science.[37]

Darwinism had deeply penetrated not only liberal but also conservative Protestant seminaries. Conservative Presbyterian John Gresham Machen, the founder of Westminster Theological Seminary, was a closet theistic evolutionist, a fact only revealed in his private letters.[38] It was surprising that conservative

---

34 Barton, 2018, p. 6.

35 Browne, 2002, p. 249.

36 Himmelfarb, 1996, pp. 280–281.

37 Ibid., p. 283.

38 North, 2010, p. 40.

seminaries had incorporated evolution into their theology, because the Westminster Confession should in theory have influenced the Churches of England and Scotland and the Presbyterian Churches worldwide.

Eugenics: evolution with its sleeves rolled up

It's worth digressing slightly to examine eugenics, the applied "science" developed by Charles Darwin's half-cousin, Francis Galton. We have already seen that animal and plant breeding was carried out in ancient times. But as the Darwinian theory became accepted, Francis Galton developed the idea of eugenics—using a range of techniques to "improve" the stock of humans, who were increasingly viewed as animals. Galton outlined the principles of eugenics in 1883. He viewed it as a science which aimed to selectively control human breeding as government-sanctioned policy. This would accelerate positive evolutionary variations in the population. Galton wanted to "further evolution, especially that of the human race".[39]

In Britain, the Eugenics Society was formed in the early 1900s to advance eugenic policy. By 1911, Charles Darwin's son, Leonard Darwin, assumed the presidency of the society; his inaugural address reveals eugenics as a reversal of Christian morality. Whereas Christ teaches that charity to the disadvantaged is a duty, something good, the eugenicist sees charity to the "unfit" as problematic; Leonard Darwin characterized assistance to the poor as a problem that led to "anti-eugenic effects".[40] Applied Christianity and applied Darwinism were in direct opposition.

American eugenicists canvassed euthanasia by public gas chambers,[41] but in practice, forced sterilization was the most radical policy they implemented. The first forced sterilization law was passed in Indiana in 1907 after the urging of the medical profession.[42]

To assist governments to implement such laws, an American eugenicist, Harry Laughlin, drafted a model law—what we would today call a template. Laughlin's 1914 Model Eugenical Sterilization Law had all the legal bells and whistles required; states merely had to tweak it for their local requirements, then enact it. The law recommended the forced sterilization of those with mental illness, feeble-mindedness, the blind, the deaf, the deformed, drunkards, epileptics, criminals, orphans, those dependent on financial support, the homeless and the poor.[43] Virginia passed a law based on Laughlin's model in 1924.

---

39 Galton, 1907, p. 220.
40 Darwin, 1911, p. 15.
41 Black, 2003.
42 Micklos & Carlson, 2000.
43 Lombardo, _.

The Nazi government in Germany borrowed from Laughlin's law in 1933 and proceeded to sterilize more than 350,000 people.[44] Nazi academia was so pleased with Laughlin's model law that they awarded him an honorary degree in 1936, citing his work in the science of racial cleansing.[45]

Harry Laughlin died in 1943, several years before the Nuremburg trials revealed the true horror of the Nazi regime to a shocked world. But Laughlin was not the only former eugenicist to have his legacy tarnished by the revelations at Nuremberg. Madison Grant, another leading American eugenicist,[46] had written a book in 1916 titled *The Passing of the Great Race.* Translated into German in 1925, the book was well received by the early Nazi movement, with Hitler reportedly announcing, "this book is my Bible".[47] One of the defendants at Nuremberg, Karl Brandt, who had served on the Nazi euthanasia program, produced a copy of the book and cited its contents in his defense.[48] One wonders whether the American prosecutors saw something vaguely treasonous in the way their compatriot, Grant, had provided "intellectual" succor to Hitler's regime—a regime that America and its allies had expended considerable blood and treasure in defeating.

Following the revelations that Nazi Germany had practiced eugenics on a large scale, the eugenicists of the Anglosphere largely—but not completely—abandoned their agenda. As they were members of the elite, with considerable influence over the media, the disgrace of the eugenics movement was simply swept under the carpet. There would be little in the way of public outcry, investigations, or reparations for the victims of applied Darwinism.[49]

Darwinism into schools

In 1968, American schoolteacher Susan Epperson won the right to teach human evolution in *Epperson v. Arkansas,* a decision that was upheld by the U.S. Supreme Court. This decision was tantamount to a proclamation that a Genesis-as-history interpretation of the early chapters of the Bible had been disproved by science. At this point secular humanists had achieved a powerful pedagogical advantage in schools throughout the U.S., which specifically meant the right to teach the secular origins account pertaining to humans *as science* within the science curriculum. Similar decisions were soon enacted by the educational bureaucracies in other Western countries. By default, secular

44 Ibid.

45 Ibid.

46 Grant happened to have served alongside Harry Laughlin on the publication committee of the 1932 International Congress of Eugenics, held in the Museum of Natural History in New York.

47 Spiro, 2009, p. xi.

48 Ibid., pp. xi–xii.

49 Some jurisdictions have provided compensation for victims of eugenics. For example, Japan's top court ruled in 2024 that forced sterilization was unconstitutional and ordered compensation payments for people affected (Ng & Nakayama, 2024).

humanism became the established religious perspective of the West. The secular state, an idea which sprung partly from Protestant biblical theology,[50] had been replaced with the secular humanist state.

Recalling that universal education was initially instituted by the Reformers so that all could read the Bible for themselves, the teaching of Darwinism in schools was highly significant. It meant that the secular humanists had, cuckoo-like, commandeered the public education system for their own ends. From that point forward, state-funded education pushed Darwinism—what Darwin referred to as the "Devil's gospel"—as a neutral scientific fact. Over time, generations of young people have been exposed to these teachings, and this has arguably been one of the most significant factors in the mass de-Christianization of the West.

A grand-scheme view

If Christianity reached a peak—politically and culturally—at the time of Thomas Aquinas, it was a situation which took seven centuries to reverse. Following the emergence of humanism in the aftermath of the Black Death, the elite progressively abandoned Christianity, but the masses retained, to a greater or lesser extent, a certain level of respect for the Bible for several more centuries.

This has changed decisively in recent years. From the perspective of young-earth creationists, the emergence of Darwinism, its official acceptance by various denominations and incorporation into theology, coupled with the compulsory education of Darwinism via state school systems, are factors in the de-Christianization of the West. Western civilization can now be truly described as post-Christian—the polar opposite of Aquinas' Europe.

Broken walls

Prior to the emergence of secular origins science, the key question that Christians sought to answer regarding creation had been: *how long ago did creation occur?* A range of learned church leaders, like Bishop Ussher[51] in Britain, applied their considerable expertise to this question and published their results.

But this would no longer do. By the late eighteenth, or early nineteenth century, merely totting up the years since creation—as important and informative as this was—had been exposed by secular origins science as an insufficient support for the Genesis-as-history view. Secular scientists had

---

50 Luther's doctrine of the two kingdoms (the spiritual and secular), the focus on individual conscience, rejection of papal authority, and the replacement of divine right with covenant theology (Calvin) led towards the secular state. It was the devout English Puritans who made the universities secular. They wished to "expunge[e] ... from the universities ... all efforts at the inculcation of religion and piety ..." The universities were to "confine themselves to the knowledge of the secular realm, above all the scientific-experimental." (Nisbet, R., 2009, p. 131).

51 We discuss aspects of Ussher's approach in Chapter 17.

called into question the historicity of Genesis. The very foundations of the faith were under attack.

Many Christians, like Philip Henry Gosse (mentioned earlier), were confounded by the new origins science. Although the decades following the publication of *Origin* were a sort of heyday for the Western church—with foreign missions abounding, high levels of church attendance and general acceptance of the Bible as a source of morality—Christians who favored a straight-forward reading of Genesis initially lacked credible answers to humanists and their Darwinian origins science.

As the 1900s progressed, the church had no well-credentialled champion to defend a straightforward reading of Genesis. Ronald Numbers notes that Harry Rimmer (1890–1952), "who reached a wider audience among American evangelicals during the second quarter of the [twentieth] century" than any other antievolutionist, was a "degreeless evangelist"[52] whom critics dismissed as "a charlatan and buffoon".[53] Another prominent antievolutionist of the era was George McCready Price, an "armchair scientist" who "felt insecure about his lack of formal scientific training and his limited familiarity with the evidence he was disputing".[54] Overall, these early defenders of the Genesis account appeared to be long on stage charisma but short on credible scientific argument.

From the standpoint of young-earth creationists, this intellectual climate was in some ways analogous to the state of Jerusalem's walls in the Old Testament following the Babylonian exile. The broken walls were a humiliation because they left the inhabitants defenseless. In the origins sphere, not only were the walls of Western Christendom broken, but the seats of influence set up by Christians—in churches, schools and universities that had become liberal—were often occupied by secular humanists. To society at large, the Christian foundations described in Genesis had been pulverized.

Elsewhere in the Old Testament, King David had asked, "When the foundations are destroyed, what can the righteous do?" (Psalm 11:3.) An answer is given in the book of Ezra: rebuild the foundations.[55]

The creationists respond

The creationists sought to defend the historicity of the Bible, particularly the early chapters of Genesis. They saw their task as analogous to what the early Christians had done when faced with the heresy of the Gnostics in the first through fourth centuries. The Gnostics had mythologized or allegorized parts

---

52 Numbers, 1992, p. 69.

53 Ibid., p. 64.

54 Ibid., p. 89.

55 Ezra 3:10–13.

of the Bible; the early church responded by composing creeds which emphasized important scriptural truths, such as the Trinity and the historicity of the birth, death and resurrection of Christ. In fact, these early Christians were directly following the example of the apostle Peter, who in 2 Peter 3:3–7, excoriated those who denied the historicity of the world-wide Flood of Genesis.

Protestant Christian laymen John Whitcomb and Henry Morris published *The Genesis Flood* in 1961, a book which sought to establish the historicity of the great deluge. This book was a breakthrough: the authors presented significant scientific evidence for Noah's Flood and encouraged many Christians concerning the trustworthiness of the Bible. In the well-credentialled Morris, whose PhD in hydraulic engineering from the University of Minnesota lent the gravitas required to undertake a scientific study on the deluge, the embryonic creationist movement finally had an advocate to be taken seriously.

This seminal book came a full century after the publication of *Origin*. Although not all its arguments have stood the test of time, *The Genesis Flood* was of historical significance because it sparked the creation science movement. And whereas many of the forerunners of the creation movement were Seventh Day Adventists, key figures in the movement would now emerge from the same protestant camp that Whitcomb and Morris belonged to. Specifically, this was *lay Protestantism*; Protestant (and Catholic) seminaries and hierarchies were mostly in favor of theistic evolution.[56]

Prior to the "creationist renaissance of the 1960s",[57] the field was dominated by solo writers and short-lived societies, such as "Religion and Science Association" and the "Deluge Geology Society" from the 1930s.[58] But the excitement surrounding *The Genesis Flood* brought enough impetus for several sustainable creation ministries to commence. These have produced increasingly sophisticated scholarly research in support of a straightforward reading of Genesis 1–11 as history. These ministries include the Institute for Creation Research, Creation Ministries International and Answers in Genesis. Like science in general, some of the research has stood the test of time and some has not. Nevertheless, the creation movement has been making steady, visible progress.

The emergence of stable creation ministries resolved the problems that hampered early efforts to defend a straightforward reading of Genesis. The scriptural geologists of the early nineteenth century, and the disastrous solo effort of Gosse, were examples of an unorganized approach which proved ineffective. Creation ministries enable networks of individuals with expertise in a variety of scientific and other disciplines to coordinate their efforts to

---

56 Elsewhere, The Kolbe Center for the Study of Creation is a Roman Catholic creationist organization that has been active for several decades. The Seventh Day Adventist Christian University espouses six-day creation.

57 Numbers, 1992, p. xi.

58 Ibid., pp. 102, 118.

produce materials that defend a biblical approach to origins research. Scientific articles and books are reviewed and assessed from a biblical worldview. Creation ministries act to coordinate a research community investigating the history, philosophy, theology, and science of origins speculations. But these ministries also provide speakers to attend local churches to communicate with lay Christians and address their questions, thereby maintaining a link with grass-roots Christianity.

In addition to corporate creation ministries, solo creationist writers have emerged. It is important for lay Christians to identify whether materials are produced by solo writers or from corporate creation ministries; up-do-date materials from a research community are likely to be more reliable, more balanced, and less prone to error or miscommunication.[59]

Creationist science

One of the criticisms levelled against the creation movement by atheists is that it relies on "God of the gaps" arguments. By this, atheists mean that if creationists encounter phenomena which they cannot explain by natural means (a "gap" in understanding), they simply declare that a Divine miracle is responsible and move on. This kind of "explanation" is held in contempt by atheists because it cuts short the investigative process.

However, the creation movement seeks to avoid these types of arguments, recognizing that they fall into the same intellectual dead-end that Philip Henry Gosse meandered into with *Omphalos*. To show how these arguments are avoided, we first categorize and describe the types of miracles found in the Bible. (We suggest the following categories for illustrative purposes, not as a final word—accepting that different Christians could reasonably disagree with some of the classifications):

Creative—*ex nihilo* (Latin for "out of nothing"): In this miracle type, God creates matter out of nothing. An example from the Bible is God's creation of physical matter out of nothing as recorded in Genesis 1:1 (and affirmed in Hebrews 11:3). Other possible examples which are less certain include the widow of Zarephath's flour and oil not running out in the days of Elijah (1 Kings 17:14–16), and the miraculous increase of the five loaves and two fishes into enough food for "about five thousand men, in addition to women and children" (Matthew 14:13–21).

Transformative—in this miracle type, God transforms existing matter into a different form through miraculous, rather than naturalistic processes. In Genesis, this includes the shaping of dust into the first man (Genesis 2:7). Elsewhere in the Old Testament, Aaron's staff was miraculously transformed

59 For the record, this book was drafted in consultation with experienced creationists.

into a snake (Exodus 7:10), and in the New Testament, Jesus miraculously transformed water into wine (John 2:1–11).

Naturalistic—In our final miracle, type, God intervenes through naturalistic processes, which may be miraculously sped up or magnified and applied to a specific time or location. In the Old Testament, this includes the great Flood, where rain and the reservoirs of the deep were opened; and the plagues of locusts, lice and flies in Egypt (Exodus 8). We could also include here the defeat of Amalek (Exodus 17:8–13). Although Amalek's army was superior to the Israelite army in natural terms, it nevertheless faced a miraculous defeat as long as Moses held up his hands. The budding of Aaron's rod (Numbers 17), and Jesus' healing the man's withered hand (Matthew 12:9–14) also contain elements which fall within this category, but because they are not "in sync" with the surroundings, (the budding, and growth of the hand were sped up with respect to other rates of change on earth), they represent a subset of naturalistic miracles.

Of these categories, the ex-nihilo and transformative suspend science; there is no naturalistic explanation for these. The creation movement accepts miracles of these types where they are stated in the Bible, but creationists generally look for explanations of naturalistic type, especially where the miracle applies to the whole creation rather than a single item, which are potentially amenable to investigation and modeling as naturalistic processes. Contrast this with Gosse's argument in *Omphalos*; the deliberate seeding of sedimentary layers with fossils, where no creatures had previously lived, would fall into either the first or second categories (depending on one's interpretation), but certainly not the third.

With creationism, the major ministries, such as the Institute for Creation Research, Creation Ministries International, and Answers in Genesis, deliberately avoid (where possible) explanations that are not amenable to scientific causation.

In Chapter 4, we quoted the National Academies' statement in *Science, Evolution and Creationism*.[60]

> Because ... appeals to the supernatural are not testable [which includes falsifiability] using the rules and processes of scientific inquiry, they cannot be a part of science.

We noted that we wholeheartedly agree with this statement *when applied to scientific evidence*. But now, after categorizing miracles, we see that naturalistic miracles which affect the entire earth at the same time and the same rate (which constitute appeals to the supernatural) *are testable in principle*, because they are naturalistic miracles. Further, in the Christian worldview, both naturalistic

60 National Academy of Sciences and Institute of Medicine, 2008, p. 39.

miracles and natural phenomena are *similar* subsets of the supernatural (see Acts 17:28)—a typical difference is that they proceed at different rates. Therefore, we strongly disagree with the above statement as a general assertion, concluding that the National Academies' statement reveals an anti-theistic bias in its presuppositions.

The creationist project places science onto a foundation of biblical presuppositions. This is an ongoing task which needs to keep abreast with new biblical and scientific knowledge as it comes to light over time.

*

Secular humanists are right to cast a wary eye on the creationist movement. The history of ideas that we have reviewed in these chapters shows that when even a single person puts forth a coherent argument, it can, over time, germinate into a large-scale movement. That was true both of Boccaccio's *Decameron*, which foreshadowed the anti-Christian brand of secular humanism, and John Wycliffe's Bible translation, the morning star of the Reformation. In our time, it has been true of *The Genesis Flood*.

And this is where we see creationism heading. The early work is now complete. The movement is growing in sophistication, even as secular humanists continue to maintain their view through doubtful rhetoric and coercion.[61] [62] The rebuilding of the foundations of the city of God, on presuppositions derived from the Bible, is moving forward.

References

Allen, G. (1888). *Charles Darwin*. London: Longmans, Green, and Co.

Barton, R. (2018). *The X Club: Power and Authority in Victorian Science*. Chicago: The University of Chicago Press.

Bergman, J. (2012). *Slaughter of the Dissidents: The Shocking Truth About Killing The Careers Of Darwin Doubters* (Vol. 1). Southworth: Leafcutter Press.

Black, E. (2003, September). The Horrifying American Roots of Nazi Eugenics. Retrieved July 5, 2024, from hnn.us/article/the-horrifying-american-roots-of-nazi-eugenics

Browne, J. (2002). *Charles Darwin: The power of place*. (Vol. 2). Princeton, NJ: Princeton University Press.

Darwin, C. (1860, August 8). To T.H. Huxley 8 August. Retrieved December 31, 2023, from darwinproject.ac.uk/letter/?docId=letters/DCP-LETT-2893.xml

Darwin, C. (1871). *Origin of Species*. New York: D. Appleton and Company.

Darwin, C. (1909). *The Origin of Species*. New York: P.F. Collier & Son.

61 "How many more cases are there in which teachers are not hired or are fired, students failed or their grades lowered, promotions denied, admission to graduate programs refused, theses rejected, books or articles denied publication, or scientific projects denied government funding simply because the evolutionary establishment does not believe in equal rights for creationists? Usually the persecution is more subtle than in Dr Bergman's case. Reasons are fabricated or frivolous reasoning is substituted. But in more cases than most of us realize, the real reason is religious bigotry" (Bergman, 2012, p. xviii).

62 When David Gelernter, a well-known Yale University professor "turned away from Charles Darwin's theory of evolution", the intellectual behavior of Darwinists was telling: "As far as they are concerned, take your life in your hands to challenge it intellectually. They will destroy you if you challenge it ... [there] is nothing approaching free speech on this topic ... I am attacking their religion. It is a big issue for them" (Prestigiacomo, 2019).

Darwin, L. (1911, June 1). Presidential Address by Major Leonard Darwin June 1st, 1911. Retrieved December 26, 2023, from wellcomecollection.org/works/vtxnz3xv/items

Desmond, A. (1994). *Huxley: the Devil's disciple*. London: Michael Jospeh Ltd.

Desmond, A. & Moore, J. (1991). *Darwin*. London: Michael Joseph.

Galton, F. (1907). *Inquiries into Human Faculty and its Development*. London: J.M. Dent & Co.

Gosse, E. (1890). *The Life of Philip Henry Gosse, F.R.S.* London: Kegan Paul, Trench, Trubner & Co. Ltd.

Gosse, E. (2004). *Father and Son: A Study of Two Temperaments*. Project Gutenberg. Retrieved November 3, 2024, from gutenberg.org/ebooks/2540

Gosse, P. H. (1998). *Omphalos: An Attempt to Untie the Geological Knot* (reprint of 1857 book). Woodbridge: Ox Bow Press.

Himmelfarb, G. (1996). *Darwin and the Darwinian Revolution*. Chicago: Elephant Paperbacks, Ivan R. Dee, Publisher.

Hoyle, F. (1999). *Mathematics of Evolution*. Memphis, TN: Acorn Enterprises.

Huxley, T. H. (1893). *Darwiniana Essays*. New York: D. Appleton and Company.

Lombardo, P. (_). Eugenics Sterilization Laws. Retrieved July 5, 2024, from eugenicsarchive.org/html/eugenics/essay8text.html

Micklos, D., & Carlson, E. (2000, Nov). Engineering American society: the lesson of eugenics. *Nature Reviews. Genetics*, 1(2), pp. 153–158.

National Academy of Sciences and Institute of Medicine. (2008). *Science, Evolution, and Creationism*. Washington: The National Academies Press.

Ng, K., & Nakayama, C. (2024, July 3). Japan top court says forced sterilisation unconstitutional. Retrieved December 23, 2024, from bbc.com/news/articles/c0krnjy72j0o

Nisbet, R. (2009). *History of the Idea of Progress*. New Brunswick: Transaction Publishers.

North, G. (2010). The Road to Dayton: Creation vs. Evolutionary Eugenics. Dallas: GaryNorth.com. Retrieved December 31, 2023, from garynorth.com/RoadtoDayton.pdf

Numbers, R. L. (1992). *The Creationists*. New York: Alfred A. Knopf.

Prestigiacomo, A. (2019). WATCH: Renowned Yale Prof Leaves Darwinism, Says Intelligent Design 'Absolutely Serious' Theory. Retrieved March 1, 2025, from dailywire.com/news/watch-renowned-yale-prof-leaves-darwinism-says-amanda-prestigiacomo

Sarfati, J. (2015, March 10). God created with functional maturity, not 'appearance of age. Retrieved January 1, 2024, from creation.com/is-apparent-age-biblical

Spiro, J. (2009). *Defending the Master Race: Conservation, Eugenics and the legacy of Madison Grant*. Lebanon, NH: University Press of New England.

Thyer, B. A. (1995). The X club and the Secret Ring: Lessons on How Behavior Analysis Can Take Over Psychology. *The Behavior Analyst*, 18, 23–31.

Turner, F. M. (1974). *Between Science and Religion: the Reaction to Scientific Naturalism in Late Victorian England*. New Haven and London: Yale University Press.

8
Terms of the Debate

When a group is maneuvered into using the vocabulary of its opposition, it argues from weakness, for it has already conceded the terms of the discussion.[1]

This observation by journalist Otto Scott is certainly true of the origins debate. The secular humanist origins account includes words that are umbrella terms, that is, they encompass several distinct concepts. A prime example is the word *evolution*. As we'll soon see, this savvy use of the term "evolution" has conferred an advantage on secular humanists, and a corresponding disadvantage on creationists. Scott went on to point out that the culprit for disadvantageous vocabulary is "a failure of analysis"; it's possible for the creationist movement to resolve these rhetorical issues, but these terms must be unpacked—very careful analysis must be undertaken. That's the task of this chapter.

In this chapter we explain the conceptualization of evolution within "fundamentalist"[2] Christian circles and, more specifically, the creation movement. We will show how key issues have been analyzed and vocabulary clarified to challenge secular humanist ideas and recapture the terms of the debate. Nevertheless, the incompatibilities in vocabulary between creationists and secular humanists means that both groups inevitably talk past each other. The simple word "evolution" has different meanings in creationist and secular humanist contexts. But the first thing we need to clarify is the difference between evolution as a causal mechanism (part of empirical science—which investigates repeated phenomena) and as a holistic origins theory (part of applied science investigating one-off events).

---

1 Scott, 1985.

2 A "fundamentalist" Christian is one who accepts the five fundamentals summarized by the Doctrinal Deliverance of 1910: that the Bible is inspired by the Holy Spirit and is without error, that Jesus was born of the virgin Mary, that Christ was offered as a sacrifice to satisfy Divine justice, that Christ rose bodily on the third day after the crucifixion, and that Christ performed miracles as recounted in the New Testament. Most evangelical, Reformed and charismatic Protestants, Orthodox Christians and the majority of Roman Catholics historically fall within this definition of "fundamentalist" Christian.

A crucial but subtle distinction

To illustrate this point, we refer back to the MH370 case study. To clarify the issue, we'll focus on one of the more outlandish scenarios—the "elopement" theory.

The expert that proposed this theory probably had reasonably sophisticated reasoning, but we will present a simplified, imaginary version to illustrate the concepts. Imagine a hypothetical MH370 investigation in which aviation experts are invited to give evidence to a panel in the same manner mentioned in Ean Higgins' book. The panel's job is to establish the most likely scenario that accounts for the MH370 disappearance. Let's say our expert is a Freudian psychologist who presupposes that most human decisions reflect underlying sexual impulses. We enter the interview room while the expert is submitting evidence for his "elopement" theory:

Expert: The first thing I want to establish is that the history of aviation is replete with examples of pilots using planes to transport themselves closer to their romantic partners.

Panel: Do you have any evidence of that?

Expert: I have several documented examples. (The expert tables supporting evidence of such, which the panel examines and finds reputable.)

Expert: The next point is that there are many instances in which pilots have bailed out of multi-engine aircraft into seas and oceans.

Panel: What evidence do you have?

Expert: Check these out. They're sufficient to prove my point. (The expert again tables supporting evidence which is reputable.)

Expert: My final point is that pilots who bailed out into oceans or seas are often picked up successfully by boat.

Panel: And your evidence is?

Expert: I have several examples. Some of these pickups occurred at night, which is when MH370 disappeared. (The expert again produces irrefutable evidence in support.)

Panel: Well, what do you think happened on the night MH370 disappeared?

Expert: Based on what I showed you, I conclude that the captain of MH370 sought to elope with his lover by bailing out of the Boeing 777 over the South China Sea. He parachuted into the sea, where he was picked up by a waiting boat chartered by his lover's family. And remember—each component of my narrative has a precedent in aviation history.

We draw attention to the conceptual shift that takes place in this discussion. The first three exchanges are in the realm of *demonstrable fact*, whereas the expert's final response is a *forensic theory*. The forensic theory fits the data, but it is only one theory among several, and it does not even sound the most likely. In fact, it sounds preposterous. It seems highly unlikely that anyone would elope in this way. Nevertheless, our expert is reluctant to jettison his idea because it is compatible with the Freudian presupposition of his worldview.

We now turn to Darwin's origins speculation and contrast it with the MH370 scenario. The following table (8.1) illustrates the two-stage progression of thought in each. The top row contains repeatable phenomena that serve as part of the support for the forensic/origins theory; all this evidence is in the realm of empirical science. Anyone who doubts that any of these phenomena occur is free to conduct experiments in the lab, or in the field—they will find that these are all demonstrable or observable.

The second row, highlighted in grey, is different. It contains forensic or origins theories; these are not categorized as empirical science—they are applied science. The theories fit the data, but fitting the data is not enough to prove that the theory *actually describes what happened*. Multiple theories could fit the data; therefore, the task of forensics and origins science is to select the most likely scenario from the range of possible scenarios, while making clear any presuppositions which affect the decision.

Table 8.1: Distinguishing causal model from forensic or origins science

Type of science	**MH370 investigation**	**Biology**	**Type of evidence**
Causal model science (or equivalent)	Demonstrated evidence of: Pilots transporting themselves to romantic partners Pilots bailing out of large aircraft into seas Successful pick up of pilots by boats, including at night	Demonstrated evidence of: Natural selection Variation Genetic drift Mutations in DNA	Empirical: repeatable phenomena
Forensic or origins science	Proposed theory: The MH370 captain eloped by bailing out of the 777 airliner into the South China Sea, where he was picked up by a boat chartered by his lover's family.	Proposed theory: All life evolved from a common ancestor via naturalistic means. This involved gradual processes of variation occurring over very long time periods	Circumstantial: from one-off past events

A couple of points need to be emphasized. First, items in the first row are not disputed by informed people, because all these are demonstrable in the lab or in the field. To reiterate, these may be taken as evidence to support forensic or origins accounts, but they certainly *do not prove that these occurred;* other accounts are possible.

However, the grey box represents a shift from empirical science to the applied science of either forensics or origin science. Here, presuppositions play an important role, and the presuppositions that investigators bring may be influenced by political or, as admitted by Lewontin (refer Chapter 4), ideological/religious reasons.

Anecdotally, lay people may not even be aware of the shift between empirical and origins science that we have with Darwinism, but rather assume that words and phrases like "natural selection" and "mutation" are synonymous with, and proof of, Darwin's origin account. But they are not. They belong to a different class of science.

Table 8.2 shows how the word *evolution* can be used in two different ways by secular origins scientists, who commonly apply the label "evolution" to mechanisms of variation as a catch-all term. The table lists two empirical causal mechanisms of variation—natural selection and mutation—but there are others, including genetic drift and so on. As the table indicates, these *mechanisms* are accepted by both creationists and secular scientists. By contrast, evolution as a holistic origins theory (shaded grey for clarity) always involves extrapolation of causal mechanisms in a manner that is very unlikely. That is, it cannot be currently observed, relies on fragmentary circumstantial (typically not empirical) evidence, and is underpinned by the presuppositions of naturalism and gradualism (the latter is inconsistent with fossil evidence, as we discuss in Part 2).

Table 8.2: The term "evolution"—encompassing observed and unobserved processes

Causal mechanisms of "evolution" (variation)	**Observed examples—accepted by creationists and secular origins scientists**	**Unobserved examples—not accepted by creationists but promoted by secular origins scientists**
Natural selection	Natural selection, as observed in the lab or nature, selects variation in the direction of increased adaptation, e.g. increased prevalence of antibiotic resistance in bacteria when exposed to antibiotics.	Natural selection, over unobserved, lengthy timescales, purportedly contributes to increased complexity and increased genetic information, e.g. apelike ancestors evolving into humans.
Mutation	Observable small-scale or large-scale mutations, e.g. polyploidy in plants (refer Macroevolution section that follows below).	An unobserved process in which many small mutations over a long period purportedly contribute to increasing complexity, e.g. the evolution of the human eye.

It is here that we see an important reason creationists reject evolution as an origins theory. Darwinism, and like theories, extrapolate causal mechanisms to an unwarranted degree and in the wrong direction (the degree and type of extrapolation accepted by secular scientists is far beyond what empirical science can support, as we explain in Part 2)—as far as is necessary to show that God is not needed to explain the diversity of life. Thus, the theory is rendered extremely unlikely, even if, like the MH370 "elopement" explanation, it seems superficially compatible with the facts.

In this chapter, and several other places in the book, we use the term "evolution" in the restricted sense of empirical causal model science. While informed creationists might not be confused by this use, we recognize that evolution is a tainted term and is generally best avoided by creationists for that reason.

Early days

Historically, Bible-believing Christians sensed serious problems with Darwinism and communicated these concerns by opposing "evolution". A classic case was the Anti-Evolution League of America, established in the 1920s, with a now-famous image shown in Figure 8.1.[3] As well-intentioned as the League was, its overly broad name left its members wide open to condemnation from the Darwinist camp for ignorance. The phrase "anti-evolution" suggested

Figure 8.1: Anti-Evolution League bookstall, 1925

a rejection not only of secular origins theories, but also empirical causal model science.

3 Licht & Liftarn, 2014.

In fact, early Christian opposition to "evolution" was oftentimes more nuanced than a total rejection implied by the Anti-Evolution League name. In the classic "anti-evolution" statute of Arkansas (1928) overturned by the U.S. Supreme Court in 1968, it was the teaching that humans had "ascended or descended from a lower order of animals"[4] which was prohibited. Evolution as empirical causal model science was not mentioned in the statute and arguably was not off limits in the classroom. This more nuanced position rejected holistic Darwinian origins theories but was silent on empirical causal model science.

Evidence for *evolution*

Charles Darwin argued in *Origin* that species were not rigidly fixed. Because the scientific consensus at that time was that species were immutable, non-Christians and many lay Christians interpreted the creation account of Genesis to imply that every plant and animal species was originally created specially by God. In *Origin*, Darwin specifically opposed this view of special creation.

Darwin discussed the concept of variation that a breeder of animals may encounter. He used the term "domestication" to describe the process of breeding animals more suited to human purposes. Darwin provided examples, such as the breeding of pigeons and other animals. He wrote:

> It is, therefore, of the highest importance to gain a clear insight into the means of modification and co-adaptation. At the commencement of my observations it seemed to me probable that a careful study of domesticated animals and of cultivated plants would offer the best chance of making out this obscure problem. Nor have I been disappointed; in this and in all other perplexing cases I have invariably found that our knowledge, imperfect though it be, of variation under domestication, afforded the best and safest clue.[5]

Today, examples of this type of change are well known. They are identified by a sub-population of a kind changing with respect to some characteristic. For example:

- Antibiotic resistance—Applying antibiotics to bacteria in the laboratory causes a selective pressure that tends to increase the proportion of antibiotic resistance among bacteria as those that are not resistant die off.[6]

4 Fortas, 1968.
5 Darwin, 1963.
6 Munita & Arias, 2016.

- Decades ago, when the chemical DDT was in widespread use to suppress mosquito populations, resistance spread over time due to reproduction of mosquitoes that had survived this artificial selection.[7]
- In the early days of penicillin production, scientists tried to increase the amount of penicillin produced by inducing mutations in the penicillin mold. This was achieved by successively applying X-rays, ultraviolet light and toxic chemicals to the mold, and mutated mold organisms that produced more penicillin were selected. This dramatically improved the production efficiency of penicillin, demonstrating a practical application of induced mutation and selection for human use.[8] It is possible that such mutations damaged a penicillin production control mechanism in the organism, leading to increased penicillin production.
- During the COVID-19 pandemic, new strains of the virus, such as Delta and Omicron, emerged over time. These changes in the strains roughly correlated with a theory of pandemics suggesting that pathogenic organisms move towards higher infectivity but lower mortality rates.[9] The emergence of different strains may be evidence of mutations combined with selective pressure.

All these examples represent changes occurring within particular species.

But Darwin went on to speculate that these changes, when coupled with *natural selection* acting over long time periods, could cause speciation—the generation of new species. Indeed, he briefly suggested in *Origin* that these processes could explain "the production of the higher animals",[10] a concept which was distinct from domestication and speciation.

Historically, the scientific community distinguished between these two concepts using the terms *microevolution* and *macroevolution*. Because these terms are still found in some contemporary school textbooks in the West, we discuss them here.

Microevolution

Microevolution refers to variation within species, as per the examples provided earlier. Another good example is dog breeding. Hundreds of breeds have been developed, with very significant variation across a range of attributes. Dogs range from the Irish Wolfhound to the Chihuahua, from the English sheepdog to the Mexican hairless, and from the docile Labrador to highly aggressive guard dog breeds. These are examples of microevolution

---

7 Livadas & Georgopoulos, 1953.

8 Adrio & Demain, 2006.

9 Methot, 2012.

10 Darwin, 1963, p. 560.

directed by humans. Darwin posited that this effect can also occur in the wild because of selective pressures. His well-known examples included the varied beak shapes and sizes of the finches on the Galapagos Islands. In each of these cases, variation occurs within species. Dogs remain dogs even if their features vary widely; finches remain finches, and so on. A feature of microevolution is that it proceeds quickly enough to produce readily observable results. A skilled dog breeder can establish a new breed in a matter of years. Breeding of the Irish wolfhound, for example, which is the tallest dog breed in existence, commenced in the early 1860s, with the breed standardized by the mid-1880s.[11]

Dog breeding illustrates another important point about these variational changes. From an ancestral dog kind, both the massive Irish Wolfhound and the diminutive Chihuahua breeds have been developed. Both breeds contain less genetic information than the ancestral kind because they were produced by selective breeding; it is probably not possible to breed an Irish Wolfhound into a Chihuahua-like-dog or vice versa. Both dog breeds are probably missing essential genetic information that the other contains, because selection of varieties leads to *loss of information,* not gain. However, there is a limit to how much information can be removed by selective breeding without making the animal infertile.

Another key feature of microevolution is that it is consistent with the Bible, which directly mentions at least two instances of this biological process. In the biblical book of Esther, King Ahasuerus utilizes horses bred for speed to carry messages across his kingdom (Esther 8:10). Clearly, the art and science of selecting and breeding certain animals for a desired characteristic—in this case speed—was known and practiced by the ancients.

A further example of microevolution, this time from the plant world, is noted by the prophet Isaiah. In Isaiah 5:4, the prophet describes disappointment at finding wild grapes when choice grapes were expected. This shows that the ancients had distinguished between wild and specially bred varieties of grapes. Whether the breeding of the grapes was carried out by progressively removing those plants which provided poor fruit, or by cross-pollinating the best plants, we cannot be sure. Either way, this example and the previous show that human-directed microevolutionary processes in both plants and animals were practiced in the ancient world and noted in the Bible. Microevolution was also involved in the process by which God transferred the high-quality specimens of Laban's flocks to Jacob (Genesis 30). In this last example, the superficial characteristic being varied was the coloration of the wool, but the important characteristics were strength and vigor, as the Bible mentions.

The Bible contains other descriptions which imply microevolution. A notable case is found in the Genesis account of the confusion of languages at

11 Flaim, 2023.

Babel. The language barriers that the people suddenly experienced caused them to coalesce into isolated populations, forming the precursors of the various ethno-linguistic groups we have today. Human ethnic groups can differ in several anatomical features, such as skull shape, skin pigmentation, presence of prominent skin folds on the upper eyelid (epicanthal folds) and so on. Clearly, microevolutionary effects operating in geographically isolated human populations have produced these well-known distinguishing features. In fact, the downstream effects of the Babel events cannot be properly understood without invoking the effects of microevolution.

Another example of an event from Genesis with microevolutionary effects is Noah's Flood, which can be described in biological terms as a worldwide near-extinction event. This had tremendous microevolutionary effects on the human population by dramatically reducing the gene pool, rendering most pre-Flood human populations completely extinct.

In microevolution, an effect known as *devolution* can and does occur. This refers to a loss of genetic information in a population which might manifest as a loss of function. By devolution we do not necessarily mean directed change, but merely loss, corruption or garbling of existing genetic information in the gene pool. Several examples of devolutionary effects have been observed. These include African elephants losing their tusks over several generations in response to the pressures of poaching. In this scenario, elephants with small or no tusks (which might result from a mutation) experience survival benefits because poachers do not target them; these characteristics are passed to their offspring. Further examples include cave fish losing eyesight,[12] and loss of flight in the white-throated rail on islands in the southwest Indian Ocean.[13] In the plant world, the hongyacha tea plant has lost the ability to produce caffeine, which is a naturally occurring pesticide.[14]

Microevolution can cause a gradual change in a trait (such as body size, coloration, or physiological traits) due to a so-called geographic gradient. For example, the geographic gradient might relate to an environmental factor like temperature or altitude. This can lead to variation within a species and results in what is known as a *cline* (a gradient of change within a species).

Macroevolution

Darwin hypothesized that microevolutionary effects, if carried on over a very long period, would lead to the second category of evolution, historically termed *macroevolution*. Macroevolution refers to the development of new species; an example is the alleged emergence of modern humans from ape-like

12 Wieland, 2000.
13 Cserhati, 2019.
14 Hughes & Carter, 2023.

ancestors. A feature of macroevolution is that it is hypothesized to occur over such long time periods that it cannot be directly observed. (Whereas historians regard 2000 years ago as ancient, macroevolutionists regard two million years ago as relatively recent.) Darwin suggested that all living organisms are descended from primordial organisms, from which different types of organisms evolved over time by the process of natural selection. In Darwin's account, natural selection implies the "demographic death" of that part of the population not selected by "nature".

Unlike microevolution, the concept of macroevolution is not described explicitly in the Bible. In fact, on superficial examination, it appears incompatible with the biblical account. Humans emerging from ape-like ancestors contradicts the plain teaching of Genesis. For this reason, in the last century or so, some Christians advocated that microevolution is acceptable, but macroevolution is incompatible with the Bible. However, this view is now considered incorrect by creationists. Several Christian intellectuals before Darwin had analyzed the account of Noah's Flood and had postulated that speciation had occurred by transmutation within kinds,[15] a belief also held by modern creationists. Therefore, creationists accept speciation[16] (which is categorized by secular scientists as a type of macroevolution).

Today it is known that some animals, and many plants, show evidence of macroevolution. In the animal world, a small deer—the muntjac—native to China, India, and parts of South-East Asia, is a case in point. The Indian muntjac (which has been introduced to the UK, where it is known as Reeves's muntjac) cannot breed successfully with the Chinese muntjac because these sub-types have different numbers of chromosomes. As a result, these deer are technically different species. What appears to have happened is that the two varieties were geographically separated at some point in the past, and a process called chromosomal fusion, which rearranges existing genetic material into fewer chromosomes, has resulted in them becoming separate species. Because of this known type of edge case, Christians that blithely deny the possibility of macroevolution fall neatly into an evolutionary "gotcha" trap. Macroevolution in animals can and does occur; the muntjac is an example. Horses, zebras and donkeys have almost certainly descended from the same ancestral kind and have also experienced chromosomal fusion.

---

15 Transmutation within kinds refers to speciation within one of the "kinds" that God created; for example, it could include speciation resulting in tigers and lions from an original feline kind. This concept was postulated by thinkers including Athanasius Kircher (1602–1680) and John Wilkins (1614–1672). (Sarfati, 2018, p. 577).

16 There are several reasons that creationists hold the belief that speciation within kinds has occurred: (1) there are examples of hybridization records showing a circle of hybridization (discussed later in this chapter) between species, such as the cat family; (2) in the case of the cat, evidence suggests that multiple species were present soon after the Flood; (3) logistics of the ark, mean that while there was sufficient space for all kinds, there was probably not enough for all known species; (4) speciation is observed to occur rapidly.

In plants, macroevolution is relatively common. Caused by an effect known as polyploidy, it occurs in flowering plant species when aberrant separation of chromosomes leads to multiple sets of chromosomes being found within each cell. This leads to instant speciation—the generation of a new flowering plant species. It is estimated that approximately half of flowering plant species have formed due to polyploidy.[17] Speciation by polyploidy is unquestionably macroevolution.

Both of these examples—chromosomal fusion in the muntjac and polyploidy in plants—result in instant speciation. That is, if a muntjac conceives a fawn which exhibits chromosomal fusion, that fawn is a new species, one which might not be able to produce fertile progeny with the parent species. (In order to breed successfully as a new species, a similar chromosomal fusion would be required in a member of the opposite sex.) Likewise, polyploidy in plants is also instant speciation.

These processes are not Darwinian macroevolution. Darwinian evolution describes speciation as occurring extremely gradually over long periods of time via adaptive changes retained by natural selection, without the types of sudden changes we see with the muntjac and plant polyploidy. Creationists generally do accept that variation and speciation can proceed via mechanisms of natural selection, mutation, genetic drift, and so on. However, creationists believe these changes can occur quickly, and result in net losses of genetic information, not gains.

An imprecise umbrella term

The existing vocabulary left some Christians in a bind. The term evolution, used by itself, contained within it both the notions of micro- and macroevolution. On the one hand, evidence consistent with microevolution of both plants and animals is written into the Bible. Macroevolution, by contrast, when used by secular humanists, contains the unbiblical concept of humans emerging from other species, which young-earth creationists do not agree with, but also the examples cited above, both of which represent incontestable, empirical evidence for macroevolution. Given these difficulties, creationists resorted to several neat sobriquets in an attempt to isolate, categorize and describe the processes that are non-biblical. Thus, we read in the creationist literature of *goo-to-you evolution*, or *molecules-to-man evolution*. These may have some use in a descriptive sense (and do correctly identify the central issue) but they do not qualify as scientific terminology. Secular humanists owned the linguistic terrain, and the power of this rhetorical advantage is hard to overstate. We have summarized these concepts in Figure 8.2. Some Christians over the years have considered the areas in grey highlight to have been

17 Patterson, 1999, p. 33.

"tainted" terms because they contain the concept of speculative macroevolution

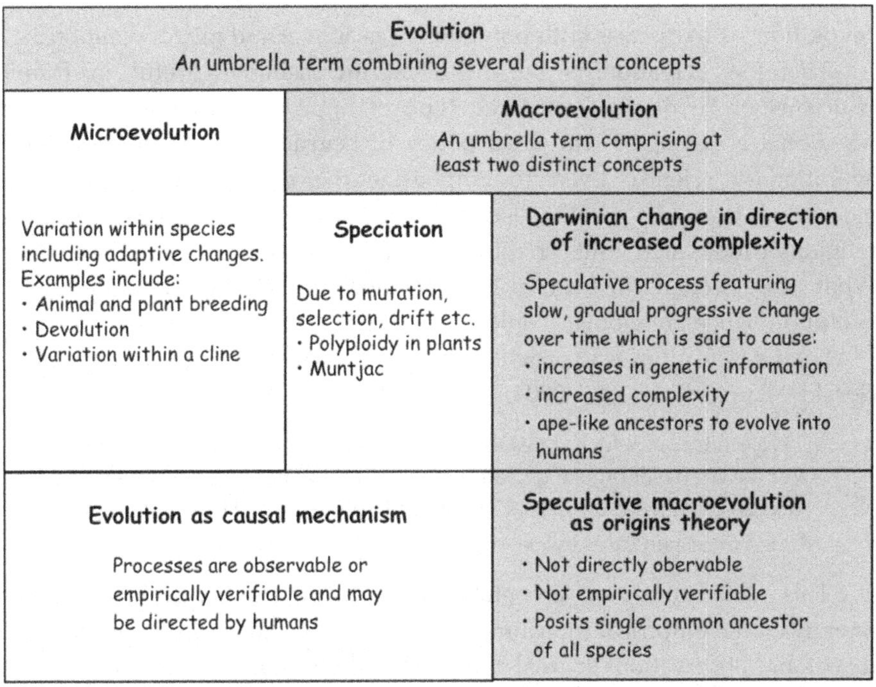

Figure 8.2: Categories of evolution according to micro/macro distinctions

as origins theory.

Evolution is an umbrella term not only because it includes the subcategories shown in Figure 8.2, but also because it subsumes both causal model science (variation by natural selection) and a holistic naturalistic origins account, as indicated in the lower section of the diagram. Darwin explained the causes by which animals in the wild adapt to their environment (natural selection), and this explanation was an example of causal model science, an empirical science. But Darwin's speculation that slight changes over very long time periods could result in the development of all known species is an example of origins science, a field of enquiry that deals in circumstantial evidence. In origins science, as in forensics, speculation carries a certain amount of weight (think of the "terrorist-hijacking-gone-wrong" theory in MH370), whereas in empirical causal model science, harder evidence is preferred. By conflating these, Darwinists can posit speculative "just-so stories" of how certain animals or behaviors evolved, and also expect these stories to be taken as "scientific".

The creation movement soon learned to critique the tainted greyed out terms—such as evolution or macroevolution—with care. These terms contain within them proven concepts (white). To dismiss macroevolution is to dismiss polyploidy, a known process which is common in plants, and to deny known examples of speciation among vertebrates involving mutation, selection and genetic drift, or chromosomal fusion (e.g. muntjac and also lizards). To dismiss "evolution" is to dismiss both polyploidy (as above) and microevolution.

Therefore, creationists began to be increasingly careful in framing criticisms of "evolution". But the type of approach discussed so far still resembled a defensive rearguard policy of begrudgingly accepting parts of evolutionary science. Even to emphasize the distinction of micro- and macroevolution was to suggest that Christians needed to put arbitrary, religiously motivated limits on the extent of evolution which could be accepted. What was needed, instead, was a scientifically defensible definition for evolution which creationists could accept.

Such a definition was provided by the British zoologist and physiologist Gerald Allan Kerkut (1927–2004), who wrote:

> There is a theory which states that many living animals can be observed over the course of time to undergo changes so that new species are formed. This can be called the "Special Theory of Evolution" and can be demonstrated in certain cases by experiments.[18]

This definition was acceptable to creationists because it explicitly referenced the empirical causal model of variation and could be supported by repeatable observations in real time. (This definition is quoted by Creation Ministries International.[19]) Having described evolution as a causal model, Kerkut continued to describe the Darwinian form of evolution:

> On the other hand there is the theory that all the living forms in the world have arisen from a single source which itself came from an inorganic form. This theory can be called the "General Theory of Evolution" and the evidence that supports it is not sufficiently strong to allow us to consider it as anything more than a working hypothesis. It is not clear whether the changes that bring about speciation are of the same nature as those that brought about the development of new phyla. The answer will be found by future experimental work and not by dogmatic assertions that the General Theory of Evolution must be correct because there is nothing else that will satisfactorily take its place.[20]

This definition is also typically acceptable to creationists because it clearly delineates Darwin's speculative extrapolation. It is Kerkut's idea of the General Theory of Evolution which creationists typically refer to by the word "evolution" or phrases "goo-to-you evolution" or "molecules-to-man

18 Kerkut, 1960 A, p. 157.
19 Kerkut, What is evolution? (Kerkut), 1960 B.
20 Kerkut, 1960 A, p. 157.

evolution". Understanding these distinctions is critical to understanding creationist literature.

The created kinds

There is another term in the origins debate, this time from the Bible, that needs careful consideration. This is the word translated as "kind" from the Hebrew word *baramin* that is used in Genesis. Modern science does not refer to the concept of kind at all, and indeed Darwin's theory referred to the origin of *species*. This results in a potential point of incongruence in the origins debate that Christians need to be aware of.

The biblical account in Genesis describes various living organisms being created "after their kinds". This applies to vegetation such as grasses, seed-yielding herbs and fruit-bearing trees, aquatic animals, birds, livestock, creeping things, animals of the earth and man. In this latter case, a single man was created, from whom a woman was made, and this original couple was the ancestral gene pool of the human species. Mankind is both a kind and a species. In fact, all the original kinds were both a kind and a species.

This leads to the question—what is a biblical kind, and how do we define it?

As mentioned earlier, a problem with identifying kinds today is that they have become blurred by variations and speciation since the creation week. Consequently, it is not a simple matter to reconstruct what the original kinds were.

One type of approach has been to consider the contemporary classification of several species, such as in the table below. This classification is known as a morphological taxonomy because it is based on physical characteristics of organisms.

Table 8.3: Morphological taxonomy

Category	Indian Muntjac	Reeves's Muntjac	Human	Chimpanzee
Kingdom:	Animalia	Animalia	Animalia	Animalia
Phylum:	Chordata	Chordata	Chordata	Chordata
Class:	Mammalia	Mammalia	Mammalia	Mammalia
Order:	Artiodactyla	Artiodactyla	Primates	Primates
Family:	Cervidae	Cervidae	Hominidae	Hominidae
Genus:	*Muntiacus*	*Muntiacus*	*Homo*	*Pan*
Species:	*M. muntjak*	*M. reevesi*	*H. sapiens*	*P. troglodytes*

The Indian Muntjac and Reeves's Muntjac, discussed above, are two distinct species, where one of these species is the product of instant speciation from the other by chromosomal fusion. Reasoning from this example, one might surmise that the biblical kind in this case may correspond to at least the genus level (*Muntiacus*), but it is difficult to be definite.

The second example is the human and chimpanzee, which the Bible tells us are different kinds. In the biological taxonomy, these are classified in the same family, but different genera.

Based on this comparison, it might be a starting consideration that the biblical kinds correspond roughly to the genus level, giving us a possible morphological definition. However, this brief analysis cannot be taken to be definitive in every case.

A problem with this descriptive definition is that it purports to discover a rule applicable to morphological taxonomy (meaning, the physical characteristics of the creature, such as whether it has a cloven hoof). However, the rule ought to be established by reference to heredity such as changes in DNA, because evidence from DNA is more significant than evidence about the physical characteristics, or morphology, of organisms. If we try to define "kind" in relation to the taxonomy above, our definition will need to be updated as new scientific information, such as new DNA evidence, comes to light.

This morphological approach to defining kinds used by some creationists also typically included a concept called the "fixity of kinds", which emphasizes limits to the amount of change within a kind which can occur. It states that one kind cannot evolve into another kind. Once again, this had the appearance of placing arbitrary or religiously motivated limits on the extent of variation which was acceptable. A new approach was sought to help explain certain facets of the Genesis account, incorporating recent advances in biology and genetics.

One reason creationists needed a new approach to understanding speciation was a conundrum associated with the global flood narrative. Creationists had found it difficult to explain how all known terrestrial species, both extant and extinct, could fit on the ark and survive for a year or more. From the outset of the creationist movement, variation and speciation were seen as important parts of the solution to that conundrum.

A different conception of kinds

In 1991, creationist writer Carl Wieland offered a new approach to understanding biblical kinds. Instead of defining kinds by morphology, Wieland suggested a definition based on biological descent: "Groups of living organisms belong in the same created kind if they have descended from the same ancestral gene pool".[21] Importantly, the phrase "created kind" restricts the definition to the biblically mentioned kinds—it does not necessarily encompass unicellular organisms such as bacteria, or even simpler structures such as viruses, neither of which are mentioned in the Bible. Wieland's definition also precludes one kind evolving into another, because they have different ancestors; ancestry cannot be changed post hoc. Hence it renders talk of "fixity of kinds" redundant, because "no new kinds can appear *by definition*".[22]

Kinds have been subject to variational, adaptive processes throughout history. Just as mankind (species) encompasses Adam and Eve and all descendants, each kind encompasses the original created members and all their descendants, without specifying any conditions or limitations on the historical processes which may have affected specific sub-populations.

But Wieland's definition raises another question: how can creationists identify shared lineages among species?

A proposed answer is the hybridization criterion. (Hybridization refers to sexual reproduction between different but related species. For example, male lions mated with female tigers result in offspring called ligers. Similarly, crossing horses and donkeys results in mules or hinnies.) This posits that different species which can hybridize belong to the same ancestral kind, even if the offspring are relatively or completely infertile.[23] The absence of empirical evidence for hybridization does not necessarily preclude two species having derived from the same ancestral kind, but successful hybridization makes it almost certain.

For example, large cats such as the lion and tiger can hybridize (into ligers and tigons) and are therefore viewed as belonging to the same ancestral kind by the hybridization criterion. Similarly, the serval and the domestic cat can hybridize, indicating derivation from the same ancestral kind.

But hybridization among cats goes much further than these two examples. By establishing all known hybridization combinations among the cat family,

21 Wieland, 1991.

22 Ibid.

23 Lightner, 2018.

creationist writer David Catchpoole presented a hybridization linkage between the domestic cat and the tiger.

Domestic cats can be crossed with the 3–9 kg margay, *Leopardus weidii*, which readily interbreeds with the 11–16 kg ocelot, *Leopardus pardalus*, which has been crossed with the puma (35–100 kg), *Puma concolor*, which has been hybridized with the leopard (30–85 kg), *Panthera pardus*, which can be crossed with the lion (120–250 kg), *Panthera leo*, which can readily interbreed with the tiger, *Panthera tigris*, with adult tigers ranging from 110 kg to a massive 320 kg.[24]

Secular humanist versus creationist views of biological history

The Wieland definition is useful because it focuses attention on an important distinction between the secular and biblical accounts. The secular account starts from a position of genetic poverty by positing an ancestral gene pool comprising a single primordial organism. From there, it is claimed that mutations and natural selection increase the amount of information, resulting in the diversity of life. On the other hand, the biblical account describes multiple ancestral gene pools originating from the original created kinds. In the case of humans, an individual man, Adam, was created, from whom the woman, Eve, was cloned (Genesis 2:21–22).

Figures 8.3[25] and 8.4[26] show the shape of variation and speciation across time within the secular and creationist paradigms respectively.

24 Catchpoole, 2015.
25 Adapted from Sarfati, *Refuting Evolution*—Chapter 2, 2019.
26 Ibid.

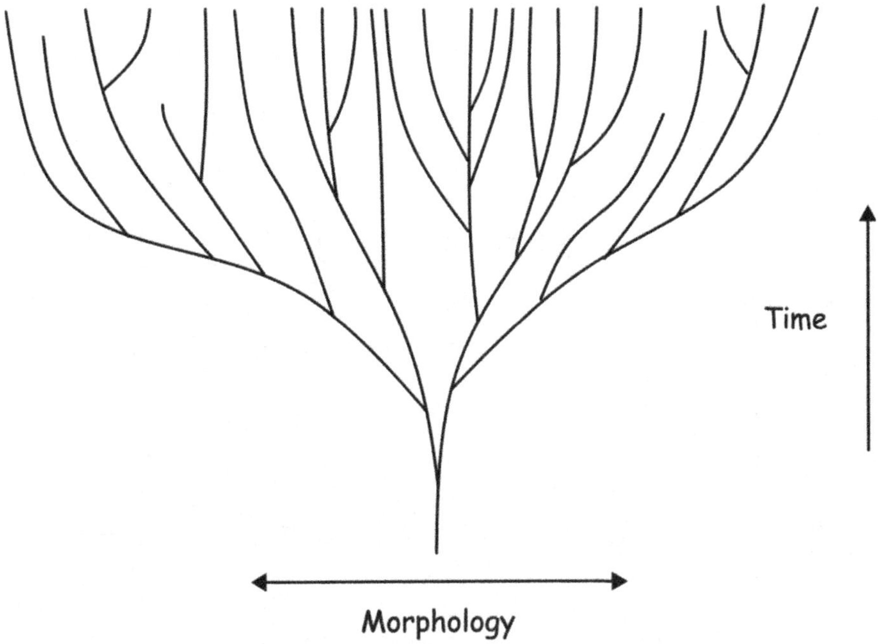

Figure 8.3: The secular tree of life

The secular origins account posits an evolutionary tree branching from a single common ancestor.

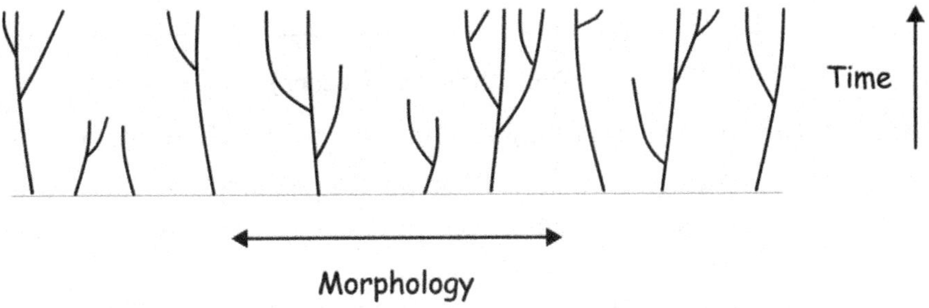

Figure 8.4: The orchard of creation

The biblical origins account is more like an orchard. Variation and speciation affect numerous kinds.[27] This variation, which results in sub-branches of kinds, could legitimately be termed *evolution*, however, creationists

27 Sarfati, *Refuting Evolution*—Chapter 2, 2019.

avoid this tainted term. Note that mankind is not represented in the diagram, as man has varieties (ethno-linguistic groups and nations) but they are not sister species; i.e. there has been no branching off of further species within the kind.

It's useful to consider the pathways of informational change suggested by the secular humanist and biblical paradigms.

The secular humanist view has several features shown in Figure 8.5 below.

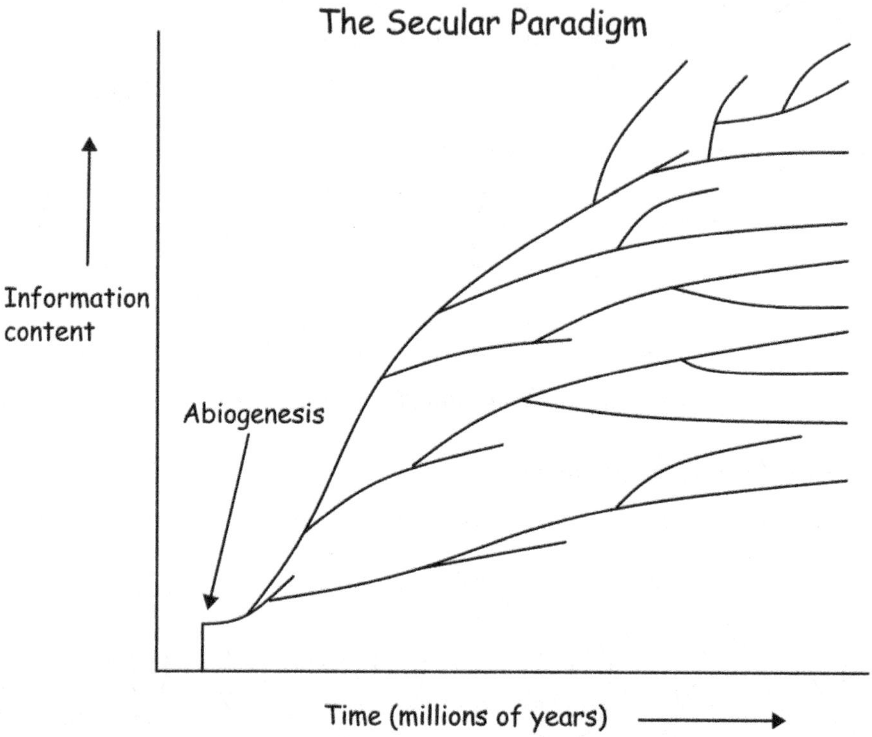

Figure 8.5: Changes in genetic information over time in the secular paradigm

- An immediate information jump, corresponding to abiogenesis, or the origin of a living cell from non-living matter, is represented in the lower left part of the diagram.
- Branching represents new varieties, species, and higher taxonomic categories such as families, phyla, orders etc.
- Upward sloping lines represent a gain of information over time.
- Some branches are angled downwards, indicating net loss of information, even though this could increase adaptation of the organism to its environment.
- Some species become extinct, indicated by a line not reaching the right-hand side of the diagram.

- Overall, the net direction of change is upward when proceeding from left to right, starting from a relatively low point on the left side of the diagram.

The upward slope of the lines is crucially important in Darwinian evolution. Increased complexity or increased information content is a necessary part of the theory, as Richard Dawkins has made clear:

> The feature of living matter that most demands explanation is that it is almost unimaginably complicated in directions that convey a powerful illusion of deliberate design.[28]

For secular humanists, the task of evolution is to explain "the illusion of deliberate design" and "unimaginable complexity" in organisms. The secular humanist explanation must detail how the universe went from zero biological information at the beginning to the vast amount of information existing today.

In the creationist scheme of variation and speciation, a 2-dimensional diagram can only show the progression of a single kind from the "orchard of life". The essential characteristics are:

- The high point of information content corresponds with the act of creation during the creation week, where God saw that all was "very good". The animal and plant kinds were created with an inbuilt genetic capacity to adapt to different environments that they encounter through history.
- Variation and speciation occur rapidly after creation due to mechanisms like genetic drift, natural selection, human selection and hybridization. All these generally cause information losses but may increase adaptation of creatures to their environments.
- The Genesis flood causes a severe genetic bottleneck for sentient land-dwelling animals.
- Rapid variation and speciation occur after the world-wide flood event, and then slow. Organisms have an inbuilt capacity for variation and adaptation; DNA was designed by God to allow variation and adaptation to environmental conditions. As selection and genetic drift occur, particular populations contain progressively less information, only a part of that contained in the parent population. Daughter populations, each of which is more specialized, have less potential to vary in other directions. A net decline in genetic information occurs. A domestic dog has less information than its mongrel ancestor, which in turn has less than the wolf, which has less than the original ancestral (created) kind.

---

28 Dawkins, 2003, p. 79.

- Speciation doesn't necessarily have to slow: certain types of mutations might actually increase the rate of speciation by introducing reproductive barriers, although these types of mutations are relatively uncommon.

The net direction of information change, following the high point at creation, is relatively flat or declining from left to right.

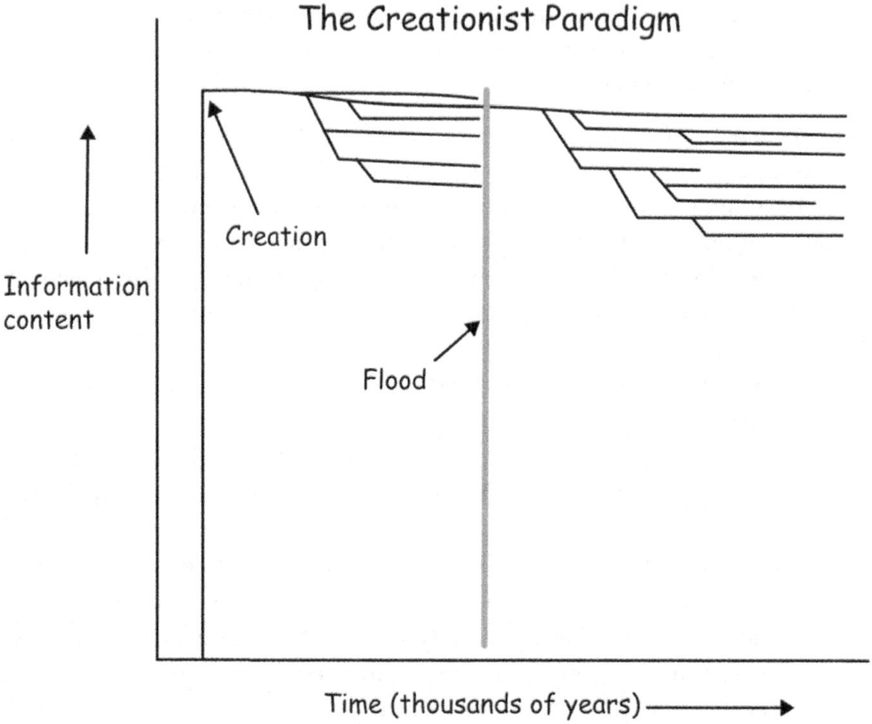

Figure 8.6: Changes in genetic information over time in the young-earth creationist paradigm

Death

Since death is such an important component of the secular humanist evolutionary explanation, we need to be clear on the biblical meaning of life as used in Genesis 1–11. For example, Genesis 1:20 describes the waters bringing forth swarms of living creatures, and the Hebrew for "living creatures" (nephesh chayyah) literally means "animate life", namely those animals which breathe[29] through the nostrils and have lifeblood.[30] In the Genesis 1–11 sense,

29 Sarna, 1989, p. 10.
30 Sarfati, 2018, p. 248.

plants do not have life, nor insects.[31] Following the same logic, microbes such as bacteria do not have life in the meaning of Genesis 1–11.

Within a short period of time, the fall of Adam and Eve occurred, and animal/human death entered the creation as part of God's curse.

Different understandings of change

We can now summarize key differences between the types of historical evolution posited by secular humanists and the speciation and variation posited by creationists.

Table 8.4: Essential characteristics of secular and creationist schemes

Feature	Secular paradigm	Creationist paradigm
Origin of life on earth	Abiogenesis	Acts of creation over a six-day period
Origin of information content of early lifeform/s	Random chance	Imparted into living organisms by the Divine Word
Initial number of living organisms	Single primordial organism	Many organisms within various kinds (plant/animal); one man and one woman
Cause of informational change since origin of earthly life	Evolutionary mechanisms	Natural selection, genetic drift and other empirically verifiable mechanisms of variation
Information gain since origin of earthly life	Extremely large	Minimal or zero among more complex animals, information loss vastly more likely
Timeframe of existence of life on earth	Billions of years	About six thousand years
Rate of change of animal varieties	Slow, gradual	Rapid, possibly decelerating for complex animals
Death	Naturally present from the outset	Present only after the fall of man (for sentient animals)

Motte and Bailey

The imprecise term "evolution" enables secular humanists to engage in a rhetorical technique called "motte and bailey". This technique takes its name from medieval strongholds that featured two defensive barriers, known as the *motte* and *bailey* respectively.[32] The outer wall, or bailey, enclosed a wide area where the inhabitants could live and work. If an attacking force breached the bailey, people would retreat to the motte, a castle built on higher ground that was cramped but enabled archers to ward off attackers. Once the attackers were rebuffed, people could return to their daily lives in the bailey.

In secular origins science, the "bailey" corresponds to common evolutionary descent from a single primordial organism—evolution as origins

31 Ibid., p. 247.
32 Sarfati, 2022.

science. Scientifically, this is unsupported, difficult to defend, and unlikely to hold up under scrutiny. Nevertheless, it is pushed as fact by many secular scientists.

When challenged, the secular scientist retreats to a more defensible, less controversial position—the "motte". In this case, the defensible position is evolutionary change as empirical causal model science (which includes mechanisms of natural selection, mutation, genetic drift etc., and may include scientific data).

After the challenge ends, the secular humanist may shift back to implying or stating outright their more extreme claim about evolution as origins science. It enables a secular humanist to make an ambitious assertion while having a safe fallback position.

Use of the motte and bailey tactic has been criticized by philosophers[33] and social commentators because it can be misleading and allows individuals to push a controversial view without fully defending it. It may also confuse opponents when the position being defended changes from bailey to motte and vice versa. A simple way to avert the use of the motte and bailey technique is to avoid umbrella terms.

Conclusion

In unpicking the term "evolution", we noted first that it is an umbrella term, and critiquing its unbiblical sub-components requires care and knowledge. We recommend that Christians move away from discussions of microevolution and macroevolution because these terms unhelpfully center the debate around the extent of evolution rather than the key issue—the type of evolution. Christians should also consider avoiding tainted terms which can lead to confusion. For example, rather than using the umbrella word "evolution", consider the use of more specific terms such as speciation, natural selection, genetic drift, and variation when indicated.

In relation to the evolution of man, the creationist position remains unaltered from the Arkansas statute mentioned earlier—evolution of humans from lower species did not occur. But this is no longer held on purely biblical grounds—it is now known that human evolution from lower animal kinds is not scientifically supportable, which we discuss more fully in Part 2.

We encourage Christians to adopt the definition for kind which is based on biological descent: "Groups of living organisms belong in the same created kind if they have descended from the same ancestral gene pool." This definition is both biblical and useable. The hybridization criterion is also useful as a concept

33 For instance, philosopher Nicholas Shackel from Cardiff University introduced the term mote and bailey in his 2005 paper "The Vacuity of Postmodernist Methodology", *Metaphilosophy*, 36(3).

to shed light on the contemporary diversity of varieties and species comprising specific created kinds.

Creationist and secular humanist conceptions of change in the origins sphere are of different types.

References

Adrio, J. L., & Demain, A. L. (2006, March). Genetic improvement of processes yeilding microbial products. *FEMS Microbiology Reviews*, 30(2), 187–214. doi: 10.1111/j.1574-6976.2005.00009.x

Catchpoole, D. (2015, October). Cats big and small. *Journal of Creation, 37*(4), pp. 34–37.

Cserhati, M. (2019, August 12). *Rails derail evolution*. Retrieved February 8, 2024, from creation.com/rails-derail-evolution

Darwin, C. (1963). *The origin of species by means of natural selection or the preservation of favoured races in the struggle for life* (Reprint of the Sixth Edition with a Preface by Sir Gavin de Beer). London: Oxford University Press.

Dawkins, R. (2003). *A Devil's Chaplain: Reflections on Hope, Lies, Science and Love*. Boston: Houghton Mifflin Company.

Flaim, D. (2023, May 10). Irish Wolfhound History: Behind the Ancient Breed of Ireland. Retrieved from akc.org/expert-advice/dog-breeds/irish-wolfhound-history-behind-the-breed/

Fortas, M. J. (1968). *Epperson v. Arkansas, 393 U.S. 97 (1968)*. Retrieved from supreme.justia.com/cases/federal/us/393/97/#F3

Hughes, J., & Carter, R. (2023, July 10). *Naturally caffeine-free tea*. Retrieved February 8, 2024, from creation.com/caffeine-free-mutation

Kerkut, G. (1960 A). *Implications of Evolution*. New York: Pergamon Press.

Kerkut, G. (1960 B). What is evolution? (Kerkut). Retrieved June 3, 2024, from creation.com/evolution-definition-kerkut

Licht, M., & Liftarn. (2014, April 13). File:Anti-EvolutionLeague.jpg. *Literary Digest, 25 July, 1925*. Wikimedia Commons. Retrieved December 2, 2023, from en.m.wikipedia.org/wiki/File:Anti-EvolutionLeague.jpg; see also: flickr.com/photos/notionscapital/9650462984/in/photolist-fGM9qQ-WomPzZ

Lightner, J. K. (2018, April). Hybridization shaking up the evolutionary Tree of Life–what does it mean for creationists? *Journal of Creation*, 32(1), pp. 10–13. Retrieved from creation.com/hybridization-and-evolution

Livadas, G., & Georgopoulos, G. (1953). Development of resistance to DDT by Anopheles sacharovi in Greece. *Bull World Health Organ.*, 8(4), 497–511. Retrieved from pubmed.ncbi.nlm.nih.gov/13066984/

Methot, P. (2012). Why do Parasites Harm their Host? On the Origin and Legacy of Theobald Smith's "Law of Declining Virulence"—1900–1980. *Hist. Phil. Life Sci.*, 34, 561–601.

Munita, J. M., & Arias, C. A. (2016, April 8). Mechanisms of Antibiotic Resistance. *Microbiology Spectrum, 4*(2), 1–24. Retrieved December 2, 2023, from journals.asm.org/doi/pdf/10.1128/microbiolspec.vmbf-0016-2015

Patterson, C. (1999). *Evolution* (Second edn.). Ithaca, New York: Cornell University Press.

Sarfati. J. (2022). Editorial: Medieval castles and modern fallacies. *Creation*, 44(3), p. 5.

Sarfati, J. (2019). *Refuting Evolution—Chapter 2*. Retrieved July 8, 2024, from creation.com/refuting-evolution-chapter-2-variation-and-natural-selection-versus-evolution

Sarfati, J. D. (2018). *The Genesis Account: A Theological, Historical, and Scientific Commentary on Genesis 1–11* (Third edn.). Powder Springs, Georgia: Creation Book Publishers.Sarna, N. M. (1989). *Genesis: The JPS Torah Commentary*. Philadelphia: The Jewish Publication Society.

Scott, O. (1985). The Conservative Counter-Revolution. *Modern Age*, 29(3), 206–211.

Wieland, C. (1991, April). Variation, information and the created kind. *Journal of Creation, 5*(1), pp. 42–47. Retrieved December 2, 2023, from creation.com/variation-information-and-the-created-kind

Wieland, C. (2000, August 9). *New eyes for blind cave fish?* Retrieved February 8, 2024, from creation.com/new-eyes-for-blind-cave-fish

Part 2
Rebuttal of Secular Origins

9

Big Bang: the collapse of infinity

For millennia, people have looked heavenward and gazed in wonder at the night sky. Where did these magnificent heavenly bodies come from? Were they divinely created? Or were they always there?

The shifting fortunes of the eternal universe

In ancient Greece, Aristotle believed in an eternal universe, an idea which was rational and addressed the existence problem. But as Christianity spread, this type of scheme fell out of favor. It was decisively rejected by Augustine in favor of the biblical account, which spoke of "the beginning". Augustine's position was reinforced centuries later by Thomas Aquinas. In the seventeenth century, Isaac Newton, who had developed a theory of gravitation, proposed an infinite number of stars spread throughout infinite space to counter the tendency of gravity to cause a finite arrangement to collapse inward.[1] Despite his theory of a universe of infinite spatial extent, Newton still held to a universe with a finite beginning in time caused by the creative act of God mentioned in Genesis.

However, Newton's cosmological speculations reopened the Western mind to an eternal universe. How could a universe that extends infinitely in all directions have only existed for a finite time? The notion seemed illogical.[2] The eternal universe concept reappeared, a concept which found favor with the increasingly de-Christianized educated elite. By virtue of its eternality, an eternal universe did not require a Creator—it had always existed. It was infinite, autonomous, aimless, and impersonal.[3] By the late 1800s, this had become the dominant view among Western scientists.

But even as this view reached dominance, discoveries in other areas of science began to undermine it. The brilliant German physicist and mathematician Rudolf Clausius (1822–1888) formulated the second law of thermodynamics, which resulted in one of the first substantial scientific arguments against the eternal universe.

---

1 It is now recognized that Newton's argument requires a perfectly uniform distribution of celestial bodies. In the real world an infinite array of stars would eventually collapse under gravity because any slight inhomogeneity in the density and distribution of celestial bodies would lead to progressive clumping due to gravitation.

2 North, 2012, p. 457, quoting Koyré.

3 Ibid., p. 457.

The second law tells us that energy systems will move towards a state of equilibrium; for example, if a cup of boiling water is poured into a partially filled jug which contains water at room temperature, heat will be transferred from the hot water to the cool, resulting in a warm mixture at equilibrium. Heat from the warm water will then transfer to the surroundings, cooling the water and warming the room ever so slightly. These commonly understood processes also play out over cosmic scales. The sun and stars transfer their heat to cooler matter in the universe, and if these processes continue long enough, the entire universe will reach equilibrium at a uniform temperature. This theoretical end state is known as "heat death" and, as the universe progresses towards that end-state, a property known as entropy increases. (It is not necessary to understand entropy in detail, other than to state that for any physical process at all, the entropy of the system and its surroundings (i.e. the universe as a whole) either stays the same or increases—it never decreases).

If the universe was eternal, the argument went, why hasn't heat transfer between the various celestial bodies levelled out the temperature uniformly? Otherwise put, why hasn't entropy reached a maximum? This "entropic paradox" was first mentioned in public by the German physicist and physician Adolf Fick (1829–1901),[4] and it was a valid query. Christian scientists such as Clausius in Germany, and William Thomson, 1st Baron Kelvin (Lord Kelvin 1824–1907) and James Clerk Maxwell (1831–1879) in Britain all believed this to be an important counterargument not only to the eternal universe, but also to the "materialistic and unchristian notion of a cyclical world".[5] (For a brief discussion of the "cyclical" view of the world, refer to Chapter 2). Clausius' second law of thermodynamics implied that the age of the universe is finite, so the universe must have had a beginning.

Nevertheless, Fick's entropic paradox was largely ignored within the emerging field of cosmology. In 1896, Swedish astronomer Carl Charlier argued the universe was eternal,[6] and later, Swedish Nobel prize winning physicist Svante Arrhenius claimed the universe was spatially and temporally infinite and self-perpetuating.[7] The early cosmologists were apparently content to reject or downplay the entropic paradox, by proposing counter-theories that diminished its applicability.

Writing in 1897, Edward Clodd summed up the scientific consensus as the nineteenth century drew to a close:

> [M]odern science knows nothing of a beginning, and, moreover, holds it to be unthinkable. In this it stands in direct opposition to the theological

---

4 Kragh, 2007, p. 104.
5 Ibid., p. 102.
6 Ibid, p. 105.
7 Ibid., p. 106.

dogma that God created the universe out of nothing; a dogma still accepted by the majority of Protestants and binding on Roman Catholics.[8]

The scientific consensus was running scared of anything that hinted at a beginning, anything that would provide support for the biblical origin account. Astrophysicist Sir Arthur Eddington (1882–1944), regarded the notion of a beginning as "philosophically monstrous".[9] This mindset characterized the scientific establishment for several decades into the twentieth century.[10] Prominent geophysicist Arthur Holmes "affirmed the eternity of the universe" in 1913, even though a finite universe was a solution to the entropic paradox.[11] Scientists of that period agreed with Clodd—a definite beginning to the universe was "hard to accept, indeed nearly unthinkable".[12]

But even as scientists such as Holmes were declaring that the universe cannot have had a beginning, time was running down on the eternal universe. In 1912, Vesto Slipher, an astronomer working at Lowell Observatory in Arizona, observed that the spectral lines of distant galaxies were "red shifted", a finding which suggested these galaxies were moving away from earth at tremendous speeds. In the late 1920s, the celebrated American astronomer Edwin Hubble built on this work and published what is now known as Hubble's Law. This explained that spectral red shifts increased the further distant a galaxy was from earth,[13] suggesting an expanding universe.[14]

On the theoretical side, Belgian Roman Catholic priest and astronomer, Georges Lemaître, had suggested that Einstein's equations were consistent with an expanding universe. By extrapolating the expansion of the universe backwards, he proposed the idea of a starting point (a primeval atom) of the universe, a point at which the expansion must have begun.

But this conclusion was resisted. A procession of scientists and others would emerge over the next several decades to discredit the idea of a beginning. Albert Einstein had tried to avoid it by adding a fudge factor (the cosmological constant) to his equations, an act he later admitted was his "biggest blunder".[15] [16] Others tried to explain away the idea of the expanding universe by reinterpreting Hubble's red shift pattern as evidence for "tired light", a speculation that light gradually loses energy across intergalactic distances,[17] although this remained a fringe theory. Elsewhere, Andrei Zhdanov, the

8 Clodd, 1897, p. 33.
9 Kragh, 2007, p. 157.
10 Ibid., p. 149.
11 Ibid., p. 150.
12 Ibid., p. 150.
13 Roos, 1994, p. 1.
14 Silk, 2009, p. 7.
15 Tretkoff, 2005.
16 O'Raifeartaigh, 2018.
17 Kragh, 2007, p. 164.

propagandist-in-chief of the Soviet Union, identified the finite-age universe as a reactionary 'fairy tale'.[18]

British astronomer Sir Fred Hoyle deserves special mention here, as he was one of the last to be convinced, reluctantly, of the idea of a beginning. Hoyle was keenly aware that an expanding universe with a fixed starting point hinted at a creator and believed "there was an unholy alliance between Christianity and the Big Bang theory".[19] [20] To Hoyle, an attractive alternative was the so-called Steady State model of 1948,[21] in which the universe was always expanding, but matter with low entropy was continually being created (thus solving the entropic paradox) with the result that the density of the universe remained constant. Crucially, this model required no beginning or end in time. It gave a contemporary sophistication to the ancient Greek idea of the eternal universe.

But all these denials were about to be decisively swept away. In the 1960s, the discovery of the cosmic microwave background radiation (CMBR) called into question the steady state model. The scientific consensus fell in with the Big Bang, which was eventually pegged as having occurred 13.8 billion years ago. Gone was Hoyle's steady-state model, and with it any last remaining scientific consensus regarding the eternal universe. Still, to the atheistic mindset, this was not something to be cheered; Stephen Weinberg, Nobel laureate in physics and advocate for Big Bang cosmology concluded

> [t]he steady state theory is philosophically the most attractive theory because it *least* resembles the account given in Genesis.[22]

Weinberg went on to express disappointment that the steady state theory was ruled out by experiment.

An even later holdout was the atheistic government of communist China. In 1972, a "heretical" paper published by Fang Lizhi on the Big Bang theory evoked outrage among Marxists because it betrayed "the true spirit of dialectical materialism and proletarian science", and a flurry of papers opposing the theory continued to be published until 1976.[23] But by that time, the notion that the universe had a beginning at a point in the finite past, a position in agreement with the opening verse of Genesis, had won wide acceptance because of scientific evidence.

And yet, the concept of the beginning continued to rankle secular scientists. This smoldering grievance burst back into flame in 1989. John Maddox, the

---

18 Ibid., p. 199.

19 Ibid., p. 195.

20 Hoyle had sought to belittle the concept by labelling it the "Big Bang" in a 1949 radio interview—and ironically, it was this epithet, coined by Hoyle himself, which stuck.

21 Kragh, 2007, p. 185.

22 Ibid., p. 195.

23 Ibid., pp. 199–200.

editor of *Nature*, wrote an editorial that year expressing his concern that the Big Bang was "philosophically unacceptable" and that it provided support for biblical creationists. Evidently, Maddox was still smarting at the concept of a beginning—more than 20 years after the scientific consensus had settled on it—and expressed his hope that the Hubble space telescope, due to be launched in 1990, would provide data overturning the Big Bang theory.[24] Of course, Maddox's wish came to nought. None of the data from the Hubble telescope to date has overturned the dreaded beginning. But this episode stands as an important reminder of the damage sustained by the secular origins account when the eternal universe lost scientific support.

From the standpoint of creationists, the Big Bang was revolutionary in two ways. First, scientific evidence now favored the overall biblical position that the universe had a beginning, as the first words of Genesis 1:1 proclaimed: *In the beginning*. The tables had been turned on Edward Clodd's 1897 statement—the "unthinkable" was confirmed, and those "Protestants" and "Roman Catholics" that had always believed the Bible on this point were proven to have been right all along.

The second point, equally noteworthy, was the loss of the eternal universe to secular origins theories. As we saw earlier, all credible origins theories must incorporate the infinite, and for millennia, the infinity that made sense to many non-Christians was an eternal universe. But this was no longer the case. With the eternal universe demoted as an explanation, a new, thorny problem—supplying a credible explanation for existence—had appeared on the secularist's ledger. From our perspective, this marks the point at which the entire naturalistic origins enterprise suffered a collapse. The emergence of life from inanimate matter had already required a new explanation following Louis Pasteur's disproving of spontaneous generation in the early 1860s; and a century later, not even the existence problem had a clear, definitive, and rational explanation.

Around 60 years have passed since the Big Bang became scientific orthodoxy. In that time, vast amounts of research, involving thousands of gifted scientists and extremely sophisticated equipment (such as the Hubble and James Webb space telescopes, and CERN's Large Hadron Collider particle accelerator) have been carried out under the banner of Big Bang cosmologies, which have developed into a class of theories. We cannot respond to that astronomical amount of output in this book, but nor do we need to. The purpose of this chapter is to equip the reader to understand Big Bang cosmology and its problems at a high level, and how it impacts the origins debate.

---

24 Maddox, 1989.

The underpinnings of Big Bang thought

There is a limit as to how close theories such as the Big Bang can converge with the Genesis account. Secular humanists bring their own presuppositions, which deny the God of the Bible.

For instance, one of these presuppositions is the cosmological principle. This states that the sample of space visible from earth is similar to all regions of the universe,[25] a proposition which is almost certainly unprovable. In effect, the cosmological principle denies that there is a special or unique place in the universe, declaring instead that all points are the same. This presupposition seems inconsistent with the Genesis account. During the creation week the earth does, at least up until the fourth day, and perhaps beyond, occupy a special place in the universe.

Another presupposition is that physical laws in space have always behaved identically throughout the universe,[26] both spatially and across time. In other words, Big Bang theories acknowledge a single epoch with respect to the laws of nature—the current one. This is the doctrine of philosophical uniformitarianism, dressed up in the language of cosmology. By contrast, the biblical account comprises several epochs: the creative epoch (the first six days of creation); post-creation, pre-fall history; post-fall, pre-flood history; the year-long flood; and post-flood history, which is the current epoch. The creative epoch is non-uniformitarian, whereas the current epoch is almost totally uniformitarian, except for a smattering of Divine miracles.

A final presuppositional difference we consider here is the order of formation of the earth, sun and stars. Big Bang narratives have the earth forming long after the stars, whereas the Genesis account has the earth formed on the first day of the creation week, and the sun, moon and stars created on the fourth day.

British-American astrophysicist Joseph Silk has cited two philosophical reasons why the uniformitarian presuppositions of the Big Bang should be accepted, which can be paraphrased as:[27]

1. If the laws throughout the cosmos were not constant, science would be irrelevant.
2. Explanations for the universe should be as simple as possible according to Occam's razor. Occam's razor is the principle that the simplest theory which accounts for the evidence ought to be chosen.

Silk's first point is not necessarily true. We will explain in Part 3 that certain aspects of the non-uniformitarian creative epoch may be amenable to scientific

---

25 Silk, 2001, p. 2.
26 Ibid., p. 2.
27 Silk, 2009, p. 2.

enquiry. So, science can be relevant in certain types of non-uniformitarian context.

Silk's second point amounts to the humanist rationale for disbelieving in the God of the Bible, regarded as an enormously complex "super-intelligence". However, we would counter that Occam's razor doesn't prove anything—it is a methodological convenience. Scientists choose the simpler of two theories that have similar explanatory power because this approach can allow for easier calculations. In this chapter, we'll show why the explanatory power of the Big Bang theory is lacking.

Before moving on, we draw the reader's attention to another similarity between modern cosmology and ancient Greek philosophy. The ancient Greeks were keenly interested in the "cause of causes",[28] or "first cause", a frequently mentioned concept they referred to as the "arche". (Pronounced "ahr-kee", this word is the root of English words such as "archaeology" and "archaic", but also carries the connotation of chief or principal, found in the words "archangel" and "archenemy"). The Greeks reasoned that knowledge of the first cause could be obtained if a sequence of causal events could be pursued backwards in time far enough. This is very much the methodology used by scientists to reason from current conditions back in time to the Big Bang, using mathematical modeling based on the theory of relativity. Both the ancient Greeks and modern secular scientists have sought this cause of causes *within nature*. But from the early days of Christianity, trinitarian thinkers such as Athanasius (c. 328–373) noted that from a Christian viewpoint, this entity was not to be found within nature, because it is the Triune God, who is separate from, and transcendent to the natural world. The cause of causes could only be apprehended through biblical revelation, a notion that was essential to understanding nature.[29] From the biblical viewpoint, the naturalistic and materialistic presuppositions underpinning the Greek arche, and the modern secular arche, were and are wrong. If space and time began in a Big Bang, then logically their causes cannot be found in space and time.

The development of the Big Bang account

The Big Bang is the popular name given to not one, but a class of related theories. These theories posit that space, physical matter (and perhaps time) began when an infinitesimally small point of near infinite temperature, called the "singularity", began to expand. The Big Bang theories seek to explain phenomena from the point at which the expansion of matter and radiation has already commenced. Importantly, no cause for the singularity can be

28 Cochrane, 2003, p. 400.
29 Ibid., pp. 400–401.

established on the basis of empirical science. Ideas that relate to the cause of the Big Bang are in the realm of origins science.

To give an idea of the amount of matter involved, we need to step past the very brief, extremely hot initial phase, to the formation of protons, which is theorized to have commenced shortly after the universe began to expand. The number of protons in the observable universe, known as the Eddington number, is estimated at 10^{80} (that is, the figure 1 with 80 zeroes after it).[30] In addition to protons, other elementary particles such as neutrons are theorized to have been present.

This super-hot, super-dense mass of particles continued an incredibly rapid expansion. At these early stages, matter existed in a form known as plasma, an opaque substance through which light was unable to pass.

From the 1970s, secular scientific explanations of the first microseconds of the Big Bang began to use concepts derived from theoretical particle physics known as "Grand Unified Theories". These suggested that matter had sprung into being from (almost) nothing by the spontaneous separation of matter and antimatter particles within a quantum field. It was even postulated that time itself began in the Big Bang. These developments moved secular cosmology to an even closer correspondence with the ex-nihilo creative act of Genesis 1.

What do scientists tell us about the ensuing expansion of matter? Evidence from the cosmic microwave background radiation (CMBR), which had been discovered in the 1960s, came into play. In the Big Bang narrative, the CMBR was believed to represent massively red-shifted primordial light emitted in the era when the plasma of particles condensed into atoms, rendering the opaque plasma transparent. This was presumed to have occurred several hundred thousand years after the Big Bang. However, the extreme degree of uniformity in the CMBR, to 1 part in 100,000—equivalent to no more than 1mm deviation in a 100-meter-long path—could not be explained initially. Random fluctuations occurring in the Big Bang ought to have caused a much greater degree of variation in the CMBR. A related issue, known as the horizon problem, was identified when calculations showed there was insufficient time for thermal equilibrium to occur across the extent of the universe at any point in time.[31] In a nutshell, the uniformity of the CMBR was unexpected and difficult to explain.

In 1981, American theoretical physicist and cosmologist Alan Guth proposed cosmic inflation as a solution to the horizon problem of Big Bang theories.[32] The idea was that the early universe expanded at an extremely rapid

30 The calculation of the Eddington number predates the Big Bang theory and is separate to it.

31 In fact, there was not even enough time for light to have travelled across the universe, given the degree of expansion (Lisle, 2023). For this reason, the creationist Jason Lisle has classed this as a "light travel problem".

32 Lisle, 2023.

rate—faster than the speed of light[33]—for a tiny fraction of the first second after the Big Bang (from 10^{-38} to 10^{-33} seconds).[34] [35] Although initially controversial (perhaps because it appeared as a just-so story to many cosmologists), inflation steadily gathered a scientific following. It appeared to have significant explanatory power (even though there was no consensus around the mechanisms by which the process started or stopped). Some early models of inflation implied that for a very brief time, the early universe underwent extremely rapid cooling as it expanded, a process called supercooling.[36]

Inflation was an additional bolt-on theory that enabled the data to better fit the Big Bang. But some conceptions of inflation, for example Guth's, also posited a *multiverse* as part of the theory. A multiverse is a potentially infinite number of universes separate from our own. This concept has divided scientists—some are decidedly uncomfortable with the concept, while others champion it. As cosmologist Michael Turner recently noted, the multiverse, which flows from inflation, makes some cosmologists question whether inflation can really make firm testable predictions, suggesting alternatives need to be considered.[37] But if a theory can't make firm, testable predictions, is it part of science?

Following the purported inflationary period, hundreds of thousands of years of continued expansion occurred. During this time, the cooling plasma began to form into various types of matter. This included the so-called baryonic atoms, the ordinary matter of the universe from which stars and planets are formed. "Dark matter", whose presence is inferred, yet cannot currently be detected, is believed to account for most of the mass in the universe. Under the influence of gravity, atoms coalesced into stars, which in turn gravitated into galaxies. Within stars, heavier atomic elements, such as metals, were produced by nuclear fusion. Some of these elements were subsequently ejected into space and were incorporated into planets.[38]

The timescales over which these events played out are held to be vast; the age of the universe given under this interpretation is 13.8 billion years.

Scientists have developed a range of jargon to describe elements of the Big Bang theory. Phrases such as *dark energy, dark matter* and *cosmic inflation* are par for the course in Big Bang cosmology, and these may lend a veneer of assurance—in the minds of laypeople at least—that scientists are on top of the secrets of the universe. After all, the process of naming phenomena is a legitimate part of descriptive science. However, there are major problems with

---

33 This does not contravene relativity's speed limit of c, since the postulated expansion is of space-time itself.

34 Silk, 2009, p. 116.

35 Hawking, 1998, p. 131.

36 Guth, 1997, p. 167.

37 Turner, 2022, pp. 17, 27–28.

38 This proposition requires Big Bang theories to contradict the order of celestial body formation in Genesis 1.

Big Bang theories. These include the dark matter problem,[39] the relatively low observed abundance of stellar lithium,[40] and the Hubble tension.[41] We leave it to the interested reader to seek further information on these issues.

The iteration of the Big Bang theory which is current at the time of publication takes the technical name *ΛCDM*, where "*Λ*" or uppercase lambda, a Greek letter, represents the number otherwise called the cosmological constant, and CDM is an acronym for *cold dark matter*. In this formulation, the *cosmological constant* has a meaning different (dark energy driving cosmic acceleration) from Einstein's fudge factor.

Other Big Bang models

Although ΛCDM is the current "standard cosmology", scientists have criticized it and proposed alternatives. For example, British mathematical physicist Roger Penrose has proposed a "conformal cyclic cosmology" (CCC), where the Big Bang represents an expansion based initially on photons[42] and gravitons (hypothetical particles travelling at the speed of light which convey gravitational force).[43] As the cosmos ages, components of the expanded universe become swallowed by massive black holes, which over extremely lengthy timescales are converted to photons by Hawking radiation, eventually evaporating and disappearing. This state serves as the precursor to a new Big Bang, one which is much "larger" in spatial extent than the prior instance. Penrose believes his version of the Big Bang successfully resolves the entropic paradox,[44] perhaps by the ever-expanding scale of successive universes, allowing entropy to increase forever. His theory also suggests that energy is eternal, amounting to a potential solution to the existence problem. However, for now, Penrose's idea has attracted few converts[45] and is not without its own set of problems.

From our perspective, Penrose's CCC theory illuminates the point that Big Bang theories are not necessarily incompatible with an eternal universe, or at least, eternal energy.

---

39 The Stephen Hawking Centre for Theoretical Cosmology, _.

40 Turner, 2022, p. 24.

41 Ibid., p. 27.

42 Photons are particles of light or other electromagnetic radiation which have which have particle- and wave-like behaviors.

43 Hurley, 2023.

44 Kumar, 2010.

45 Cartlidge, 2018.

Borde-Guth-Vilenkin Theorem

The Borde-Guth-Vilenkin (BGV) Theorem was proposed by Arvind Borde, Alan Guth and Alexander Vilenkin in 2003. It addresses the nature of the universe's expansion and its implications for the beginning of the universe.

The key conclusion of the BGV theorem is that any universe, which has, on average, been expanding throughout its history, cannot be infinite in the past but must have a past spacetime boundary,[46] which is cosmologist speak for a beginning. This is significant because it provides a strong indication that classical (non-quantum) cosmology cannot be extended to past infinity and that the initial singularity (or something akin to it) is an unavoidable feature of our universe.

The theorem is formulated under fairly general conditions, primarily that the average Hubble expansion rate is positive. This avoids assumptions about specific materials, energy conditions, or gravitational theories, making it broadly applicable within the framework of general relativity. The theorem uses concepts from geodesic[47] behavior to argue that any universe with an average expansion cannot be past-eternal and must therefore reach a boundary or singularity as we trace back in time.

This conclusion is compatible with the standard model of cosmology, which suggests that the universe began with the Big Bang, a state of extremely high density and temperature. The BGV theorem does not, however, specify the physical conditions at the beginning of the universe, nor does it provide a mechanism for the Big Bang, or the subsequent expansion. Instead, it sets a significant theoretical constraint, indicating that any cosmological model involving an always-expanding universe (like the standard Big Bang model) must incorporate a beginning.

Thus, on current evidence of an expanding universe, and if the Borde-Guth-Vilenkin theorem is accepted, then the universe had a beginning, and space and time themselves began at that singularity. The movement of the secular scientific account towards Genesis is unmistakable, although the presuppositional gulf between the two cannot be bridged.

Shifting infinities

Unlike Christianity, which involves one personal infinity, the Creator God, the secular cosmology comprising the Big Bang theory is associated with three infinite or quasi-infinite quantities to explain the cosmos:

46 Vilenkin, 2015.

47 Geodesic refers to the shortest ("straight line") path between two points in curved space-time. Movement of an object under the influence of gravity alone, such as a planet orbiting a star, or an object being dropped to the ground, are examples of geodesic behavior.

1. An implicit infinity to explain existence. This has been required since the scientific consensus shifted to a universe with a beginning, but the identity of this infinity has not yet been clarified by secular scientific consensus.
2. The initial singularity, which is nearly infinite in density and temperature.
3. A quasi-infinite amount of time since the beginning of the universe—13.8 billion years. Often referred to as "deep time" this quasi-infinity is, psychologically at least, very important within the secular humanist worldview. Part of the deep time concept is the age of the earth, estimated at 4.5 billion years from radiometric dating. (For further information on radiometric dating, refer to Appendix B.)

For the minority of cosmologists who accept the multiverse we add a fourth infinity or quasi-infinity:

4. A plethora of hypothetical universes comprising the multiverse, which, depending on the particular version of the multiverse, may be finite or infinite.

Note that all these infinities are impersonal and relate to philosophically uniformitarian processes, which are essential to a purely naturalistic, and essentially atheistic, worldview. If biblical monotheism (the one God of the Bible) is rejected, then it becomes essential to substitute a multiplicity of impersonal infinities or quasi-infinities.

In the following discussion, we will outline how (1) and (4) have impacted the origins debate.

Consider the apocryphal riposte by the French mathematician Pierre-Simon Laplace (1749–1827) who allegedly discussed his theory of the universe with Napoleon. Napoleon, so the story goes, expressed surprise that Laplace's theory contained no mention of God. "I have no need for that hypothesis," shot back Laplace, in a display of chutzpah which has delighted and amused secular humanists down through the years.

In fact, Laplace may have had a valid point. When his discussion with Napoleon supposedly took place (late eighteenth century), the idea that the universe operated deterministically under conditions described by philosophical uniformitarianism was becoming ascendant. A mechanistic universe supplied Laplace with the presuppositions he needed to propose answers to important questions of the period. These presuppositions, combined with Newtonian mechanics, would morph over time into the eternal universe, which, crucially, would provide a then-credible solution to the problem of existence.[48] In Laplace's mind, God was not needed to explain the operations of

48 The problem of information was almost non-existent; the discovery of DNA and the informational basis of life was many decades into the future.

an eternal mechanistic world, once immutable laws were in place as axioms. His reasoning makes a certain amount of sense.

But science has moved on since Napoleon conferred with Laplace. The eternal universe, the one infinity that seemed to cover everything, is increasingly untenable in most Big Bang interpretations. Evidently, that fact seems to be lost on those who continue to quote Laplace's riposte. If not God, then what infinity do they propose? It's a question that requires an answer.

The multiverse has also been used to explain the problem known as the *fine-tuning of the universe*, which is akin to a type of non-biological information found in matter. Fine-tuning refers to the fact that the universe appears to be precisely calibrated to support life. Is this merely an astonishing coincidence? Or is it all part of God's design, as a creationist would suggest? Hear Stephen Hawking's comment on the fine-tuning problem:

> The laws of science, as we know them at present, contain many fundamental numbers, like the size of the electric charge of the electron and the ratio of the masses of the proton and the electron. ... The remarkable fact is that the values of these numbers seem to have been very finely adjusted to make possible the development of life.[49]

Sir Fred Hoyle, who, as we saw above steadfastly resisted any implication of a beginning, also commented on the fine-tuning issue. His comment includes the now famous question as to whether "a superintellect had monkeyed with physics":

> From 1953 onward, Fowler and I have been intrigued by the remarkable relation of the 7.65 MeV energy level in the nucleus of ^{12}C to the 7.12 MeV level in ^{16}O. If you wanted to produce carbon and oxygen in roughly equal quantities by stellar nucleosynthesis, these are just the two levels you would have to fix, and your fixing would have to be just about where these levels are actually found to be. Is that another put-up, artificial job? ... I am inclined to think so. A common sense interpretation of the facts suggests that a superintellect has monkeyed with physics, as well as with chemistry and biology, and that there are no blind forces worth speaking about in nature. The numbers one calculates from the facts seen to me so overwhelming as to put this conclusion almost beyond question.[50]

For the secular humanist, a superintellect "monkeying with physics" cuts awfully close to a transcendent God, which is anathema. This is where the multiverse comes riding to the rescue: if we imagine an infinite number of separate universes with different, randomly distributed fundamental numbers, then the problem appears to go away. We just happen to be in a random universe with the right numbers—a concept secular humanists call the *anthropic*

---

49 Hawking, 1998, p. 129.
50 Hoyle, 1981.

principle. We cannot be in a universe with the wrong fundamental numbers, because we would not be here to know about it.

In fact, the multiverse does not explain a key problem. Why does a universe such as ours have stable physical regularities? Why are the "laws of nature" not chaotic?[51] This is the problem of coherence that we discussed in Chapter 2; here, secular scientists assume that the laws of each universe are relatively stable and coherent, without explaining why.

Of course, the multiverse is not a scientific theory, at least not according to Popper's demarcation criterion. Infinite universes that cannot be detected or observed cannot be falsified.[52] The multiverse, and the anthropic principle, are more in the domain of metaphysical or philosophical ideas, more dogma than science. This has led to a rift in the scientific community, pitting multiverse stalwarts such as Michio Kaku against naysayers like Sabine Hossenfelder.[53] We agree with the naysayers—the multiverse is nothing more than a series of imaginary worlds dreamed up to solve the problem that arises by denying a Creator. But unlike the eternal universe concept, in which the observable universe was simply believed to have had no beginning, the multiverse cannot be observed at all. It is completely adrift from the domain of empirical science.

Secular miracles

But we can say more about the multiverse. We regard it as one of the *secular miracles* that is required to render secular origins accounts plausible. We define a secular miracle as an event for which secularists invoke some of the attributes of God (such as infiniteness or purpose) to render plausible phenomena which are otherwise impossible in a subjective probabilistic sense (see Appendix C for an explanation of the latter term). So how does this apply to the multiverse?

The secular miracle in view is the extraordinary fine-tuning of the universe to support life. To explain this fine-tuning, the multiverse posits an infinite number of universes with randomly distributed material properties, such that it becomes possible for at least one of these universes to have all necessary constants at just the right levels.

Infinity, by definition, is not part of observable nature. The estimated number of protons in the visible universe, 10^{80}, is very large, but finite. According to secular science, the universe has been in existence for 13.8 billion years, which again, is a very lengthy period, but finite. Infinity is not scientifically discernible in nature. The multiverse is not empirically testable. It

51 Feynman, 1998, pp. 22–23.

52 Some multiverse proposals incorporate potentially observable signatures, but no decisive detection has occurred.

53 The Institute of Art and Ideas, 2022.

is not part of science; it is a device used to render the reality of fine-tuning unimportant.

However, God has infinite qualities (refer Chapter 2). Therefore, the multiverse, as an explanatory construct, sneaks an attribute of God into what is purportedly a scientific discussion.

In sum, the Big Bang theory posits a beginning, and the multiverse invokes a supra-scientific miracle, insofar as it invokes infinity, one of the attributes of God.

A beginning, and a miracle. To the creationist, that sounds familiar. But with presuppositions that rule out the existence of God, Big Bang theories must run into difficulties sooner rather than later.

Trouble in the ranks

Although ΛCDM is the reigning cosmological theory among Western cosmologists at the present time, there have been undercurrents of discontent. In 2004, a letter written by Eric Lerner, an American science writer and plasma researcher, was published as "Bucking the Big Bang".[54] More than 30 scientists co-signed Lerner's letter. The letter complained that the "dominance [of the big bang] rests more on funding decisions than on the scientific method ...". Furthermore, it noted that the modern Big Bang theory depends increasingly on unproven concepts—such as inflation, dark matter, and dark energy—that have never been directly observed. These hypothetical constructs are necessary to prevent serious divergences between what astronomers see and what the Big Bang model predicts. In any other branch of physics, repeatedly introducing new, unverified elements to resolve discrepancies between theory and evidence would not be considered acceptable and would likely prompt significant doubts about the theory's credibility.

Lerner explained further that the Big Bang theory depends heavily on these ad hoc corrections to remain viable, and even with these corrections, it has not produced any precise predictions that were later confirmed by observation. What its proponents call successes are mainly its capacity to explain existing observations after the fact, by continually adding adjustments—much like the outdated geocentric model of Ptolemy, which required increasingly complex layers of epicycles to match the data.

What are we to make of such claims about mainstream cosmology? Lerner has been dismissed as a "crackpot" by physicist and science communicator Sean Carroll,[55] but the issue is whether his points are valid. And indeed, the thrust of Lerner's complaints has been repeated more recently by American

54 Lerner, 2004.
55 Vinod, 2022.

theoretical cosmologist Michael S. Turner, co-author of *The Early Universe* (1990) and a proponent of the ΛCDM Big Bang theory, who admitted:

> We could be on the wrong track, with dark matter, dark energy and inflation being the modern equivalents of Ptolemy's epicycles. … Too much works for us to be completely off track, but there could be an unexpected Kuhnian shift ahead.[56]

Lerner's original complaint has therefore been recently substantiated by a ΛCDM paradigm proponent. Furthermore, these additional hypotheses of inflation, cold dark matter and dark energy are challenged by Occam's razor. Fred Hoyle's and Chandra Wickramasinghe's interpretation of Occam's razor led them to

> [b]e suspicious of a theory if more and more hypotheses are needed to support it as more facts become available, or as new considerations are brought to bear.[57]

Hoyle's and Wickramasinghe's warning is in line with the complaint of Eric Lerner, and the admission by Michael Turner about the ΛCDM Big Bang model.

Lerner's 2004 letter included a parting shot at the institutional environment that gave succor to Big Bang theories:

> Those who doubt the big bang fear that saying so will cost them their funding.
>
> … discordant data on red shifts, lithium and helium abundances, and galaxy distribution, among other topics, are ignored or ridiculed. This reflects a growing dogmatic mindset that is alien to the spirit of free scientific enquiry.
>
> … Funding [in cosmology] comes from only a few sources, and all the peer-review committees that control them are dominated by supporters of the big bang. As a result, the dominance of the big bang within the field has become self-sustaining, irrespective of the scientific validity of the theory.

These last comments reinforce what has been discussed earlier in this book about the way science, and in particular origins science, really works. The fear that scientists and researchers experience when challenging entrenched dogmas and theories is real and likely results in self-censorship.

The collapse of the naturalistic origins account

The loss of the eternal universe was a seismic development in modern science, and yet few seem to have recognized the flashing red light on the

---

56 Turner, 2022, p. 26.

57 Hoyle & Wickramasinghe, 1981, p. 135.

secular humanist origins dashboard. Or at least, few have attempted to address the issue. In 1997, Alan H. Guth's book *The Inflationary Universe* touched on this topic in seeking an explanation for how the universe could come from nothing. However, *A Universe From Nothing* by physicist Lawrence M. Krauss, published in 2012, is, as far as we know, the first book solely dedicated to this issue.

Resolving this issue is essential to the secular origins enterprise. The notion that the universe could arise from nothing appears irrational, a departure from the ancient Greek notion that "out of nothing comes nothing".[58] It is a glaring point of weakness for uniformitarians. Richard Dawkins was confounded on this point during a televised debate with Cardinal George Pell in 2012, when a call-in questioner asked him to explain, in layman's terms, how the universe came from nothing. (Interested readers may like to watch the whole debate on YouTube.[59]) The audience was treated to the sight of one of the most celebrated science communicators of our age defending his interpretation of the meaning of "nothing". The argument about the origin of the universe had effectively moved from the domain of science to philosophy. In our view, this was an excellent demonstration in real time of the truth of our epigraph, Isaiah 44:25: " ... who confounds the wise and makes their knowledge foolishness."

In our view, Krauss's book does not live up to its billing, as it fails to identify the infinity which explains existence. Toward the end of the book Krauss also asks rhetorically whether the question of existence is worth investigating, suggesting that perhaps it is not interesting after all, or is less interesting than other scientific questions.[60] He responds to his inability to generate a rational answer by adopting a rhetorical stance that diminishes the question.

Richard Dawkins' afterword compares Krauss's book to Darwin's *Origin of Species*, a comment which was ridiculed by reviewers. If this ridicule was based on Krauss's work discussing a less important issue than Darwin's, then we disagree. An origins account must explain both existence and information; Darwin attempted to explain biological information, and Krauss is attempting to explain existence. Both are essential. However, comparing Krauss to Darwin betrays a lack of understanding on Dawkins' part of the processes of science. Darwin was the capstone to millennia of evolutionary thought. Krauss, by contrast, is one of the earliest writers to put the new problem of existence for secular humanists on the agenda. This is the kind of role Copernicus played in the Copernican revolution, but Krauss cannot be classed as a secular

---

58 Clodd, 1897, p. 32.

59 See youtube.com/watch?v=0HI_nqppIM4

60 "Why is there something rather than nothing? Ultimately, this question may be no more significant or profound than asking why some flowers are red and some are blue. ... Either way, what is really useful is not pondering this question, but rather participating in the exciting voyage of discovery that may reveal specifically how the universe in which we live evolved and is evolving and the processes that ultimately operationally govern our existence" (Krauss, 2012, p. 178).

Copernicus on this issue—he has raised a problem without providing a semblance of a solution. Copernicus' ideas needed to be built on and refined, but at least he provided a starting point.

Nevertheless, Krauss deserves credit for flagging a serious problem faced by the secular origins enterprise. But we also note that Krauss has unwittingly highlighted another fallacy that permeates the debate—that the more science progresses, the closer humanity will get to a complete naturalistic understanding of origins. We claim the opposite—increasing scientific knowledge works against uniformitarian origins. And cosmology is one of two great examples of this, the other being abiogenesis, which we cover in the next chapter.

Secular humanist cosmology: where to from here?

From the perspective of a Bible believer, the secular humanist origins enterprise is like an MH370 investigator who insists that the plane disappeared in the South China Sea. For a while, this seems to make sense. But as more data comes in, it becomes increasingly apparent that the theory is untenable.

This is where we see uniformitarian cosmology heading. For more than 2000 years, the concept of the eternal universe had ticked a lot of boxes in both the existence and information problem categories for non-Christians. Illustrious figures like Aristotle and Einstein had warmed to it. But as more data has come in, this longstanding concept has been debunked. The Big Bang, which shares a beginning in common with the biblical account, has moved origins science into closer alignment with Genesis.

Will future discoveries allow secular science to regain its footing? Our hunch is that it won't. Perhaps fatal flaws in the presuppositional base of the Big Bang will prevent decisive progress, or perhaps further scientific progress will highlight problems underlying the theory.

It is tautological to state that the observable universe is finite. But by using this tautology we can demonstrate again the fact that modern Big Bang cosmologies contain elements that are non-scientific. The following syllogism borrows its second proposition from the one cited in Part 1:

1. All credible origins accounts must directly incorporate the concept of an entity that in one or more characteristics is infinite (for example, an entity with no beginning).
2. The observable natural world is finite.
3. All credible origins accounts must incorporate concepts outside of the observable natural world—and thus outside of empirical science.

Neither the arche proposed by the ancient Greeks, nor the modern secular humanist equivalent, the Big Bang, can give us the complete picture of the origin of the universe.

Conclusion

In this chapter we saw that by the late 1800s and into the first half of the 1900s, prominent secular scientists promoted the concept of the eternal universe because it least resembled the Genesis account and solved the existence problem. Those who held to the biblical scheme were ridiculed. But the eternal universe became untenable as evidence accrued for a beginning. Key problems for the eternal universe were the entropic paradox, evidence suggesting that the universe was expanding, and details from the cosmic microwave background radiation incompatible with a steady state universe. The consensus shifted to the Big Bang despite the remonstrations of several notable secular scientists, who believed the theory had moved too close to Genesis.

Big Bang theories are underpinned by secular presuppositions incompatible with the Bible. These include philosophical naturalism, philosophical uniformitarianism (the cosmological principle) and Occam's razor used as a philosophy (rather than a useful heuristic). When these cosmologies contain infinities, proponents insist that the infinity is impersonal. The speculative multiverse which is invoked to address the issue of fine-tuning needs to be recognized for what it is—philosophy, not science.

Although secular cosmology has moved closer to Genesis in important ways, it cannot completely converge while the presuppositions of scientists remain antagonistic to a personal Creator. But in our view, those very presuppositions will prevent secular humanist cosmology from converging on a settled origins account.

References

Cartlidge, E. (2018, August 21). New evidence for cyclic universe claimed by Roger Penrose and colleagues. Retrieved May 18, 2024, from physicsworld.com/a/new-evidence-for-cyclic-universe-claimed-by-roger-penrose-and-colleagues/

Clodd, E. (1897). *Pioneers of Evolution: From Thales to Huxley*. New York: D. Appleton and Company.

Cochrane, C. N. (1940, 2003). *Christianity and Classical Culture*. Indianapolis: Liberty Fund, Inc.

Feynman, R. P. (1998). *The Meaning Of It All: Thoughts of a Citizen-Scientist* (Kindle edn). New York: Basic Books.

Guth, A. H. (1997). *The Inflationary Universe: The Quest for a New Theory of Cosmic Origins*. New York: Helix Books.

Hawking, S. (1998). *A Brief History of Time* (10th anniversary edn). New York: Bantam Books.

Hoyle, F. (1981, November). The Universe: Past and Present Reflections. Engineering & Science, 8–12. Retrieved April 1, 2024, from calteches.library.caltech.edu/3312/1/Hoyle.pdf

Hoyle, F., & Wickramasinghe, C. (1981). *Evolution from Space: A Theory of Cosmic Creationism*. New York: Simon and Schuster, Inc.

Hurley, S. (2023, July 1). A Cyclical Universe. Retrieved May 17, 2024, from explainingscience.org/2023/07/01/a-cyclical-universe/

Kragh, H. S. (2007). *Conceptions of Cosmos: From Myths to the Accelerating Universe: A History of Cosmology*. Oxford: Oxford University Press.

Krauss, L. M. (2012). *A Universe from Nothing*. London: Simon & Schuster.

Kumar, M. (2010, October 16). Cycle of Time: An Extraordinary New View of the Universe by Roger Penrose—review. Retrieved May 17, 2024, from theguardian.com/books/2010/oct/16/cycles-time-roger-penrose-review

Lerner, E. (2004, May 22). An Open Letter to the Scientific Community. Retrieved April 6, 2024, from cosmology.info/org/open-letter-on-cosmology.html

Lisle, J. (2023, December 9). Light-travel time: a problem for the big bang. Retrieved April 1, 2024, from creation.com/light-travel

Maddox, J. (1989, August 10). Down with the Big Bang. *Nature*, 340: 425.

North, G. (2012). *Sovereignty and Dominion: An Economic Commentary on Genesis* (Vol. 2). Dallas: Five Point Press.

O'Raifeartaigh, C. (2018, October 30). Investigating the legend of Einstein's "biggest blunder". Retrieved April 6, 2024, from pubs.aip.org/physicstoday/online/30082/Investigating-the-legend-of-Einstein-s-biggest

Roos, M. (1994). *Introduction to Cosmology*. New York: John Wiley and Sons.

Silk, J. (2001). *The Big Bang* (3rd edn). New York: Henry Holt and Company, LLC.

Silk, J. (2009). *Horizons of Cosmology: Exploring Worlds Seen and Unseen*. West Conshohocken: Templeton Press.

The Institute of Art and Ideas. (2022, November 4). Should we abandon the multiverse theory? Sabine Hossenfelder, Roger Penrose, Michio Kaku. Retrieved September 30, 2023, from youtube.com/watch?v=W39kfrxOSHg

The Stephen Hawking Centre for Theoretical Cosmology. (_). The Origins of the Universe: the Big Bang—Shortcomings of the Standard Cosmology. Retrieved October 4, 2021, from ctc.cam.ac.uk/outreach/origins/big_bang_five.php

Tretkoff, E. (2005, July). This Month in Physics History: Einstein's Biggest Blunder. Retrieved April 6, 2024, from aps.org/publications/apsnews/200507/history.cfm

Turner, M. S. (2022). The Road to Precision Cosmology. *Annual Review of Nuclear and Particle Science*, 72: 1–35.

Vilenkin, A. (2015, October). The Beginning of the Universe. Retrieved April 18, 2024, from web.archive.org/web/20231227193200/https://inference-review.com/article/the-beginning-of-the-universe

Vinod, K. (2022, August 29). Explained: The Deal With Eric Lerner Saying the Big Bang Didn't Happen. Retrieved April 6, 2024, from science.thewire.in/the-sciences/eric-lerner-big-bang-jwst/

10
Abiogenesis: a secular miracle

We saw in Part 1 that the ancient Greeks believed that life could emerge from non-living matter via a process termed *spontaneous generation*. People had noticed that maggots soon turned up on dead flesh, or small water creatures appeared in ponds, and surmised that these life forms generated themselves spontaneously. This theory of spontaneous generation was largely taken for granted, and, for well over a thousand years—from the ancient Greeks who developed it, through to the mid-nineteenth century—it was never decisively challenged.

From spontaneous generation to abiogenesis

This finally changed when Louis Pasteur, the French biochemist, carried out his landmark "swan-necked flask" experiments in the late 1850s and early 1860s. These experiments demonstrated that bacteria did not grow in meat broth that had been boiled and protected from settling airborne particles, unlike separate samples that were open to the entry of such particles. This breakthrough discovery was widely accepted and became the basis of certain processes such as the sterilization of medical equipment. Although some die-hards continued to believe in spontaneous generation after Pasteur's experiments, the issue is no longer contentious today. Sterile medical equipment typically has expiry dates measured in years.

Nevertheless, an explanation for the origin of life remains an essential component of the secular origins account, as Darwinian evolution is entirely dependent on life that already exists. Darwin attempted to explain the emergence of *species*, not the emergence of life. With spontaneous generation disproved, where to for secular scientists?

The answer is *abiogenesis*, a term coined by Thomas Huxley. This somewhat clumsy word comprises three Greek roots. Most readers will recognize *bio*, meaning *life*, and *genesis*, meaning *beginning*. The prefix *a*, which appears in words like atypical and apolitical, means *not*. Therefore, the term abiogenesis refers to the beginning of life from matter that is not living (otherwise stated as nonliving matter).[1] The main distinction between spontaneous generation and

1 In contrast, the term biogenesis refers to the development of life from preexisting life, such as occurs in the processes of living cells.

abiogenesis is that the former held that life constantly emerged, whereas proponents of abiogenesis posit that the emergence of a living organism from non-living matter is an exceedingly unlikely event that may only have happened once. It is also important to distinguish abiogenesis from Darwinian evolution. Until there is a basic life-form in place, which is capable of independent replication, natural selection cannot occur. So, theories of abiogenesis cannot include natural selection. That statement is generally accepted by all sides of the debate; Theodosius Dobzhansky, a founder of the modern Neo-Darwinian synthesis, stated that "pre-biological natural selection is a contradiction in terms".[2]

Exactly how this primordial life form might have come into existence is subject to ongoing research, but the field of abiogenesis has yielded few hard answers. To get a feel for the difficulties involved, consider some of the remarks that commentators have made about abiogenesis. Professor Paul Davies, a well-credentialed thinker who has lectured and written widely on origin-of-life research, introduces readers to the issues in his 2020 book *The Demon in the Machine*. Davies poses three key questions: when, where, and how did life begin? Davies regards the simplest of these questions as when; he cites evidence for 3.5 billion years ago. More difficult is where; Davies suggests a variety of locations, including other planets, comets, or deep ocean vents in the Earth. But the most difficult is how life began. According to Davies, the simple truth is that no one knows how life first started—and it's even more challenging than that: we don't even know how to begin calculating the likelihood of it occurring.[3] [4] One major characteristic of life is its ability to handle digital information through a mathematical coding system. Any credible explanation for the emergence of life must also account for how this system of digital information processing—and crucially, the code itself—came into existence. Davies conceded that this remains an incredibly difficult challenge.[5]

Despite the extreme difficulty of the problem, Davies dismisses the concept of a chemical 'ratchet' which acts to "lock in the gains and limit the losses while the system hangs out for the next step … . A chemical soup doesn't know it's

---

2 Meyer, 2013, p. viii.

3 Davies, 2020, p. 172.

4 We disagree with Davies' statement that nobody has been able to calculate the probability for abiogenesis. The statement is true in a tautological sense; the probability of abiogenesis cannot be calculated without knowing a mechanism, and no mechanism is known for abiogenesis. However, the statement is misleading because we do not need to know the precise probability of abiogenesis to assess its likelihood. All we need is an upper bound estimate; Hoyle and Wickramasinghe published such an estimate in 1981 (Hoyle & Wickramasinghe, 1981, p. 24) and Eugene Koonin published an estimate for the RNA world hypothesis in 2007. Both showed abiogenesis to be impossible (with respect to a probabilistic "impossibility" cutoff that we discuss later in the chapter). Davies mentions none of this; instead, he (arguably) acts as a propagandist. He promotes the theory of abiogenesis (which is debunked by probabilistic arguments) because it is foundational to his ideology. This demonstrates his faith that some future discovery will render abiogenesis plausible. Within a biblical worldview, however, holding to the "fact" of abiogenesis, as Davies does, is anti-science.

5 Davies, 2020, p. 174.

trying to make life … . Scenarios in which chemistry 'strives' towards life are patently absurd".[6]

We agree with Davies on this point. The proposition that matter can purposefully bring forth life, such that life is, in the words of Christian de Duve, "a cosmic imperative", is a smuggling-in of God-like attributes of foreknowledge and purpose—a secular miracle.

Another commentator, philosopher Thomas Nagel, whom we encountered in Part 1, has stated that a "material account of how we and our fellow organisms came to exist [is] hard to believe" and that it "flies in the face of common sense".[7] Nagel goes on to advance teleological speculations—which Professor Davies dismissed—including "a cosmic predisposition to the formation of life, consciousness, and the value that is inseparable from them".[8] As an atheist, Nagel hastens to add that this explanation would be "an alternative to a miracle"[9]—which is an important admission; and indeed, we concur that abiogenesis is a secular miracle.

If these experts are correct, abiogenesis, the foundation which underpins evolution, remains not only an unsolved problem, but one that currently appears intractable. Nevertheless, secular scientists have assembled a superstructure of theory on top of abiogenesis. The Last Universal Common Ancestor (LUCA) supposedly refers to the most recent organism from which all organisms now living on earth have descended. It is said to be a pivotal concept in evolutionary biology, suggesting that all life shares a common genetic heritage. LUCA is supposedly preceded by FUCA, the First Universal Common Ancestor. But before scientists rush on ahead to discussions of LUCA and FUCA, wouldn't it be more logical to resolve the issues with abiogenesis?

The mathematics of abiogenesis

If it really is so difficult to calculate the probability of abiogenesis happening, as Davies suggests, then we need to frame our calculations carefully. What are the chances of matter randomly organizing itself into structures that are useful to life? Otherwise stated, what are the chances of information arising from randomness? To get a handle on this, we need to understand several mathematical concepts.

Exponential growth

An Indian folk tale narrates the story of a sage who developed the game of chess and explained it to a king. The king was so impressed with the game that

---

6 Ibid., p. 173.
7 Nagel, 2012, p. 5.
8 Ibid., p. 123.
9 Ibid., p. 124.

he asked the man to name his reward. Rather than asking for gold or silver from the king's treasuries, the sage asked for a quantity of rice to be measured out in an intriguing way. Starting with a single grain of rice on the first square of the chessboard, the sage stipulated that the amount of rice should be doubled on each successive square of the board. Thus, two grains of rice would go on the second square, four grains on the third square, eight on the fourth square, and so on, up to the sixty-fourth and final square on the board. Caught in the moment, the king agreed to this but was astonished that the sage should only ask for a few cups of rice despite the riches he could have requested.

The king called his servants to measure out the rice. On the first few squares, the servants easily placed the correct number of grains, but towards the end of the first row, a large coterie of servants were needed—each square took thousands of grains. Soon enough, the servants admitted defeat. They could not possibly count out the vast amount of rice required.

If the king had remembered his training in mathematics, he would never have agreed to the sage's request, which invokes the remarkable phenomenon of exponential growth. Despite the insignificant quantity of rice at the start, by the 32nd square, almost 4.3 billion grains would sit atop the first half of the board. This amount of rice occupies just over 90 cubic meters and would fit neatly into the standard trailer of a 22-wheeler truck. Continuing to the second half of the board, however, the amount of rice is not merely doubled. The volume of rice increases exponentially; the 4.3 billion figure we saw earlier applies again, but this time it is multiplied by the volume of rice in the 22-wheeler, which checks out at about 400 cubic kilometers. This is sufficient to cover the entirety of New York State to a depth of 2.8 meters (over 9 feet) or the land area of England to just over three meters. This prodigious quantity of rice vastly exceeds annual world rice output. It would take well over 2000 years at current rates of production to harvest this quantity.

Let's consider another example of exponential increase, this time related to information arising from random processes. We take the first three words of the Bible—*In the beginning*—and investigate a simple question: if a blindfolded person had a large supply of shuffled letter cards (including blank cards which would serve as spaces), how long would it take, if the person places cards at random, before these first three words were spelled out? We assume, of course, that placing the same letter multiple times is allowed.

To calculate this, we first need to ascertain how many different cards are required. We need a set of upper-case letters (26 cards), a set of lower-case letters (another 26 cards), plus a blank card for spaces. All up, this makes 53 different cards.

Next, we consider how many cards we need to place in each attempt. This is straightforward; there are two letters in the word "In", three letters in the word "the" and nine letters in "beginning", plus a couple of spaces to separate the words. This adds to 16 places in all.

We now assume our blindfolded card-placer puts cards down at the rate of one per second. The rules of the game require the person to put down a card in each of the 16 spaces, and then, if *In the beginning* is not spelt out, immediately commence another attempt with a supply of freshly shuffled cards. Each attempt would take 16 seconds, and the game continues until *In the beginning* is spelt out correctly. We want a sense of how long it would take, on average, to spell out these three words correctly.

The answer is readily calculated. There are 53 possibilities for the first place, 53 possibilities for the second, and so on for each of the 16 places. The total number of possibilities is represented mathematically by 53^{16}, which means 53 multiplied by itself 16 times. This works out to a very large number—210 million times larger than the number of grains of rice on the chessboard from our previous example. In fact, the number is so large that, as we saw with the rice, the environment lacks the capacity to supply it. If we had not one, but *a billion blindfolded players*, each putting down 1 card per second, for 13.8 billion years (the reputed age of the universe given by secular science) there would be less than a one percent chance that we would ever see the words *In the beginning* spelt out—a result which is just as surprisingly impressive as the rice on chessboard example.

Tapping out the universe in six chessboards

In a final example, we'll consider a calculation presented by Jonathan Sarfati, a biblical origins researcher and writer, who sought to provide an estimate of the number of atomic interactions that have ever occurred since the beginning of the universe. By atomic interactions, Sarfati was referring to the jostling of atoms that occurs in matter. For example, if helium gas were magnified to the extent that we could see the atoms in motion, we would see atoms colliding with each other at an extremely rapid pace.[10] Sarfati estimated that on average, each atom collides with the atoms that surround it at a rate of one trillion times per second (that is, 10^{12} times per second in mathematical notation). Then, using the secular age of the universe (13.8 billion years), Sarfati presented the following calculation:[11]

1. 10^{80} atoms in the visible universe [Sarfati could have used a lower number here; 10^{80} refers to the number of *protons* in the visible universe. Since many atoms have more than one proton, Sarfati is being lenient.]

---

10 This atomic jostling also occurs in liquids and solids; it occurs wherever matter is above the temperature of absolute zero (known as zero degrees Kelvin), the temperature at which such atomic movement stops.
11 Sarfati, 2008, p. 159.

2. 10^{12} atomic interactions per second [As we saw above, Sarfati estimates that each of the 10^{80} atoms interacts with surrounding atoms at an average rate of 10^{12} (a trillion times) per second.]
3. 10^{18} seconds in the universe, according to secular science [In the name of simplified notation, Sarfati is generous in his time estimate; 10^{18} seconds is over 31 billion years, more than double the secular age of the universe, currently given as 13.8 billion years.]
4. Multiplying these three figures together yields: Only 10^{110} interactions possible

We denote this figure, 10^{110}, as the *Sarfati Limit,* as it represents the maximum number of atomic interactions that have taken place in the observable universe (or, more correctly, could take place if the universe existed for 31 billion years). In Figure 10.1, we have presented these numbers by extending the rice-on-chessboard example onto multiple chessboards. The sage's challenge begins as per the original example, then continues progressively onto subsequent boards.

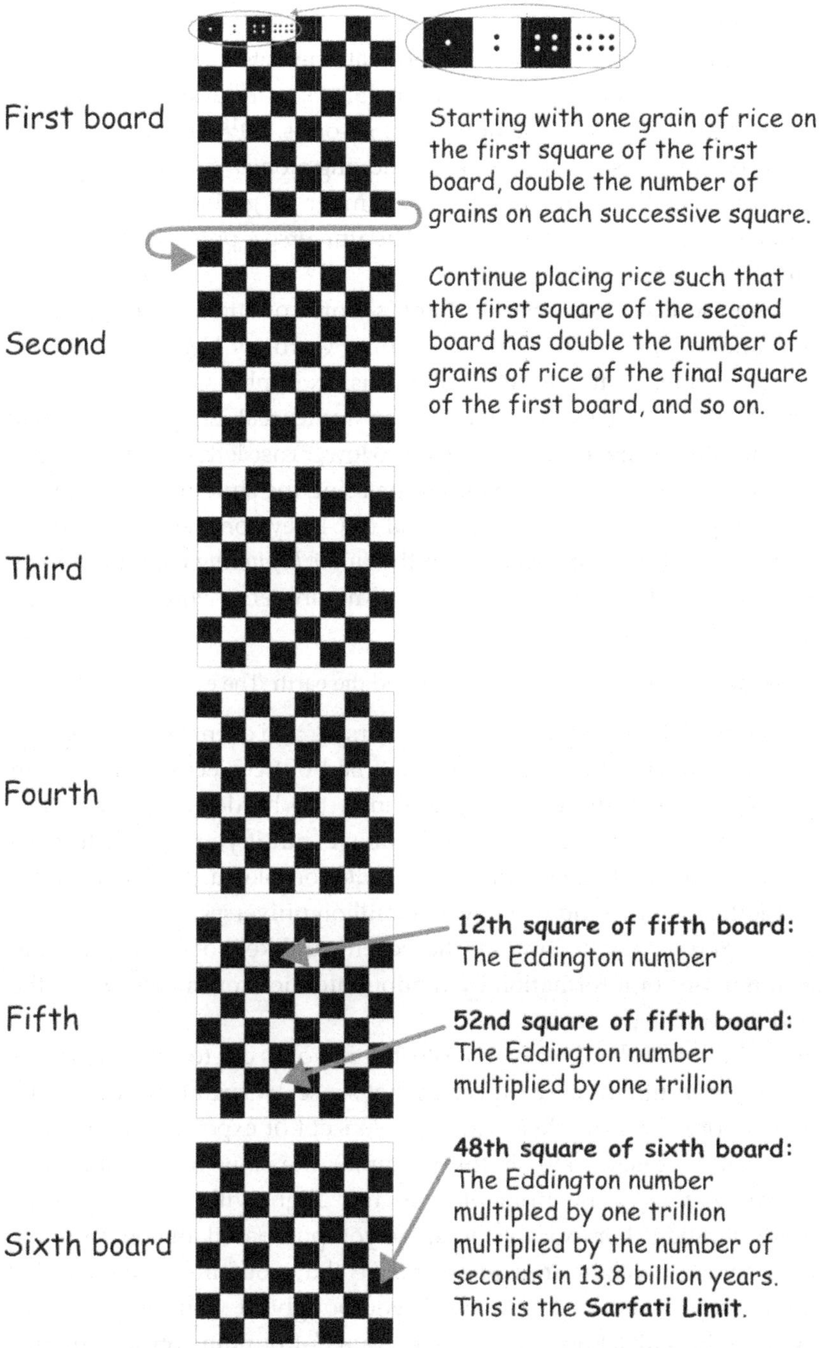

Figure 10.1: How to represent the atomic interactions of the universe in six chessboards

On the first square of the second board, we place double the number of grains of rice that we had on the final square of the first board, and so on.

The Eddington number, 10^{80}, is reached on the 12th square of the fifth chessboard. To multiply that by the number of atomic interactions per second, one trillion, we move along 40 squares, to the 52nd square of the same board. Finally, we multiply all this by the number of seconds, 10^{18}, that the universe has existed (according to secular science) by moving along a further 60 squares, to the 48th square of the sixth board. The number of grains of rice *on that individual square* equals the Sarfati Limit.[12] The universe's potential for atomic interactions is exhausted at that point.

Out of interest, if we had a Sarfati Limit's-worth of attempts to lay down letter cards randomly, how much of the opening words of Genesis would we expect to see written out correctly? For this example, we need to add punctuation to the keyboard so that we can write real English sentences. Assuming a simplified keyboard comprising 26 lower case letters, 26 upper case letters (to do away with the complexities of pressing the shift key), a space bar and 11 items of punctuation, we have a total of 64 keys on the board.[13] If we play the game according to the rules from the *In the beginning* example above, here's what we would mathematically expect the universe would provide us with (MEV translation):

> In the beginning God created the heavens and the earth. The e

That's Genesis 1:1 and the first snippet of verse 2, and even then, you would only have a 67% chance of seeing the above typed out correctly. Note that in these types of exercises, the more you advance, the harder it is to advance further. A single universe might furnish the above, but if you wanted to have that same 67% chance of seeing the word *earth* completed, by advancing a further four letters, you would need over 16 million universes.

What this exercise demonstrates is the extreme unlikelihood of assembling a significant amount of information by random atomic movements, given the finiteness of the visible universe.

The Sarfati Limit, 10^{110}, is easy to write, but it works out to be vastly larger than most people would initially expect. In that sense, we are all rather like the Indian king of yore; we struggle to grasp the concept of exponential numbers. As the exponent increases ("110" in this example), the value of the expression increases exponentially. The principles do not align with our day-to-day experience of numbers. For example, imagine you purchased an item from the store that is sold by weight. You might notice that your item weighed 510 grams. If you were asked later how much you bought, it would be perfectly reasonable to reply "half a kilogram" or "500 grams". Rounding off such figures

---

12 The number of grains on the 48th square of the sixth board actually exceeds the Sarfati Limit by 50%, but the previous square has less than the Sarfati Limit, so we go with the 48th square.

13 This is less than the number of keys on the usual keyboards provided with computers; a laptop keyboard usually has around 80 keys, and a standalone keyboard has over 100 keys.

is a normal and indeed practical part of the way we speak. But now consider a number in exponent form—namely 10^{510}. In this case, if you were asked about this number, it would be inappropriate, even misleading, to round off the exponent and reply "about 10^{500}". As big as 10^{500} is, it's only *one ten billionth* of 10^{510}. In this example, a seemingly slight change in the exponent results in an outsize increase in the number.

The concept of exponential numbers is at the heart of our discussion of abiogenesis, because the size of the environment required to generate information randomly increases exponentially for each additional bit of information. Of course, we won't be talking about assembling a few words. The amount of information in the DNA of living organisms varies, but it can easily exceed the information contained in multiple sets of encyclopedias. The question arises: how did the vast amount of information in DNA come into existence? For the Christian, the explanation is that God imparted this information into His creation via the *logos*, the spoken Word. For secular humanists, who do not believe in a super-intelligent God orchestrating the act of creation, random processes are generally invoked as a key part of the explanation.

In this chapter we will examine how good these explanations are. We will use the insights provided by James F. Coppedge,[14] Fred Hoyle & Chandra Wickramasinghe[15] and John Lennox[16] via a "back of the envelope" style calculation using conservative assumptions. This will give us a feel for the numbers and the issues confronting abiogenesis. We will also discuss some implications for the secular scientific worldview. But before we get to that we need to briefly explain the connection between probabilities and exponential numbers.

The probability of impossibility

When the French mathematicians Blaise Pascal and Pierre de Fermat corresponded about probability in 1654, they discussed probability as the ratio of favorable outcomes to total possible outcomes in games of chance. This was the original definition of probability and remains the simplest, so we will use it here. Games of chance still offer excellent illustrations of the concepts we need to cover, so we briefly discuss Powerball,[17] a lottery that is conducted in various countries.

---

14 James Coppedge's *Evolution: Possible or Impossible? Molecular Biology and the Laws of Chance in Nontechnical Language* was first published in 1973. With a PhD in theology and a master's degree in chemistry (UCLA), his book contains probabilistic critiques of abiogenesis and evolution as origins theories and was most recently re-published in 2002.

15 See Introduction, Chapter 1 and Chapter 2 of *Evolution from Space* (Hoyle & Wickramasinghe, 1981).

16 See Chapter 7 ("The origin of life") of *God's Undertaker* (Lennox, 2007).

17 As Christians, we caution against gambling and games of chance, which can be destructive to individuals and families.

In the U.S. Powerball, a participant selects five white ball numbers (which range from 1 to 69), and one red Powerball number (which ranges from 1 to 26). To win the jackpot, all numbers on the entry must match the balls which are drawn. There are 292,201,338 possible combinations of numbers, which means that the probability of selecting the correct numbers is about 1 in 292 million. We can write this probability as exactly $\frac{1}{292{,}201{,}338}$ or a rounded approximation as 3.4×10^{-9}, which is more compact to print. The negative exponent in the second form tells us that 10^9 belongs in the denominator. (In other words, 10^{-9} is the same as $\frac{1}{10^9}$.)

Suppose instead we wish to find the probability of winning the U.S. Powerball jackpot twice in a row, from only two entries. This calculates as $(3.4 \times 10^{-9})^2 = 1.2 \times 10^{-17}$. Because the negative exponent is now "17", we are exponentially closer to zero than at first. This probability is extremely low, and as far as we know, this outcome has never been recorded in U.S. Powerball history—although there are some individuals that have notched up multiple wins at lower prize levels.

The probability of winning three U.S. Powerball jackpots in a row from only three entries is 4×10^{-26}, an extraordinarily tiny number. As the number of consecutive Powerball jackpot wins increases, the probability of it occurring approaches closer and closer to zero. It never reaches exactly zero, no matter how large the denominator becomes. But this raises a question: at what point do we regard a highly unlikely event as actually impossible? For instance, if no one has ever won three U.S. Powerball jackpots from three attempts, is it actually possible? We know that it is *theoretically* possible, with a chance calculated as 4×10^{-26}—which is above zero—but would it ever occur in the real world?

At least one scientist has considered this kind of question and suggested a very low probability that would be regarded by an informed person as impossible in the real world. Lee Spetner, an American Israeli scientist who taught information and communication theory at Johns Hopkins University, investigated the relationship between information and the theory of evolution and suggested that any probability less than 10^{-45} is close enough to zero to be considered impossible in the real world.[18] The probability of 10^{-45} is about 47 times *less likely* than one person winning the U.S. Powerball jackpot five times from five separate entries. In the real world, of course, the authorities would not accept an outcome where a person won five jackpots off five attempts. The lotto and law enforcement agencies would almost certainly conclude that fraud was involved. Payouts would be withheld by the lotto pending an exhaustive investigation. This indicates that the incredibly low chances of winning five

---

18 Spetner, 1999, p. 94.

Powerball jackpots in a row is effectively regarded by the authorities as impossible; Spetner's impossibility cutoff is, as we said, 20 million times less likely again.

The figure of 10^{-45} that Spetner advanced is for all practical purposes indistinguishable from a probability of zero. For the rest of this book, a probability of 10^{-45}or less will be taken to be probabilistically impossible (or alternatively, a secular miracle). We can also define a probability cutoff of 10^{-90} as *impossibly impossible* because it is the probability that one impossible event is followed by another impossible event. (10^{-90} is equal to $10^{-45} \times 10^{-45}$).

Now let's combine all three considerations—the typing game on the simplified 64-character keyboard; a Sarfati Limit's worth of random keyboard strokes; and Spetner's impossibility criterion—to ascertain how much of Genesis it is actually possible to assemble. Here it is:

> In the beginning God created the heavens and the earth. The earth was without form, a

It is impossible, given the constraints of the exercise, and the universe, to add the letters "nd" to the "a", making the complete word "and". For the record, this is 85 characters, and, given our keyboard which had 64 keys, the above text string works out to about 64 bytes of information. At that point, the visible universe is expected to be tapped out: it's impossible to proceed any further.

When we consider that the DNA of even the simplest organisms can easily contain *hundreds of thousands* of bytes of information, the extreme difficulties faced by abiogenesis researchers become apparent. In essence, the visible universe cannot furnish enough atomic interactions to randomly generate that amount of information; abiogenesis is left looking completely implausible. That is why it has been described as the hardest problem in biology.[19]

A putative candidate for abiogenesis

Living cells are the basic building blocks of all living creatures on earth, whether single- celled, such as bacteria and yeasts, or multicellular organisms—like plants, animals and people.[20] The simplest organism, therefore, comprises a single cell.

A cell is somewhat analogous to a self-sufficient city. Think for a moment about what a city needs to function. It needs a ubiquitous form of power that residents and businesses can plug into. For the city to maintain itself, and grow, it needs businesses that convert raw materials into building materials—

---

19 Koonin, 2007.

20 A separate category, viruses, are organism-like forms containing some genetic material. However, viruses must hijack living cells to reproduce, because they lack the essential cellular machinery needed for independent metabolism and reproduction.

sawmills, brickworks, ironworks and the like. It also needs factories that create machines that the power generation and building material suppliers use to make their products. Naturally, moving all these raw materials and products around requires an efficient transport network, and an effective communication system.

For the analogy to be effective, our imaginary city must be the kind of metropolis that can set up other cities like itself. It would do this by dividing a portion of the workforce, raw materials, equipment and know-how and sending these to a separate location where another city of the same kind can be rebuilt.

What are the minimum requirements for a city like this to operate? We will only consider two functional requirements: an energy source and factories that produce machines used within the city. For the sake of simplicity, we'll ignore transport and communication.

In the cell, the machines correspond to proteins, which are a kind of wonder material, analogous to Lego™. They comprise twenty different types of amino acids which are strung together like beads on a necklace; this string is in turn folded into three dimensional structures like a pliable wire. The specific order of the amino acids determines the final shape that the protein will take when folded. Proteins can act as enzymes, speeding up molecular processes, and also serve as building materials, to give structure.

Now, let's consider the energy source. In cities, energy sources historically have been coal or diesel. Both are energy-rich substances that can be burned to produce electricity. In a cell, one of the most ubiquitous high-energy molecules is glucose. In a similar way, cells break down glucose, releasing energy (similar to our city burning coal) which is stored in a molecule called ATP (adenine triphosphate). This is analogous to an electricity supply in a city in that this molecule is used almost everywhere in the cell to provide energy.

The breakdown of glucose is performed in a multi-step process known as the Krebs cycle.[21] The Krebs cycle has multiple steps, rather like a production line. A production line consists of successive stations, each equipped with a unique machine to perform a specific task. Machines must be used in order to perform the work fast enough and at a sufficient level of quality for the factory to be sustainable. Similarly, the Krebs production line in the cell is equipped with molecular machines, known as enzymes, which are types of protein able to speed up the production line sufficiently to sustain life. (Without these machines, there is no production line and the processes occur too slowly to support life.)

---

21 The Krebs cycle is named after Sir Hans Adolf Krebs, a German-born British biochemist who discovered the cycle in 1937.

The Krebs cycle takes in a glucose molecule that has been cut in half (this preliminary step, known as the glycolytic pathway, involves ten different enzymes, but we won't cover it in any further detail here) then adjusts the molecule further in eight successive steps, where each step is facilitated by one type of enzyme. The output of the cycle is several high energy molecules which will ultimately be used as a ubiquitous cellular energy source—along with carbon dioxide, which is respired out of the organism.

In sum, metabolizing glucose—and only that—takes ten enzymes to cut the molecule in half, and another eight enzymes to operate the Krebs cycle.

The second process we consider is the manufacture of machines in our city. This is important, because if our city replicates itself, the new city will need its own copy of each of the machines. Similarly, in the cell, there must be a mechanism for producing proteins that act as enzymes. The proteins are produced to blueprints, and the blueprints for proteins are stored as DNA.

In a real city, talented engineers in factories could, if needed, reverse engineer existing machines to make new blueprints which themselves can later be used to make copies of the machines. However, cells are unable to reverse engineer proteins to make the DNA blueprint. In a cell, information flows from the DNA blueprint to proteins (via an intermediate molecule called messenger RNA, or mRNA for short), but not in the opposite direction. This principle is so important that biologists refer to it as the central dogma of molecular biology. The upshot of this for our simple cell: it is not enough to randomly assemble functional proteins; corresponding DNA blueprints for similar functional proteins (enzymes) must also be randomly assembled, as they cannot be reverse engineered.

Although single-celled organisms such as bacteria are among the simplest, they are still enormously complex. Externally, bacteria are bounded by a cell membrane composed of various types of molecules. Internally, the cytoplasm is a gel-like fluid containing DNA, inclusion bodies, signaling proteins, transport molecules, flagellar motors (in some bacteria), enzymes, salts, and other protein-manufacturing machines (ribosomes). The cytoplasm also contains cytoskeletal filaments which are important in cell division and determination of cell shape. The filaments assist with chemotaxis, the movement of the bacterium in response to external chemical stimuli. (Chemotaxis enables bacteria to move towards food molecules and away from toxins.)

Secular biologists may not consider bacteria to be the first life form, but they would generally agree that the first life was some form of single-celled organism. They further agree that living cells require several minimum features: a membrane, enzymes which assist the metabolism of energy-containing molecules, information storing and function-controlling systems, and molecular machinery to reproduce (make copies of themselves).

The question of abiogenesis can be rephrased as: *How did a single-celled organism come about?* But we can refine this to a narrower query: What is the probability that random processes can generate a minimum set of proteins and associated DNA required to support life in a cell? We will answer this question in the following back-of-the-envelope analysis, making sure to tilt the table in favor of abiogenesis at every decision point.[22] That way, it is more difficult for a critic to refute our argument.

Minimum requirements

Scientists have investigated the minimum number of protein-encoding genes required for life across a range of single-celled organisms. By altering particular genes to inactivate them and then observing whether the organism can survive and reproduce, the essential genes can be identified. If a gene can be inactivated without causing a failure of reproduction, it is non-essential; the rest are essential. Among bacteria with the lowest number of essential genes is *Bacillus subtilis*, a hay bacillus which is found in the intestines of cattle; scientists have identified 271 essential genes.[23] Well-known American expert in evolutionary and computational biology, Eugene Koonin, estimated a "bare-bones" gene set at 150 genes, but conceded that it was questionable "whether such a minimalist cell could survive under any realistic conditions".[24 25] A very conservative approach would be to reduce the minimal number of genes to 100, but for our back-of-the-envelope analysis, we will reduce the number to ten to allow for the possibility that putative early life forms were far simpler than those we know today.

Recall that the Krebs cycle, which metabolizes glucose fragments, takes eight enzymes to complete. This was also preceded by the 10 enzymes required for glycolysis, for a total of 18 enzymes. This shows how conservative we are in specifying ten enzymes—we haven't even considered reproduction or the DNA blueprint production of proteins.

A critic could counter that a putative first organism may not have used glucose as its energy source and therefore may not have required the Krebs cycle. However, alternate metabolic pathways require multiple enzymes. For example, the Calvin cycle, which is a process in photosynthesis, is typically said to involve ten enzymes.

---

22 Mathematically, this is known as calculating an upper bound of the probability of abiogenesis (a type of Fermi estimate). The probability of abiogenesis is less than or equal to the upper bound. If the upper bound probability is impossible, then abiogenesis is disproven (that is, $P(abiogenesis) \leq P(upper\ limit\ abiogenesis)$, so if $P(upper\ limit\ abiogenesis) \leq P(impossible)$ then abiogenesis is probabilistically impossible, and is disproven).

23 Kobayashi, Ehrlich, Albertini, ..., & Ogasawara, 2003.

24 Koonin, 2000, p. 113.

25 The organism must also encode structural RNA for ribosomes, but we will ignore this extra information requirement in this chapter to simplify our calculations.

Furthermore, DNA replication involves five or more enzymes. Hence, a minimal set of ten enzymes is very conservative, and would hardly cover the requirements for replication and metabolism, let alone other critical cellular functions that are essential to life, such as those afforded by structural proteins.

The next task in specifying our putative first life form is to decide upon the length of proteins. This also tells us the length of the DNA blueprints that are needed to inform assembly of the proteins. To help readers visualize how we approach this step, we present a snippet of DNA in Figure 10.2.

DNA takes the form of a multi-rung double helix in which each rung is known as a base pair. The letters A, C, G and T on the base pairs stand for adenine, cytosine, guanine and thymine respectively—these are chemical compounds known as nucleobases. The figure shows a number of base pairs flattened out, and each base pair triple drawn distinctly from adjacent triples for illustrative purposes. The lower strand carries the genetic information, and

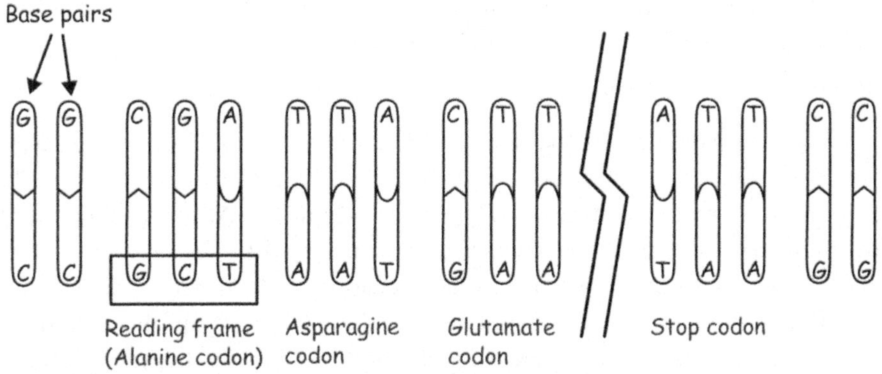

Figure 10.2: Assembly of protein using the DNA blueprint

the upper strand is a complement of this. Notice that the base pairs always comprise either A paired with T, or C paired with G.[26]

In the diagram, the box indicating the reading frame is analogous to the reading head of a ribosome. The reading frame encompasses three bases, which together comprise a codon. The reading frame shows a GCT codon, which codes for the amino acid alanine (refer Appendix C for a complete listing of codon sequences and their corresponding amino acids). The ribosome processes the triplet and reaches out into the cell cytoplasm, locates an alanine amino acid molecule (which, for the record, is attached to a special form of RNA

26 For simplicity, we will ignore messenger RNA, and our description will imply that the ribosome works directly on the information strand of DNA. In reality, messenger RNA is used as an intermediate molecule to transfer information from DNA to the ribosome. That's too complicated for our purpose, which is only to illustrate the fact that proteins are formed using the information in DNA.

called transfer RNA), and chemically links it to the protein under formation. The ribosome then moves down three bases. The codon at this point is AAT, which codes for the amino acid asparagine. The ribosome grabs an asparagine amino acid and links it to the chain. In this case, the asparagine will be linked to the alanine from the previous step. The ribosome continues down the chain in this manner; the next codon is GAA, which will result in the amino acid glutamate being linked to the asparagine molecule.

This process ends after all the amino acids that comprise the protein have been strung together in order. The DNA contains a stop codon further down, TAA, which tells the ribosome to "cast off" the protein chain. The cell could apply further processing steps, such as folding up the chain into a three-dimensional shape, trimming the chain, or some chemical modification, but at this point, it is virtually ready for use in the cell.

Scientists have established that the average protein contains about 400 amino acids.[27] In our back-of-the-envelope calculation, we will assume a lower protein length of one hundred amino acids, which is simpler to assemble, and therefore tilts the table in favor of abiogenesis.

Minimum information in a protein

There are some genes that are found across a wide variety of organisms. We have already met the Krebs cycle; the genes that code for enzymes used in this cycle will be found in the DNA of most organisms. Although there might be slight variation across the Krebs cycle enzymes in different organisms, there are particular amino acids in certain locations which are virtually non-negotiable—they must be present in the right location for the enzyme to function. These particular amino acids might be essential to the protein to fold correctly or have a special function at what is known as the "active site" of the enzyme. The number of non-negotiable amino acids could be as low as a dozen in a simple enzyme, but in more complex enzymes, several dozen are likely.

For our calculation, we will set the number of these virtually non-negotiable amino acids at ten. But we will allow one hundred different variations of the non-negotiables; it may be possible for a different amino acid with similar chemical properties to be substituted in some locations. Our model will not put any constraints on the other 90 amino acids in each protein; we will assume they can be random without affecting function. This again is highly favorable to abiogenesis, because in reality, some of these types of substitutions are likely harmful.

27 Mushegian & Koonin, 1996, p. 10273.

Calculating the probability of abiogenesis

Having set these parameters, we can now calculate the probability of randomly assembling, on earth, a complete set of proteins and DNA segments essential for life.

The steps we follow to calculate this probability are outlined below. They include an estimate of the maximum amount of constituent material, such as carbon, that is available for protein and DNA assembly; we refer to this as the "primordial sludge". (The full mathematical workings and explanations are provided in Appendix C, including a brief justification of our probabilistic approach):

1. Calculate the probability of assembling ten proteins essential to life randomly. This works out to 7.9×10^{-111}, considerably less likely than an impossible impossibility.
2. Calculate the probability of assembling ten DNA segments corresponding to the proteins (or their variants). This works out to 8.1×10^{-97}, also less likely than our impossible impossibility cutoff.
3. Calculate the maximum number of chances (or "dice rolls") available to assemble the essential proteins and DNA segments, given the amount of "primordial sludge" and time available (half a billion years according to contemporary secular science), and the time taken to assemble or reassemble the molecules. The maximum number of chances works out to 3×10^{54}.
4. We combine the above values to yield an overall probability of randomly assembling a minimalistic set of molecules required for life. The overall probability of assembling a minimalistic set of molecules required for life on earth is 1.9×10^{-152}.

Recalling that Lee Spetner's cutoff of probabilistic impossibility was 10^{-45}, it is clear that abiogenesis cannot have occurred on earth. It is less probable than three impossible events (events whose probability is 10^{-45}) occurring in succession.

However, this calculation is only concerned with assembling some of the components of a dead cell. We did not factor in how to bring these components to life; intuitively, this is an enormous jump. Nor did we deal with the problem of forming other parts of the cell, such as the membrane. We are faced with the fact that our probability for abiogenesis is an enormous overestimate, even though it is already less possible than an impossibly impossible event.

The problem of stereoisomers

Our assessment of the probability of abiogenesis occurring on earth is wildly optimistic, and not simply because we tilted the table at every turn to favor the odds and ignored complexities like membrane formation. There is an additional problem, related to what are known as stereoisomers, that we should also take into account.

All amino acids found in living organisms (except for glycine) come in two forms, which are essentially mirror images of each other, called the levo (left) and dextro (right) amino acids. (Think of a pair of fitted gloves; on first inspection both gloves look identical, but in fact they are mirror images of the other; the left-handed glove does not fit a person's right hand, and vice versa.) These two "handed" forms of amino acids are called stereoisomers. When amino acids are produced by conditions which scientists assume may have been present on earth in the distant past, they form in approximately equal amounts of left- and right-handed amino-acids. This was demonstrated by the famous Miller-Urey experiment.[28] But the protein in all living things is made up only of the levo amino acids.

Consider what this means in practical terms for the formation of the amino acid polymers. Imagine a vat of left-handed amino acids, and a separate vat of right-handed amino acids. Dye the first vat red, the second blue, then pour them together to produce a purple-colored mixture. For any protein suitable for life to assemble by random processes, only the red-dyed left-handed amino acids can be permitted to join together. But how is it possible to filter them out from the mixture?

It certainly cannot be achieved by random processes. The chance of forming ten proteins composed of one hundred levo amino acids from a 50:50 mixture of levo and dextro forms, even assuming the proteins are composed of fifty percent glycine which has no handedness, is $2^{-10\times100\times0.5} = 3.1 \times 10^{-151}$, which is well under Spetner's impossibility threshold. If so much as a single right-handed amino acid joins the chain, it may be rendered useless for life. Importantly, left-handed and right-handed amino acids have identical chemical properties, that is, either form will just as readily join an amino acid chain as the other. Using intelligently designed processes, modern-day biochemists can synthesize a biologically active protein, but they must use purely left-handed batches of each amino acid—which, outside of living things, are not found in nature. Random processes always produce a near 50:50 mixture.[29]

Secular scientists raised the possibility that clay minerals might help filter out levo stereoisomers, but later investigations showed that this was not an adequate solution. Another possibility was that a hypothetical polarized light source in space might have led to an excess of 'handedness' in amino acids

28 The Miller-Urey experiment of 1952, widely cited in biology textbooks, showed that some of the amino acids needed for proteins could be produced from inorganic constituents in environments that were thought to resemble that of the primordial earth. However, the amino acids were a mix of levo and dextro—with no known mechanism for filtering out the dextro type.

29 Since the LH and RH forms readily convert to each other with equal frequency, any batch of amino acids with even a small imbalance of one form or the other will immediately trend towards the 50:50 ('racemic') state.

which were then transported to earth on meteorites. This idea, too, led nowhere.

No reasonable scientifically verified explanation has been proposed to account for how this "optical purity", as it's called, may have occurred in abiogenesis. It represents another major hurdle for theories of abiogenesis, and one we can factor into our probability assessment qualitatively. Basically, if the conditions to generate levo amino acids in sufficient purity to permit random assembly of the proteins needed for life have proven impossible to simulate up to now, then the fraction of carbon atoms supposedly exposed to such a process on earth must be miniscule. It cannot have been a widespread process, or it would already have been discovered. In this case, our calculated probability for abiogenesis, impossible as it was, is far too high. Abiogenesis is much less likely than we calculated.

The spatial and temporal proximity problems

There are further issues for abiogenesis. Arthur Chadwick, a research professor of biology and geology, identified a proximity problem for abiogenesis—the problem of spatial and temporal proximity.[30] The crux of Chadwick's proximity problem is that the required molecules must be in the right place (together) at the right time (simultaneously). If this problem, with both its components, were applied to our earlier calculation of the probability of ten proteins and DNA molecules forming spontaneously somewhere on earth, it would render it far more unlikely. Within the vast, global amount of sludge we posited, one protein molecule could have formed at the bottom of the Mariana Trench, another off the coast of North America, and so on. But for life to form, all these molecules need to form in almost exactly the same location—within a few millimeters of each other at most. If we had stipulated this constraint, it would have lowered the probability significantly. The other part of Chadwick's proximity problem relates to timing. It is no good for one molecule required for life to form at the start of the abiogenesis window, and another hundreds of millions of years later. For simplicity, we did not specify these constraints earlier, but if we had, it would have decreased the probability enormously.

Chadwick's proximity problem amounts to a catch-22 situation for abiogenesis. As the time frame and spatial extent are increased to add extra rolls of the abiogenesis dice, making the calculation more favorable for abiogenesis, the greater potential separation of the assembled components in space, and in time, work against the formation of a living cell.

Another aspect of the temporal component of the proximity problem is that to increase the chances of forming the molecules, we said (in Appendix C) that

30 Chadwick, 2005.

the molecules were very unstable and disintegrated immediately back to their constituent monomers. Once again, if we had added a constraint that the molecules had to be somewhat stable, to persist long enough for several proteins and DNA molecules to exist simultaneously, then we would have dramatically reduced the number of rolls of the dice, and hence the probability of abiogenesis happening.

Approaches to the abiogenesis problem

RNA World: a leading secular theory

Francis Crick, who co-discovered the chemical structure of DNA, speculated that life could have formed on earth via a mechanism involving RNA. Crick speculated that RNA could have served three functions simultaneously: storing information as genetic material; catalyzing biochemical reactions; and performing self-replication.[31] This idea, which is known as the *RNA world hypothesis*, has never been observed or scientifically verified. However, the RNA world concept is considered a leading contender in abiogenesis explanations. Nevertheless, Crick conceded:

> An honest man, armed with all the knowledge available to us now, could only state that in some sense, the origin of life appears at the moment to be almost a miracle, so many are the conditions which would have had to be satisfied to get it going.[32]

Evolutionary expert Eugene Koonin has acknowledged that the RNA world hypothesis is incredibly unlikely. Koonin published a back-of-the-envelope estimate of the probability of abiogenesis via the RNA world scenario across the expanse of the observable universe, concluding that the chance of abiogenesis was less than 10^{-1018} (using conservative assumptions, but ignoring the proximity problem). This is unimaginably less likely than odds we calculated for assembling 10 primitive proteins and some DNA using our own very conservative assumptions. Because he considered abiogenesis practically impossible, Koonin invoked infinity via the cosmic multiverse (or a spatially infinite universe) and the anthropic principle in order to guarantee—intellectually, at least—the emergence of life by abiogenesis. In other words, no matter how improbable it might be for abiogenesis to occur in one universe, among an infinite number of universes, abiogenesis is bound to happen once, or even multiple times.

Interestingly, Koonin cautioned that the infinite multiverse could be extended to allow the emergence by chance of *all organisms without Darwinian*

---

31 Initially, the RNA replication is supposed to be slow, error prone, non-enzymatic copying; later replication is supposed to be enhanced once ribosome-like RNA molecules have evolved to catalyze it.

32 Crick, 1981, p. 88.

evolution, although he was alarmed by the possibility of this extension because it rendered evolution superfluous:

> A crucial aspect of the framework developed here is brought about by a disturbing (almost nightmarish) but inevitable question: in the infinitely redundant world of ... [many worlds in one], why is biological evolution, and in particular, Darwinian selection relevant at all?[33]

Koonin's thinking here is fascinating. If an infinity is used to render the RNA world theory plausible, there is no logical reason that that infinity should stop at a single-celled organism. It could just as plausibly produce dogs, cats and trees, and indeed all of the biological complexity on earth. Koonin had reasoned his way to an infinity of the sort that enables virtually any secular miracle to occur. However, Koonin argued that he would only invoke this infinity to explain abiogenesis. This retained Darwinian evolution as the explanation of all downstream diversity and complexity. This was primarily for psychological, rather than logical reasons—the absence of Darwin was "nightmarish".

Until direct observations support the "RNA world" theory, it is on precisely the same level as secular humanists regard the Bible—a just-so story—a fairy tale.

Panspermia

In addition to his RNA world speculations, Crick came up with an even more fanciful theory—directed panspermia.[34] This proposed that intelligent aliens had sent living materials via space travel to Earth. This theory implicitly invoked another form of impersonal quasi-infinity to help deal with the improbability of abiogenesis. The idea is that there are billions or trillions of earth-like planets in the cosmos and life could have begun on any of them. Even though the probability of life occurring on earth is minuscule, if you have enough earths, so the argument went, then the probability that life began on one of them increases. This is a way of increasing the numerator on the probability fraction to counter the enormous denominator.

Hoyle and Wickramasinghe, also proponents of panspermia, although of an undirected type,[35] suggested that this hypothesis has been repressed by the scientific community because of concerns that the promulgation of these theories "might start filling the churches again".[36] The other aspect to the theory is that by directing the problem of abiogenesis to another, unknown planetary system, it is possible that quite different conditions might have prevailed there.

---

33 Koonin, 2007.

34 Crick, 1976.

35 Directed panspermia suggests an advanced alien civilization has distributed life to earth via spaceships, whereas undirected panspermia posits the distribution of life to earth via natural means such as by comets.

36 Hoyle & Wickramasinghe, 1981, p. 3.

Since conditions on earth are not favorable for abiogenesis to have occurred in a probabilistic sense, perhaps somewhere else offers better, unique conditions.

But in any case, what is the probability that abiogenesis has occurred somewhere else in the observable universe? According to theoretical astrophysicist and science communicator Ethan Siegel, there are up to 10^{30} starless planets in the universe. These are unlikely to be able to support life. However, there are estimated to be 10^{25} planets that orbit stars.[37] If we assume that all of these are habitable, we can calculate the probability of forming ten proteins and the related DNA molecules by random means on these planets. Of course, it is unlikely that all of these could support life; some planets are gaseous, instead of rocky, and some do not orbit an appropriate distance from an appropriate star. Nevertheless, once again we tilt the table in favor of abiogenesis by assuming that all these planets are habitable, and that on average they contain ten times as much accessible carbon as earth. We also increase the timeframe to the purported age of the universe—13.8 billion years. This raises the probability from our earlier, highly conservative figure of 1.9×10^{-152} to 5.3×10^{-125} which is still impossibly impossible according to our definition.

Even if we were to tilt the table still further by invoking some unknown property of dark matter as increasing the probability of abiogenesis by a factor of a trillion, there is no material shift in the result. The chance would then be 5.3×10^{-113} which is still impossibly impossible. Assuming a large number of universes doesn't help much either: if there are a trillion trillion trillion universes (each which mirrors our own) in the multiverse, the probability increases to of 5.3×10^{-77}, still well beyond Spetner's impossibility cutoff of 10^{-45}. But the proximity problem now includes a time span of 13.8 billion years instead of half a billion years, and a spatial extent of a trillion trillion trillion universes instead of just one. Which, as we showed earlier, makes the problem worse for abiogenesis.

Nevertheless, secular scientists still make statements that imply that life could have formed elsewhere. For example, Edward J. Weiler, a former NASA Director, said that:

> We now know the number of stars in the universe is something like one followed by 23 zeros. Given that number, how arrogant to think ours is the only sun with a planet that supports life, and that it's the only solar system with intelligent life.[38]

The layperson, confronted with Weiler's comment, might well take it at face value, and accept that belief in the biblical account is "arrogant". But by combining an understanding of probability and the minimum requirements of

37 Siegel, 2022.
38 Davies, 2010, p. 24.

even the simplest cell, the glaring light of mathematical clarity shows up Weiler's remark as ignorant. Forget the number of life-friendly planets; every interaction of amino acids and DNA nucleotides that can ever have occurred since the visible universe is said to have begun gets us nowhere near the requirements for the simplest cell. That conclusion holds even if we expand the scope of our inquiry to include a trillion trillion trillion universes. In short, an infinity is required—an infinity that nature cannot supply and empirical science cannot detect.

The concept of directed panspermia (or any panspermia) is futile. The probabilistic problem of abiogenesis applies across the visible universe, the multiverse and across all purported time. The large number of life-friendly planets cited by Siegel makes no material difference.

Hoyle and Wickramasinghe's comment stands: "No matter how large the environment one considers, life cannot have had a random beginning".[39]

Secular miracles: time and purpose

George Wald, a Nobel prize-winning scientist, commented on the improbability of abiogenesis, invoking the 'miracle' of time to render it more plausible.[40] He admitted that spontaneous generation was scientifically impossible, so he removed the question from the domain of science by making it a presupposition of his worldview. To resolve this contradiction in logic, Wald sought to redefine the meaning of "impossible":

> The important point is that since the origin of life belongs in the category of at-least-once-phenomena, time is on its side. However improbable we regard this event, or any of the steps which it involves, given enough time it will almost certainly happen at least once. And for life as we know it, with its capacity for growth and reproduction, once may be enough.
>
> Time is in fact the hero of the plot. The time with which we have to deal is of the order of two billion years. What we regard as impossible on the basis of human experience is meaningless here. Given so much time, the "impossible" becomes possible, the possible probable, and the probable virtually certain. One has only to wait: time itself performs the miracles.

In his reasoning, Wald moves sequentially from "impossible" to "possible", to "probable", to "virtually certain" based on the "miracle" of time, conceding, in the process, that abiogenesis is indeed a secular miracle.

This comports with philosopher Nagel's conclusion. In Nagel's case, he invoked material purposivism.[41] Whatever God-like attribute is chosen matters

39 Hoyle & Wickramasinghe, 1981, p. 148.
40 Wald, 1955, pp. 9, 12.
41 Nagel, 2012, p. 124.

little—secular scientists end up resorting to a secular miracle to explain abiogenesis.

Other escape hatches

Fred Hoyle and Chandra Wickramasinghe canvassed several means of escaping the logic of the type of argument we have described and rejected them all.

One such historical argument was to suggest that the number of essential enzymes required may be much lower than a hundred. Hoyle and Wickramasinghe pointed out that even if the number of enzymes is reduced to single digits (our analysis posited ten), the likelihood of generating these enzymes by random processes remains infinitesimally small.[42] And it is highly doubtful that a self-reproducing organism could exist with less than ten distinct proteins.

Another key escape hatch is to suggest there are a multitude of possible alternative chemical systems which could support life. The argument then suggests that it was only necessary for abiogenesis to have occurred in one such system, after which life transferred to the current chemical system.

Hoyle and Wickramasinghe point out that all known life-forms are carbon based—hence a non-carbon-based abiogenic system is speculative (another just-so story). The property of carbon which renders it useful for life is the ability for each atom to bond four ways—two of the bonds can be used to link other carbon atoms in a head-to-toe chain, leaving two additional bonding points available, one on each side of the chain. These can bond to a variety of different atoms or molecules, including other carbon chains. This enables enormous structural diversity in biological molecules to be generated, with resultant versatility. For example, if the carbon chain bonds with hydrogen atoms along both sides, the resulting molecule is like part of a "fat" molecule, that is, non-water soluble. Silicon is often compared to carbon as a basis for life, as it also allows for four-way bonding, but complex chains built of silicon tend to be less stable. It cannot form molecules of the same complexity that we observe in living cells. Like a poor man's carbon, it has some tantalizing properties, but ultimately it just doesn't have the life-friendly properties that carbon has.

Almost all known terrestrial life-forms use glucose as a fuel. As we said earlier, this requires a set of complex enzymes; their joint actions are known as the Krebs cycle.[43] Although there is a simpler system of glucose metabolism, namely the glycolytic pathway, it still involves about ten enzymes. But our analysis did not depend upon the specifics of the chemical system of metabolism. If one were to suggest that there were once alternative life-

---

42 Hoyle & Wickramasinghe, 1981, p. 26.

43 Ibid., p. 28.

supporting chemical systems which did not require enzymes, it would be to invoke a very implausible "just-so" story. It is true that some methanogens and nitrifying bacteria have metabolisms based on molecules other than glucose, but they still require a multiplicity of enzymes for metabolism. No specific non-enzymatic alternate systems are known to exist. It is very likely that if such systems did exist, they would already have been discovered.

Irrespective of an alternative chemical system, a randomly assembled organism must contend with the kind of probability problem we outlined, in combination with Chadwick's proximity problem. Increase the probability of random assembly of essential precursor molecules by making the prebiotic "soup" larger, or the time available longer, and the proximity problems are exacerbated, meaning that the precursor molecules, even if assembled, are likely too far apart to combine in a microscopic primordial cell.

Sir Fred Hoyle and Chandra Wickramasinghe concluded that since assembling a minimal set of enzymes for life cannot be random, it necessarily required intelligence.[44] They were able to deduce a characteristic of that intelligence:

> To be consistent logically, we have to say that the intelligence which assembled the enzymes did not itself contain them. This is tantamount to arguing that carbonaceous life was invented by a non-carbonaceous intelligence .... .[45]

As non-Christians (and in the case of Hoyle, as an outspoken atheist) they hastened to add "... which by no means need be *God,* however." (Emphasis in original.)

There are two other approaches which we will consider briefly. The first is to avoid the hard problem of how a digital information management system, including the ability to replicate, emerged. Instead, scientists focus on "earlier" stages in abiogenesis: simple metabolic networks in the "Metabolism-First Hypothesis"; assembling lipid molecules into compartments in the "Lipid World Hypothesis"; or assembling simple proteins in the Amyloid Hypothesis. None of these theories explain the emergence of information-bearing DNA or RNA.

The second approach starts by recognizing that none of the abiogenesis hypotheses are successful or likely. Therefore, proponents suggest that several hypotheses should be combined together—for example, the Lipid World and the RNA World. However, elementary probability theory tells us that combining the probability of two very unlikely events such that both must occur, results in a lower probability than either event by itself. If we accept Koonin's probability calculation, which shows that the RNA world is

---

44 Ibid., p. 150.

45 Ibid., p. 139.

probabilistically impossible, then combining the RNA world theory with any other system also renders the combination impossible.

Forcing their hands: the response of secularists

The conundrum of the emergence of life has forced secular humanists into a state of cognitive dissonance. In principle, scientists such as George Wald agree that spontaneous generation has been disproved. In practice, they believe that life emerged from non-living matter, at least once. Wald argued that a quasi-infinity of several billion years was sufficient for abiogenesis.

In realizing that the RNA world hypothesis was essentially impossible, Koonin retreated to a philosophical scheme invoking an infinite multiverse to render virtually any event possible, no matter how improbable. Unlike Wald's quasi-infinity, Koonin invoked an undetectable "infinity of the gaps". One hundred and fifty years after Gosse, it is the secular origins proponents that are now openly promoting schemes whose purpose appears to be to cut short the investigative process. In other words, the debate has flipped in this respect: creationists (or at least the large, organized ministries) have moved on from science-stopping arguments, whereas secularists have been dragged into them. "Infinity of the gaps" arguments are outside of science and provide no predictive ability.

One of the most prominent atheists of the twentieth century, philosopher Antony Flew, made the intellectual move to theism as scientific advancements unhelpful to abiogenesis progressively accumulated. Flew invoked an unidentified divine infinity as a solution to the problem. In effect, Flew no longer believed in abiogenesis.

The teleological approach, espoused by Thomas Nagel, acknowledges the intractable problem of abiogenesis, but shortcuts it by moving to purposivism. This is a non-scientific philosophical device which implies life simply had to form. This approach is arguably a move in the direction of pantheism—a belief in some sort of god suffusing the entire material realm.

Richard Dawkins, the outspoken atheist, made a comment on the chances of randomly forming protein molecules that applies to abiogenesis (even though Dawkins wasn't referring to abiogenesis here):

> You don't need to be a mathematician or a physicist to calculate that an eye or a haemoglobin molecule would take from here to infinity to self-assemble by sheer higgledy-piggledy luck.[46]

If luck cannot generate the proteins comprising hemoglobin, then neither can it generate the essential proteins and DNA for abiogenesis to occur. In effect, Dawkins has acknowledged the impossibility of abiogenesis—in the

46 Dawkins, 2016, Chapter 3.

absence of an appropriate infinity—even though secularists like him tend not to go on the public record with direct disavowals of abiogenesis. Although he has not identified an infinity, his worldview leads him to expect that abiogenesis will be solved in the future.

The responses of these non-Christian thinkers are the telltale signs that the field of abiogenesis has become trapped in a web of intellectual futility. Abiogenesis is revealed as a dogma which lacks scientific support.

Conclusion

Since Louis Pasteur disproved spontaneous generation in the 1860s, the immense complexity of even the "simplest" life forms has been increasingly revealed by scientific investigation. The discovery of the structure of DNA in the 1950s revealed the chemical requirements of life as so complex that random processes cannot ever have assembled more than a few of the basic constituents of life. Generating information by random processes becomes exponentially more unlikely as it progresses. Modern scientific progress in molecular biology has reinforced Louis Pasteur's finding that life does not arise spontaneously.

Nevertheless, secular humanists have posited several escape hatches which purportedly increase the odds of the emergence of life. The problem with each of these is that they are unsubstantiated by evidence. In other words, they are no more than just-so stories in the guise of scientific language. Some explanations effectively amount to the argument that there is virtually no minimum informational requirement for life-like behaviors,[47] a concept which flies in the face of established science. Others invoke infinity in a manner which is not subject to empirical science.[48]

The probabilistic analysis presented in this chapter, and fleshed out in Appendix C, does not account for the dynamic, moment-by-moment information management processes that actually take place in living cells. Instead, our analysis set a far lower bar—the probability for the random assembly of several important components of a lifeless cell. Even this, however, is impossible.

Bible believers can, and should, hold a far more rational view: *advances in molecular biology have disproved abiogenesis*. This in turn undercuts the foundations of any materialistic or atheistic theory of evolution, because without an adequate explanation for the origin of life, the next step up on the ladder—secular conceptions of biological evolution—are set adrift. Indeed, the impossibility of abiogenesis is a showstopper for evolution as a holistic origins account.

---

47 For example, the original RNA world hypothesis.

48 Both George Wald and Eugene Koonin invoked a quasi-infinity or infinity to argue that any event, even if highly improbable, is certain to occur.

However, this is far from the only problem confronting holistic theories of evolution, as we will discover in the next three chapters.

References

Chadwick, A. V. (2005). Abiogenic Origin of Life: A Theory in Crisis. Retrieved April 24, 2024, from origins.swau.edu/papers/life/chadwick/default.html

Crick, F. (1976, November 5). Directed Panspermia. Retrieved April 22, 2024, from profiles.nlm.nih.gov/101584582X375

Crick, F. (1981). *Life Itself: Its Origin and Nature*. New York: Simon and Schuster.

Davies, P. (2010). *The Eerie Silence: Renewing Our Search for Alien Intelligence*. New York: Houghton Mifflin Harcourt.

Davies, P. (2020). *The Demon in the Machine: How hidden webs of information are solving the mystery of life*. London: Penguin Press.

Dawkins, R. (2016). *Climbing Mount Improbable* (20th Anniversary edn—eBook). London: Penguin Random House UK.

Hoyle, F., & Wickramasinghe, C. (1981). *Evolution from Space: A Theory of Cosmic Creationism*. New York: Simon and Schuster, Inc.

Kobayashi, K., Ehrlich, S., Albertini, G., ..., & Ogasawara, N. (2003). Essential Bacillus subtilis genes. *PNAS*, 100(8), 4678–4683.

Koonin, E. V. (2000). How Many Genes Can Make A Cell: The Minimal-Gene-Set Concept. *Annual Review of Genomics and Human Genetics*, 1, 99–116.

Koonin, E. V. (2007). The cosmological model of eternal inflation and the transition from chance to biological evolution in the history of life. *Biology Direct*, 2, 15–36. doi: 10.1186/1745-6150-2-15

Lennox, J. C. (2007). *God's Undertaker: Has Science Buried God?* Oxford: Lion Hudson Plc.

Meyer, S. C. (2013). *Darwin's Doubt: The Explosive Origin of Animal Life and the Case for Intelligent Design*. New York: HarperCollins Publishers.

Mushegian, A. R., & Koonin, E. V. (1996). A minimal gene set for cellular life derived by comparison of complete bacterial genomes. *Proceedings of the National Academy of Sciences of the United States of America*, 93, 10268–10273.

Nagel, T. (2012). *Mind and Cosmos: Why the Materialist Neo-Darwinian Conception of Nature is Almost Certainly False*. New York: Oxford University Press.

Sarfati, J. (2008). *By Design*. Powder Springs GA: Creation Book Publishers.

Siegel, E. (2022, January 17). How many planets are there in the Universe? Retrieved June 29, 2025 from bigthink.com/starts-with-a-bang/planets-universe/

Spetner, L. (1998). *Not by Chance! Shattering the Modern Theory of Evolution*. New York: The Judaica Press, Inc.

Wald, G. (1955). The Origin of Life. In D. Flanagan (ed.), *The Physics and Chemistry of Life*. A Scientific American Book. New York: Simon and Schuster.

11

The fossil record: contradicting Darwin's gradualism

In the previous chapter we showed the impossibility of life forming by random processes. Abiogenesis—foundational to a materialistic or atheistic worldview—is indistinguishable from a secular miracle. This means that theories of evolutionary descent take as their starting point an initial condition of dogma masquerading as science.

Darwin's celebrated position in origins science relates to his purported explanation of biological diversity. Starting with a single or several primordial life forms, all higher life forms, Darwin hypothesized, have gradually evolved, with natural selection being "the main but not the exclusive means of modification".[1] One of the hallmarks of Darwin's scheme was the gradual nature of these changes, a point which he emphasized in *Origin*:

> As natural selection acts solely by accumulating slight, successive, favourable variations, it can produce no great or sudden modifications; it can act only by short and slow steps. Hence, the canon of "Natura non facit saltum," which every fresh addition to our knowledge tends to confirm, is on this theory intelligible.[2]

The Latin phrase *natura non facit saltum*—nature does not make jumps—identified Darwin as a Huttonian, convinced of philosophical uniformitarianism, which is synonymous with naturalistic gradualism. Darwin believed that evolution did not occur in jumps with intervening periods of quiescence, but instead by a continuous succession of small, slow steps. If this were so, progressive change should generally be visible in lengthy sequences of fossil or animal specimens, but abrupt changes should scarcely, if ever, be present. We need to consider the fossil evidence from the perspective of the secular worldview, to determine if it is consistent with Darwin's and his successor's claims. For this reason, we will mostly quote secular scientists and scholars in this chapter, sometimes at length so that readers may engage fully with the source material.

1 Darwin, 1909, p. 519.
2 Ibid., p. 510.

In this chapter, we will show that the fossil record has *never* supported Darwinian gradualism. In fact, there are specific details in the fossil record that are highly problematic for Darwinism.

Pre-Darwinian investigations

At the close of the eighteenth century, Napoleon Bonaparte commissioned a French scientific expedition to Egypt, taking with him a large company of the most accomplished French naturalists. One of these men, Geoffroy Saint-Hilaire, discovered mummified animals in the tombs and temples of Upper and Lower Egypt.[3] Saint-Hilaire took these artefacts—"cats, ibises, birds of prey, dogs, monkeys, crocodiles and the head of an ox"[4] back to France, where the Secretary of the French Institute, Baron Cuvier, investigated them. Cuvier wanted to test the evolutionary hypothesis that animals are subject to gradual changes in their structure over time, and he now had in his possession a range of specimens that were estimated at three to four thousand years old. Cuvier undertook his investigation by

> comparing the skeletons and such other parts of the mummies as were in a fitting state of preservation, with the corresponding parts of the representatives of the same species now living in Egypt. He arrived at the conviction that no appreciable change had taken place in these animals in the course of this considerable lapse of time, and the justice of his conclusion is not disputed.[5]

Cuvier's investigations convinced him of the "the fixity, not only of species, but of varieties".[6] But his evolutionary opponents were nonplussed. In the first place, they argued, a four-thousand-year time span may have been insufficient for appreciable changes to manifest themselves. Secondly, there was no evidence that the geography and overall conditions of Egypt had changed to any significant extent over the 4000 years; consequently, the animals in question did not experience any great evolutionary impetus towards change.[7]

In fact, the length of time is not the decisive factor in evolution—what counts is the number of generations. Cats, for example, have a generation time as short as one to two years, so over a four-thousand-year period, up to 4,000 generations could have occurred. This number is considerably greater than that achieved with the modern-day staple of live experimentation, the fruit fly. With a generation time of ten to twelve days, experiments since the 1960s would have produced around 2,000 generations. Cuvier's mummified cats certainly

---

3 Cuvier, 1827, p. 111.
4 Ibid.
5 Huxley, 2013.
6 Osborn, 1894, p. 211.
7 Huxley, 2013.

represented a large number of generations for a vertebrate, and we note Cuvier's conclusion that no appreciable change had taken place. Baron Cuvier's evidence does not support Darwin's scheme of unrelenting, slow, gradual evolution adding up to large-scale changes, but neither does it decisively refute it. What is needed, therefore, from the secular perspective, is evidence on even longer timescales, and such evidence can only, for all practical purposes, come from the fossil record.

Change in the fossil record

When we turn to the evidence contained in the fossil record, Darwin's scheme of slow, gradual change receives minimal support. We will present a succession of quotations from experts and scientists that show that the fossil record has never definitively supported Darwin's description of evolution—even when interpreted according to the tenets of philosophical uniformitarian geology.

Our starting point is the pre-Darwinian era. Baron Cuvier, mentioned above, was a French naturalist whom Darwin himself listed in *Origin* as one of the "most eminent palaeontologists" of his time. In his *Essay on the Theory of the Earth* (1827), Cuvier touched on the notion of gradual evolutionary changes that had been championed by the likes of Erasmus Darwin, and dismissed it on the basis that no intermediate forms in the fossil record had ever been found:

> ...to those especially who believe in the possibility of indefinite alteration of forms in organised bodies; and who think that, during a succession of ages, and by repeated changes of habitudes, all the species might be changed into one another, or might result from a single species.
>
> Yet to these persons an answer may be given from their own system. If the species have changed by degrees, we ought to find traces of these gradual modifications. Thus, between the palæotheria and our present species, we should be able to discover some intermediate forms; and yet no such discovery has ever been made.[8]

Thirty-two years after these words were published, *Origin* was released to a triumphant reception. The intervening decades brought no change to the lack of intermediate forms that Cuvier had noted in 1827. Darwin was therefore forced to acknowledge the problem in *Origin*:

> ... the number of intermediate varieties, which have formerly existed on the earth, should be truly enormous. Why then is not every geological formation and every stratum full of such intermediate links? Geology assuredly does not reveal any such graduated organic chain; and this, perhaps

8 Cuvier, 1827, p. 103.

> is the most obvious and gravest objection that can be urged against my theory.[9]

Following the publication of *Origin* and its increasing acceptance by the scientific community, paleontologists began searching for fossil evidence in support of Darwin's origins theory. One such person was the American Othniel Marsh (1831–1899), president of the National Academy of Sciences. Marsh and his team were very successful in locating and retrieving fossils on the North American continent—a particularly celebrated discovery was evidence which purported to show the evolution of the horse. Marsh showed his equine collection to Thomas Huxley, who was traveling in the United States at the time. Huxley, in turn, delivered a series of lectures in New York in September 1876 which cited the evidence turned up by Marsh. The third of these lectures, titled *The Demonstrative Evidence of Evolution*, was based on equine fossils that Marsh had shown Huxley at Yale. Huxley told his audience:

> Professor Marsh's kindness has enabled me to put before you a diagram, every figure in which is an actual representation of some specimen which is to be seen at Yale at this present time (Fig. 9).[10]

The figure Huxley referred to, which we do not reproduce here, presented a selection of bones and teeth from the modern horse through five successive "evolutionary" antecedents. Huxley continued:

> The succession of forms which he [Marsh] has brought together carries us from the top to the bottom of the Tertiaries. Firstly, there is the true horse. Next we have the American Pliocene form of the horse (*Pliohippus*); in the conformation of its limbs it presents some very slight deviations from the ordinary horse, and the crowns of the grinding teeth are shorter. Then comes the *Protohippus*, which represents the European *Hipparion*, having one large digit and two small ones on each foot, and the general characters of the fore-arm and leg to which I have referred... Next, in the backward order in time, is the *Miohippus*, which corresponds pretty nearly with the *Anchitherium* of Europe. It presents three complete toes – one large median and two smaller lateral ones; and there is a rudiment of that digit, which answers to the little finger of the human hand.
>
> The European record of the pedigree of the horse stops here; in the American Tertiaries, on the contrary, the series of ancestral equine forms is continued into the Eocene formations. An older Miocene form, termed *Mesohippus*, has three toes in front, with a large splint-like rudiment representing the little finger; and three toes behind. The radius and ulna, the tibia and the fibula, are distinct, and the short crowned molar teeth are anchitherioid in pattern.

---

9 Darwin, 1860, p. 280.

10 Huxley, 2013.

Huxley concluded his review of the diagram with the following comment:

> But the most important discovery of all is the *Orohippus*, which comes from the Eocene formation, and is the oldest member of the equine series, as yet known. Here we find four complete toes on the front-limb, three toes on the hind-limb, a well-developed ulna, a well-developed fibula, and short-crowned grinders of simple pattern.
>
> Thus, thanks to these important researches, it has become evident that, so far as our present knowledge extends, the history of the horse-type is exactly and precisely that which could have been predicted from a knowledge of the principles of evolution.[11]

As far as we are aware, this supposed evolutionary history of the horse represented one of the high-water marks of fossil evidence for Darwinian evolution in mammals.[12] But unknown to Huxley and his audience, however, was future research which would undermine the conclusions Huxley presented in New York. We'll cite that below, but by the time of our next quote, 1886, the mood regarding fossil evidence for Darwinian evolutionary processes had again darkened.

The year 1886 was four years after Darwin's death, and 27 years after the publication of *Origin*. The challenge of finding fossil evidence in support of Darwin's origins theory had been handed down to a new generation of researchers. The American educator Herbert William Conn (1859–1917) published his *Evolution of Today*[13] at this point. Conn discussed the lack of fossil evidence for Darwinian evolution in a section of his book titled *Imperfection of the Record*:[14]

> It has been clearly recognized by all scientists, that it is absolutely impossible to reconcile paleontology with evolution unless it be assumed, to start with, that the history of animals which we get from fossils is extremely imperfect … .
>
> But no one ever pretended to claim that any such a series could be found. Paleontology at the time of the appearance of Darwin's book showed very few such connecting links; while today many more are known, they are really as nothing compared with the countless numbers which must have existed. We know enough of fossils today to cause us to abandon hope of ever making out such a complete history as the theory calls for. Perceiving the great lack of these intermediate stages, some naturalists have been led to modify the first conception of evolution. Many hold the position today that development in the past has not always been slow and gradual, but by

11 Ibid.

12 Creationists consider that these same fossils are "consistent with considerable variation within a kind, even to the extent of new species" (Statham, 2024).

13 Conn, 1886.

14 Ibid., pp. 90–94.

> alternate periods of rapid development and comparative rest. This saltatory evolution, as it is called, does not demand such a long series of intermediate links. It assumes that under certain unknown circumstances species are very rapidly modified, and that a few generations only are required to produce much change … .
>
> But even with this modified and more favorable form of evolution in mind, the record of paleontology cannot be regarded otherwise than as remarkably imperfect. Its imperfection is indeed a necessity for evolution … . Indeed recent years would indicate that this imperfection is even greater than assumed by Darwin.
>
> Having reached this conception of the imperfection of the record, the evolutionist seems to have assumed an unassailable position. For whatever connecting links chance to be found, they are immediately claimed as positive arguments, while the difficulties arising from absence of links are readily answered by appeal to the imperfection of the record … .

Conn's quote is significant for several reasons. First, he mentions the concept of "saltatory evolution". The unusual word *saltatory*, meaning progress by leaps rather than gradual steps,[15] was appropriate as the fossil record seemed to suggest that the slow, gradual changes predicted by Darwin did not occur; rather, the fossil record apparently comprised lengthy periods with no change, followed by sudden periods of dramatic changes, then a return to "equilibrium" for another extended period. This conception is better known to us today by Stephen Jay Gould's terminology, *punctuated equilibrium*. Whatever terminology is used, the fact that this idea was in circulation by 1886 shows that Darwin's theory, as published, was under considerable strain. The problem that Cuvier raised in the 1820s was not going away—it was being confirmed by an additional 60 years of fieldwork.

Not only does saltatory evolution describe the fossil record more accurately than Darwinian evolution, it represents an additional danger for Darwinian processes because it shortens the amount of time available for large scale evolutionary change (refer Figure 11.1). The graph on the left-hand side shows the Darwinian conception of gradual change. Over a long period, large evolutionary change occurs gradually and constantly, represented by the smooth and sustained uptrend of the graph. The graph on the right-hand side shows the saltatory evolution mentioned by Conn. As time progresses, no appreciable change is apparent for an extended period. Then, without warning, large scale evolutionary changes occur suddenly over a relatively short timeframe. (A well-known example of this effect is termed the "Cambrian explosion", which we discuss in more detail below.) The elapsed time during

15 Saltatory is derived from the Latin word saltum, mentioned above in Darwin's quote "natura non facit saltum"—nature does not make jumps.

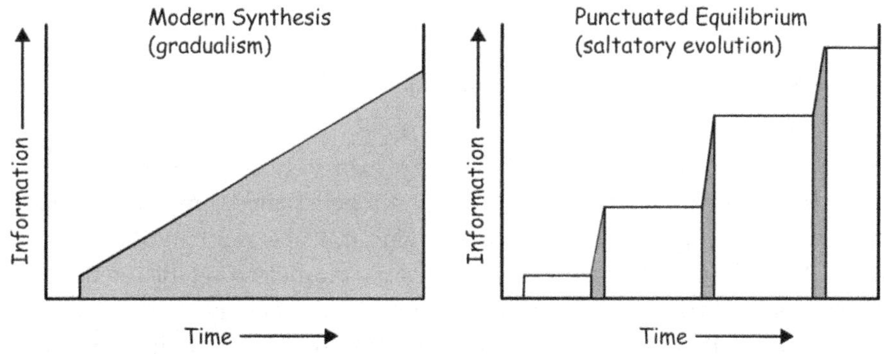

Figure 11.1: Contrasting graphs of active evolutionary change vs time

this supposed period of intense evolutionary activity is indicated on the graph; it is vastly less than in Darwin's gradual evolutionary scheme. In Conn's saltatory evolution scenario, can time really be the hero of the plot?

The second significant point Conn makes concerns the "unassailable position" of the Darwinist. If any transitional forms are found, trumpet that news from the rooftops. If links cannot be found, blame this on the extreme imperfection in the fossil record. In effect, Darwinian theory was moving in the direction of unfalsifiability by 1886. What Darwin had described as "the most obvious and gravest objection that can be urged against my theory" was now being swept under the carpet by his disciples.

For context, the lack of progress in evolutionary forensics contrasted sharply with other scientific and technological advances of the day. Emil Fischer discovered sugars, proteins and enzymes in 1884; Louis Pasteur developed and successfully trialed a rabies vaccine the following year; and in 1888 Heinrich Hertz discovered radio waves.

We now move forward to the 1930s. Two sources, relating to snakes and plants, highlight the lack of fossil evidence hampering Darwinian evolutionary studies of these species. First, fossil snakes of North America:[16]

> Such information as has accumulated concerning the extinct snakes of North America is the result of spasmodic and uncorrelated studies, a research that was made difficult by the extreme rarity and the usually fragmentary state of the available specimens. The scarcity of snake remains in the fossil record, aside from their small size, may perhaps be attributed to the fact that with comparatively few exceptions they are terrestrial in habit, and for that reason their skeletons were less apt to be preserved as fossils.

Then, plant biology:[17]

---

16 Gilmore, 1938, p. 1.

17 Godwin, 1939, p. 182.

> The steps of transition between the algae and the land flora are yet unknown, the fossil record having yielded no clear connecting links, though in one geological age (the Silurian), only algae have been found as fossil plants, and in the next (the Devonian), undoubted land plants have been identified.

Both above sources were published approximately 80 years after *Origin*, and yet the imperfection of the record was still being cited. Empirical science and technology were by this stage progressing in leaps and bounds. In 1935, Charles Richter established the well-known scale for measuring the intensity of earthquakes; in 1938, Hans Krebs had identified the energy cycle of life. In transport, scheduled airline services now linked London to far-flung areas of the British Empire. This seemed impressive at the time, but perhaps more significant were flights of prototype jet-powered aircraft in Germany (in 1939) and Britain (1941), which anticipated the jet age.

We now move forward to the 1960s, a full century after *Origin*. Errol White (1901–1985), then president of the Linnaean Society of London, remarked that he would be reluctant to try to prove organic evolution in a courtroom. Based on his work with fossil fishes, White noted that evolutionary trends are generally observable, but the specific connections between species are often missing or unreliable. He acknowledged that the mechanisms behind evolution remained unclear, and warned that blindly declaring "Darwin is God and I am his prophet" does nothing to advance scientific understanding.[18] White commented that a century after Darwin, "we still do not know the mechanics of evolution". He pointed out that the general trend in the fossil record was devoid of the specific evidence that Darwin said was necessary.

Meanwhile, technological advances in other fields continued apace. The motor car had not been developed in Darwin's day, but here, in the 1960s, men would at last land on the moon. The fossil record remained moribund for the Darwinists—providing an excellent example of the disconnect between forensic and empirical sciences.

We move forward to the 1970s, at which time Dr Colin Patterson was serving as senior paleontologist at the British Museum of Natural History. Patterson was asked why he did not include a photograph of a transitional fossil in his 1978 book *Evolution*.[19] Patterson confirmed that there was an absence of clear examples of evolutionary transitions in his book. He clarified that if he were aware of any incontrovertible examples—whether from the fossil record or living organisms—he would have certainly included them. Responding to the suggestion that an artist might illustrate these transitions, Patterson

---

18 Patterson, 1999, pp. 120–121.

19 "Creationist Luther Sunderland wrote to Dr Patterson inquiring why he had not shown one single photograph of a transitional fossil in his book. Patterson then wrote back with [this] amazing confession which was reproduced, in its entirety, in Sunderland's book *Darwin's Enigma*" (Bates, 2008).

questioned what the artist would base the illustration on, noting that he could not provide clear examples with any honesty. Relying on artistic imagination would risk misleading the reader.

Patterson further admitted that it was difficult to dispute the claim made by Stephen Jay Gould that transitional fossils were lacking. To the suggestion that Patterson at least "include a photo of the fossil each organism evolved from," he replied frankly—"I will lay it on the line—there is not one such fossil for which one could make a watertight argument".[20]

In the same decade, prominent American paleontologist Stephen Jay Gould of Harvard, (whom Patterson had cited) acknowledged:

> The extreme rarity of transitional forms in the fossil record persists as the trade secret of paleontology ... [Gradualism] was never "seen" in the rocks.[21]

This is an interesting admission. It strongly suggests that a part of the scientific establishment was knowingly running an ideologically based propaganda exercise, aimed at misleading the public.

By 1979, paleontologist David Raup of the Field Museum of Natural History noted the ongoing difficulties with the fossil record in general, and the problem with the Marsh/Huxley evidence for equine evolution in particular. In an article, Raup explained that more than a century had passed since Darwin's time, and in those 120 years, scientific understanding of the fossil record had increased enormously. (At the time of Raup's article, scientists had documented around 250,000 fossil species.) Yet despite this vast increase in data, the overall picture remained largely unchanged. According to Raup, evolutionary change in the fossil record still appeared unexpectedly abrupt and uneven. Raup noted that paradoxically, there were even fewer clear examples of gradual evolutionary transitions than were available in Darwin's era. Some of the textbook cases formerly cited as strong evidence of Darwinian evolution—like the gradual development of the horse in North America—had since been revised or even abandoned. With more detailed fossil data, what once appeared as a smooth, linear progression now seemed far more complex and discontinuous. This meant that the core difficulty Darwin acknowledged in explaining the fossil record (in 1859) remained unresolved in the 1970s. Although the fossil record does reveal signs of change, it often lacks the gradual patterns one would expect if natural selection were consistently shaping species over time.[22]

---

20 Bates, 2008.
21 Gould, 1977.
22 Raup, 1979, p. 25.

Raup had corrected the record—what Marsh and Huxley had presented in the 1870s as evidence for Darwinian evolution turned out to have been unwarranted.

With that, we move into the 1980s, when paleontologist Niles Eldredge, a colleague of Stephen Jay Gould, described his observations of the non-Darwinian signs in the fossil record. Earlier in his career, Eldredge had attempted to document clear examples of the slow, steady, directional change that had long been assumed to exist. However, he found that the evidence told a different story. Rather than showing continuous transformation, species typically appeared suddenly in the fossil record and then "tend[ed] not to change very much at all … often for millions of years". The fossils showed a persistent and unexpected resistance to evolutionary change.[23]

Eldredge further admitted:

> We paleontologists have said that the history of life supports [the story of gradual adaptive change] all the while really knowing that it does not.[24]

By the 1990s, with the turn of the millennium in view, the situation had still not changed. Fellow of the Royal Society and Professor of Evolutionary Palaeobiology at the University of Cambridge, Simon Conway Morris said:

> … biologists are much more impressed by the discreteness of organic form, and the general absence of intermediates.[25]

Robert Wesson, a political scientist who promoted a concept that chaos theory might help explain evolution, provided another frank assessment:

> It is not surprising that large evolutionary innovations are not well understood. None has ever been observed, and we have no idea whether any may be in progress. There is no good fossil record of any.[26]

An area of considerable interest is the supposed evolution of humans. One of the earliest discoveries in this regard was the remains of a Neandertal in 1856, but post-Darwin, hominid research accelerated in the early twentieth century. Several important sites for hominid research (the study of humans and their putative extinct relatives) were in Africa; the well-known "Lucy" specimen was discovered in 1974 in Ethiopia. A South African-based researcher, Lee Berger, shed light on the difficulties of hominid paleontology in his 2000 book *In the footsteps of Eve*.[27] In Berger's experience, research is thwarted by the fragmentary nature of hominid remains, the difficulties in interpretation of finds, and the frustrating possibility that future discoveries may overturn current beliefs.

---

23 Eldredge, 1996, p. 3.
24 Ibid., p. 144.
25 Conway Morris, 1998, p. 4.
26 Wesson, 1993, p. 206.
27 Berger, 2000.

Paleoanthropology is a contentious area of research in which scholars frequently disagree, and consensus is rare.

For example, Berger noted that secular scientists recognise at least 17 distinct hominid species that have lived in Africa over the past "five million years". However, not all of these species are considered by secular scientists as part of the direct human lineage, and no conclusive evidence yet identifies which one purportedly led to the emergence of the genus *Homo*. Frustratingly for Berger and his colleagues, each new fossil find tends to raise just as many questions as it resolves about early hominids and their purported evolutionary links. In fact, future discoveries may be just as likely to reveal previously unrecognised gaps in understanding as they are to fill existing ones—highlighting the dynamic and uncertain nature of reconstructions of early human history by secular scientists.[28]

Berger commented that there are often personal rivalries in the field of paleoanthropology because it is a highly interpretive domain of science. A single fossil find can define a researcher's career, with professional standing either cemented or shattered during the ensuing debates. In many cases, financial support tends to follow prominent individuals rather than academic organisations.[29]

Berger goes on to describe the features of hominid fossil remains, their discovery and re-assemblage. He states that finding an intact skull from "an early hominid is almost a myth". More often, researchers are left with fragments that must be meticulously reconstructed—and even a minor variation in how these pieces are aligned can dramatically alter interpretations. This is because brain size is used as a key indicator in assessing intelligence and human-like traits in "hominids [from] ... the distant past".

The discovery of any complete hominid bone is so uncommon that it is viewed as a remarkable event. For example, a single fossilised lower jaw—dated to four million years ago and found at Lothagam in Kenya—is all that remains of one of "our earliest ancestors". But who, asks Berger, "would be brave or foolish enough to try and name a genus or species on this paltry evidence"?[30]

While Berger and his colleagues struggled to make sense of the rare fragmentary remains they were discovering in the early 2000s, the Big Science era had begun. The multinational, multi-billion-dollar Human Genome project was well underway, and, in a technological example of international cooperation, the International Space Station had been launched a couple of years prior. The contrast with 140 years of painstaking paleontological effort

28 Ibid., pp. 29–30.
29 Ibid., p. 52.
30 Ibid., pp. 123–124.

was telling. Experts confessed that little to no progress had been made in this aspect of the secular origins enterprise.

This left Darwinism in a worse state than in 1859. At the time *Origin* was published, Darwin held out hope that future paleontological discoveries would overturn Cuvier's 1820s pronouncement and support his theory of gradual modification. The reality was the opposite. As the decades proceeded, Darwin's great hope of the 1850s had been progressively extinguished.

The "Cambrian explosion"

From the earliest days of Darwin's theory of descent, what is now often referred to as the "Cambrian explosion" has been problematic. In a relatively short period of time in the geological record (when interpreted according to uniformitarian science), an enormous diversity of relatively complex multicellular organisms appear to have sprung into existence without any clear precursors. How short a period? About ten million years, according to recent secular science, which is little more than 0.2% of the purported length of earth's history.[31] Scientists of Darwin's day were all too aware of this sudden increase in diversity, which forced Darwin to address the issue in *Origin*:

> Nevertheless, the difficulty of assigning any good reason for the absence of vast piles of strata rich in fossils beneath the Cambrian system is very great. ... The case at present must remain inexplicable; and may be truly urged as a valid argument against the views here entertained.[32]

If anything, the situation has deteriorated for theories of evolutionary descent since that time. Further discoveries of very high-quality Cambrian fossils in China around Chengjiang began about 1984. The data from Chengjiang confirmed that the Cambrian explosion occurred in a shorter period than previously thought.[33] Organisms with different body plans appeared, for example, shellfish, worms, jellyfish, starfish and finned fish. Some of these animals had complex eyes, such as the compound eyes of the extinct trilobites and *Anomalocaris*.[34] How did these evolve so quickly?

Darwin had suggested that large changes accumulate gradually over long time periods from a succession of small changes, but in the Cambrian explosion the exact opposite appears to occur. Initially there are enormous, "rapid" changes—entirely new body plans apparently arise—followed by relatively minor changes (such as speciation). The initial large changes resulted in what are called disparate body plans. Clearly, a starfish is entirely different in shape and structure to a worm or a finned fish. But these large differences in design

31 Meyer, 2013, p. 72.
32 Darwin, 1909, p. 361.
33 Meyer, 2013, p. 51.
34 Statham, 2018.

occurred, according to an evolutionary interpretation of the fossil evidence, much too quickly, and too early.

In the Cambrian explosion, representatives of the majority of the thirty-six or so different animal phyla have been identified, including the chordates—animals with a notochord, though some only during embryonic life.[35] As Darwin admitted, the Cambrian explosion is a valid argument against his conception of evolutionary descent, a concession with which we agree.

Diarthrognathus broomi

We now turn to an apparent transitional form discussed by the prominent American biologist Jerry A. Coyne. In his 2005 article "The case against intelligent design", Coyne cites a fossil described in 1958—*Diarthrognathus broomi*, a small, mammal-like reptile—as one of the finest examples of evolutionary transition ever discovered.[36]

Diarthrognathus literally means two (di) jointed (arthro) jaw (gnathus), and Coyne makes the point that the fossil specimen demonstrates that one part of its jaw joint is similar to that of a mammal and another similar to that found in reptiles. This is a transitional form, Coyne asserts, a part-way point between the jaw of a mammal and that of a reptile. The evolutionary narrative holds that a reptilian ancestor began to develop characteristics of a mammalian jaw joint on top of the reptilian joint, a demonstration of linkage in the progress from reptiles towards mammalian forms.

But, as we saw with MH370, there could be any number of alternative narratives that fit the forensic data. Darwinists like Coyne will interpret apparent transitional forms as evidence in favor of evolutionary descent because it fits their worldview. But alternative explanations should be considered. The heuristic of Occam's razor suggests that simpler explanations are preferable if they provide an adequate explanation. Also, empirically verified mechanisms are preferable to those which are unverified.

In the first place, Coyne seems to assume a direction of evolutionary descent from reptiles to mammals via *Diarthrognathus broomi*. But Darwinian evolution does not have a direction—it merely proceeds via natural selection towards forms more adapted for a particular environment. Who is to say that mammals were more adapted than reptiles in the particular (unknown) environment in which *Diarthrognathus broomi* lived? Could the direction of change have been in the opposite direction, from a mammal toward a reptile? Fossil remains cannot answer this question clearly because they only show a

---

35 This phylum incorporates the subphylum containing vertebrates, creatures which additionally have a backbone (which humans also have).

36 Coyne, 2005.

snapshot in time. Coyne appears to have merely assumed an evolutionary direction that affirms his worldview.

So what might a better interpretation be? There are several possibilities.

The first we will consider is *convergent evolution,* which is said to occur when species from different evolutionary branches develop similar traits or adaptations in response to similar environmental challenges or ecological niches, despite having a distant purported common ancestor. For example, bats have an ultrasonic echolocation system, as do dolphins and toothed whales.[37] But bats are not believed by secular scientists to be the precursors of dolphins, or vice versa. Instead, convergent evolution is invoked. The similarity arises not because these species inherited the trait from a common ancestor, but because they faced similar selective pressures that supposedly led to the independent evolution of similar solutions. So, bats evolved their echolocation system, and so did dolphins, and some birds. It was an independent evolution which happened to proceed in a similar direction.

According to evolutionists, convergence is found throughout the evolutionary tree, and there are thousands of examples.[38] The South American and African electric fish are both said to have formed their electrosensory systems independently. Of the aquatic animals which generate a high voltage to kill prey, the eels, rays, and catfish are said to have evolved the capacity separately.[39] Eyes are said to have evolved independently three or four times. A gene which controls eye development in insects and vertebrates, which is 94% similar between these organisms, is also suggested as an example of convergent evolution.[40] Other examples are the wings of bats, birds, and insects: these species supposedly developed wings separately for flight, adapting to similar ecological roles despite their very different evolutionary histories. With respect to the streamlined body shapes of sharks (fish) and dolphins (mammals), convergent evolution is invoked to explain the independent adaptation of drag-minimizing body shapes for efficient swimming.

Returning to the example of *Diarthrognathus broomi,* we posit that instead of labelling it a transitional form, Coyne (and his Darwinist colleagues) could have better interpreted it as an example of convergent evolution. This is a simpler explanation because convergent evolution is a single development from an existing form, whereas a transitional form involves development from an existing form, followed by further development to another form. The transitional explanation invokes a larger, and therefore less likely development process, instead of a more limited one. By Occam's razor, the convergent evolution explanation should be accepted until disproved. In fact, the

37 Spetner, 1998, p. 111.

38 Ibid.

39 Ibid.

40 Ibid., p. 112.

convergent evolution explanation potentially applies to any purported transitional form.

While convergent evolution offers an explanation for apparent transitional forms, there may be another empirically demonstrable alternative. Various animals and plants are known to develop anatomical changes stimulated by their environment. This is referred to as phenotypic plasticity, a phenomenon that has been observed in plants and animals for more than a century.[41] A well-known example is the Bichir fish (genus Polypterus), an amphibious fish with functional lungs. When the fish is raised exclusively on land, it holds its head further off the ground and its forelimbs closer to the midline, resulting in changes to musculature and bone structure. Significantly, if one of these fish were raised in an aquatic environment, another in a terrestrial environment, and both were fossilized, the pair would exhibit apparent transitional changes. However, this appearance of transition has nothing to do with evolution, because it happens immediately, and without genetic alteration.

The plasticity phenomenon is not rare. Certain snails grow a thicker, more protective shell in the presence of crabs.[42] Dung beetles and spadefoot toads are other examples.[43]

How many fossils with the appearance of transitional forms are in fact reflective of phenotypic plasticity? We do not know yet, but it makes more sense to invoke this empirically verified phenomenon than the transitional forms of evolutionary descent, which as we have seen earlier, are regarded by many secular scientists as doubtful, and certainly not proven.

---

41 Ibid., p. 199.

42 Ibid., p. 200.

43 Buranyi, 2022.

A final possible explanation is that *Diarthrognathus broomi* and similar apparent transitional forms represent cases of genetic mutation leading to anatomical anomalies. In humans, several types of these variations are well known: syndactyly, where fingers or toes are fused together; Marfan syndrome, where individuals are tall and thin and may have anatomical abnormalities affecting the heart; Down syndrome, which results in characteristic facial features and congenital anomalies affecting the heart and skeleton; and cleft lip, cleft palate or clubfoot, which have a genetic basis. Some apparent transitional forms, therefore, could be due to genetic or congenital malformations. These types of abnormality could well be associated with reduced survival or reproductive success (as they may be in humans) which would also explain the rarity of apparent transitional forms. The congenitally or genetically malformed will not reproduce effectively if they die off quickly.

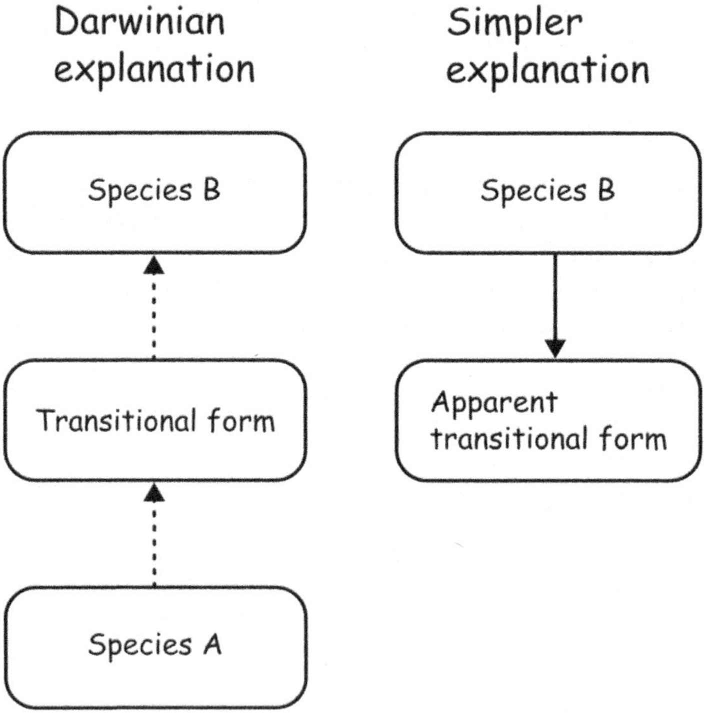

Figure 11.2: Alternate explanations for apparent transitional forms

Coyne's explanation for *Diarthrognathus broomi* is shown (Figure 11.2) at left involving multiple upward evolutionary steps. Phenotypic plasticity, convergent evolution, devolution, or congenital malformation is illustrated at right. When explaining circumstantial evidence, a range of options should be considered. It seems that Coyne has discounted explanations incompatible with his worldview.

Summing up the fossil evidence

We conclude our overview of the fossil evidence by returning to Darwin himself. In Chapter 9 of *Origin*, Darwin stated:

> The several difficulties here discussed, namely—that, though we find in our geological formations many links between the species which now exist and which formerly existed, we do not find infinitely numerous fine transitional forms closely joining them all together;—the sudden manner in which several groups of species first appear in our European formations;—the almost entire absence, as at present known, of formations rich in fossils beneath the Cambrian strata,—are all undoubtedly of the most serious nature. We see this in the fact that the most eminent palaeontologists, namely, Cuvier, Agassiz, Barrande, Pictet, Falconer, E. Forbes, &c., and all our greatest geologists, as Lyell, Murchison, Sedgwick, &c., have unanimously, often vehemently, maintained the immutability of species.[44]

The top paleontologists of Darwin's day were right: Darwinian evolution did not occur. Darwinians seek transitional forms bridging "lower" to "higher" life forms, but because natural selection and the other causal mechanisms of change have not operated in the manner Darwin predicted, they do not find them. If the evidence supports any theory of speciation in the secular scientific paradigm, it must be of the rapid, saltatory, or punctuated equilibrium type rather than Darwin's gradual version of evolution.

This is why the 160-year history of fossil finds has not materially altered the picture since Darwin published. The fossil record does not support gradualistic evolution over vast timescales because this hypothetical process did not occur. What has changed is that Darwin's initial caution against "this most obvious objection" is either ignored or explained away in a demonstration of Conn's "unassailable position"—and perhaps more importantly, Lewontin's "patent absurdities".

Scientists Sir Fred Hoyle and Chandra Wickramasinghe were among those who felt professional pressure from the Lewontin types. Hoyle and Wickramasinghe noted that they had received subtle warning (and at times explicit cautions) from colleagues that their anti-Darwinian stance was largely unacceptable within the scientific community. They, in turn, were concerned that little value was placed on empirical evidence, and how much sway was held by long-standing biases.[45]

They observed that many popular accounts continued to claim that the Darwinian theory of evolution is strongly supported by the fossil record. Yet, the more technically rigorous the writing, the "weaker the claims that are

---

44 Darwin, 1909, p. 363.

45 Hoyle & Wickramasinghe, 1981, pp. 147–148.

made". Even in the best academic texts, shortcomings in the fossil record were typically attributed to the record's incompleteness. However, Hoyle and Wickramasinghe argued that a deeper examination of the geological literature reveals a different picture. From a Darwinian perspective, the fossil record is profoundly problematic—not because geologists have failed in their work, but because the gradual evolutionary sequences demanded by the theory simply do not appear to exist:

> Although palaeontologists have recognised this truth for a century or more, they have not been able, in spite of their status as the acknowledged experts in the field, to make much of an impression on the consensus opinion.[46]

We draw particular attention to the final sentence of this quote. Readers will recall (from Chapter 7) that Darwin first marketed his theory by seeking men of influence who could "drag after them those naturalists who have too firmly fixed in their heads that a species is an entity". In other words, Darwin wanted real scientific discussion, of the sort that had been published by Cuvier, to be replaced by a softly coercive process in favor of his theory. And now, more than a century later, Hoyle and Wickramasinghe tell us that the same kind of coercive process is still in place. Paleontologists, the "acknowledged experts in the field", who recognized the insufficiency of the fossil record to support Darwinism were effectively sidelined; they were unable to shift the discourse towards a more scientifically sound basis. The widely held belief that Darwin's theory was based on scientific principles is a delusion. The reality is that Darwinism as a holistic origins theory was established (and is maintained) by ideologues in the face of more cautious and thoughtful scientific opinion.

Conclusion and outlook

Darwin was wrong about slow, gradual evolution. Even according to more recent secular science based on a philosophical uniformitarian interpretation of the fossil record, changes in animals such as speciation must have occurred relatively quickly. Creationists have come to the same conclusion, but not because of the fossil record, which we regard as largely a snapshot from Noah's flood. Before Darwin argued against the fixity of species, Athanasius Kircher (1602–1680) and John Wilkins (1614–1672) had theorized that transmutations of species (which today would generally be referred to as speciation) had occurred after the Genesis flood. They argued that this must have occurred rapidly, a view held by many creationists today.

But with respect to Darwin's theory of evolutionary descent as an explanation for the origin of all species, we can say that the fossil record is

46 Ibid.

largely irrelevant. At most, the fossil record demonstrates that living organisms have changed over time. It does not reveal how those changes occurred, nor does it provide direct evidence that "later life forms" actually descended from "earlier" ones. Claiming that such descent took place is a conclusion based on interpretation—it relies on the theoretical framework of evolution. Therefore, it is not valid to argue that the fossils support the theory unless one first assumes the theory is already true, which is begging the question (if not circular reasoning).[47]

In our view, this is exactly the trap that Darwin, and more recently, Jerry Coyne, fell into. In their minds the fossil record proves evolution because they are already convinced of the theory of evolutionary descent from a single primordial organism. But the fossil record amounts only to circumstantial evidence. It in no way proves evolution has occurred and indeed, cannot do so. As Patterson admitted, "Gould and the American Museum people are hard to contradict when they say there are no transitional fossils".[48]

With that, we take our leave of the fossil evidence and turn to the molecular biology of the evolutionary mechanism. It turns out that *natural selection*, almost an article of evolutionary faith, is in fact only part of a far larger picture—a telling point whose significance is largely, if not completely, glossed over in grade-school level biology texts.

References

Bates, G. (2008, February 18). That quote!—about the missing transitional fossils. Retrieved April 24, 2024, from creation.com/that-quote-about-the-missing-transitional-fossils

Berger, L. R. (2000). *In the footsteps of Eve*. Washington D.C. : Adventure Press, National Geographic.

Buranyi, S. (2022, June 28). Do we need a new theory of evolution? Retrieved July 6, 2023, from theguardian.com/science/2022/jun/28/do-we-need-a-new-theory-of-evolution

Conn, H. W. (1886). *Evolution of today*. New York: G. B. Putnam's Sons.

Conway Morris, S. (1998). *The Crucible of Creation: The Burgess Shale and the Rise of Animals*. Oxford: Oxford University Press.

Coyne, J. A. (2005, August 31). The Case Against Intelligent Design. Retrieved April 28, 2024, from edge.org/conversation/jerry_a_coyne-the-case-against-intelligent-design

Cuvier, G. (1827). *Essay on the theory of the earth*. Edinburgh: William Blackwood.

Darwin, C. (1860). *On the Origin of Species by Means of Natural Selection, or the Preservation of Favoured Races in the Struggle for Life*. London: John Murray.

Darwin, C. (1909). *The Origin of Species*. New York: P.F. Collier & Son.

Eldredge, N. (1996). *Reinventing Darwin: The Great Evolutionary Debate*. London: Phoenix, Orion Books Ltd.

Gilmore, C. W. (1938). *Fossil snakes of North America*. Geological Society of America.

Godwin, H. (1939). *Plant biology*. Cambridge: Cambridge University Press.

Gould, S. (1977). Evolution's Erratic Pace. *Natural History*, 86(5), 12–16.

Hoyle, F., & Wickramasinghe, C. (1981). *Evolution from Space: A Theory of Cosmic Creationism*. New York: Simon and Schuster, Inc.

Huxley, T. H. (2013, January 20). Lectures on Evolution. In *T. H. Huxley, Science and Hebrew Tradition*. Project Gutenberg. Retrieved November 16, 2024, from gutenberg.org/files/2629/2629-h/2629-h.htm

47 Spetner, 1998, p. 77.

48 Bates, 2008.

Meyer, S. C. (2013). *Darwin's Doubt: The Explosive Origin of Animal Life and the Case for Intelligent Design*. New York: HarperCollins.

Osborn, H. F. (1894). *From the Greeks to Darwin: An Outline of the Development of the Evolution Idea*. New York: Macmillan and Co.

Patterson, C. (1999). *Evolution* (Second edn). New York: Comstock Publishing Associates.

Raup, D. M. (1979). Conflicts between Darwin and Paleontology. *Field Museum of Natural History Bulletin*, 50(1), pp. 22–29. Retrieved April 27, 2024 from biodiversitylibrary.org/partpdf/376173

Spetner, L. M. (1999). *Not By Chance! Shattering the Modern Theory of Evolution*. New York: Judaica Press, Inc.

Statham, D. (2018, November 21). The Cambrian explosion—The fossils point to creation, not evolution. Retrieved from creation.com/cambrian-explosion

Wesson, R. (1993). *Beyond Natural Selection*. London: The MIT Press.

12
Mechanisms of variation: the consensus begins to fragment

We saw in Part 1 that Darwin's theory was like the Cutty Sark of origins theories. It took the ancient Greek ideas and brought them right up to modernity, but, crucially, did not escape from the original paradigm. It was still tied to ancient Greek thinking. Nevertheless, from the vantage point of 1859, Darwin seemed to have proposed a mechanism for evolution, natural selection, that ostensibly did away with the need for a creator. Variation buttressed by natural selection could do it all.

But that was 1859. The Cutty Sark is no longer state of the art—what about Darwinism? Was it updated in the same way that shipping moved ahead into new paradigms? Was there an equivalent of a steamship, or even more relevant, a modern ocean liner?

The particulars have changed, but the overall scheme has remained as it was. The secular scheme is still a two-step process comprising the emergence of life followed by evolution. Yes, Darwinian evolution has since been updated, but these updates are akin to moving from the Cutty Sark to a maxi yacht or a foiling monohull.

Ironically, these updates ended up causing even larger problems for holistic evolutionary origins theories. That is not something modern science textbooks would concede, but we'll investigate some of these problems in this chapter.

The slump

Darwin's *Origin* provided evidence that varieties could develop and suggested that speciation occurs. His proposed mechanism was natural selection working on naturally occurring variations.

When extrapolating his causal model science to a holistic account explaining all the diversity of life on earth, the force of Darwin's argument rested almost entirely on the sufficiency of his proposed mechanisms. The fossil record provided no specific support, but instead provided various strong evidence against holistic Darwinism, which Darwin himself admitted (as we saw in the previous chapter).

Darwin's theory involved heredity, but the science of genetics would not begin for several decades after *Origin* was published,[1] so his theory was incomplete. Darwin sought to fill this gap by proposing a theory of inheritance called "pangenesis", which he described in the 27th chapter of his 1868 book *The Variation of Animals and Plants Under Domestication*.

The core idea of pangenesis was that the egg and sperm received tiny migratory particles from all the different parts of the body. This theory was Lamarckian in that it implied that characteristics acquired during life were inherited. It also implied blending inheritance, the idea that the characteristics of offspring were the average of their parents. This was difficult to accept because—to cite the most obvious example—children are typically either male or female, not an average in-between sex. In an 1867 review of *Origin*, Scottish engineer Fleeming Jenkin showed that blending inheritance would quickly render natural selection inoperative.[2] Jenkin's argument caused many of Darwin's contemporaries to doubt the idea that natural selection could drive evolution,[3] even though Darwin's idea of common evolutionary descent of all species from a primordial single-celled organism remained widely accepted.[4]

In fact, the episode with Jenkin was symptomatic of a broader trend: following initial enthusiasm for Darwin's theory, the theory was increasingly viewed as overstated. Indeed, the period 1880–1920 has been referred to as "the eclipse", a time when Darwin's causal theory of natural selection went into decline. For example, the German naturalist Eberhard Dennert, writing in 1904, stated that "Darwinism will soon be a thing of the past, a matter of history; that we even now stand at its death-bed".[5] In Dennert's view, the holistic evolutionary theory of common descent was "almost universally recognized today by naturalists as a working hypothesis", while the bulk of naturalists rejected the validity of natural selection.[6]

If Dennert was right, then Darwin and his "bodyguard" had been successful in delivering a worldview shift. However, Darwinism as causal model science had become as obsolete as the Cutty Sark. Once considered effective, its mechanism no longer seemed to fit the facts; it was fast becoming a historical oddity. But after the rediscovery of Mendel's papers on the laws of inheritance around 1900, followed by further advances in genetics and statistics through the 1930s and early 1940s, Darwin was about to be reinvented by

---

1 Gregor Mendel, an Augustinian monk, published his important experimental findings relating to pea plants in 1866 ("Versuche uber Pflanzen-Hybriden" or "Experiments in plant hybridization"), less than a decade after Darwin's *Origin*, but the scientific establishment did not recognise the significance of Mendel's work for several decades.

2 Dawkins, 2003.

3 Scott & Branch, 2009.

4 Ibid.

5 Dennert, 1904, Introduction.

6 Ibid, p. 146.

population geneticists, and make a striking return. (We continue with our shipping analogy to correlate scientific and technological advances.)

From tea clipper to maxi yacht: Neo-Darwinism

By the early 1940s a new Darwinian scheme had emerged, this time incorporating concepts from a variety of disciplines, everything from Mendelian inheritance to statistical modelling. Biologist Julian Huxley's book, *Evolution: The Modern Synthesis,* presented a re-imagined Darwinism which synthesized Darwin's original ideas with population genetics. This latter field built on the work of the English statistician and biologist Ronald Fisher. The correct technical term for this theory is Modern Synthesis, but the older term, Neo-Darwinism, is also commonly applied.[7]

In addition to Fisher, the geneticists J.B.S. Haldane and Sewell Wright were key architects of statistical genetics, which was one of the foundations of Neo-Darwinism.[8] These scientists expressed some of their arguments with such brevity that it was difficult for biologists to be sure what they meant. The Neo-Darwinians also used complex mathematics, which some biologists found daunting. Other parts of their literature lacked clarity. The whole theory was "exceptionally difficult to follow ... [striking] a chill in almost any biologist who came across it".[9] In the early days, rank-and-file biologists struggled to understand the mathematical foundations of Neo-Darwinism, but accepted the general principles.

Neo-Darwinism was promoted as a unifying theory across the entire field of biology, and prominent biologists such as Theodosius Dobzhansky and Ernst Mayr emerged as champions. The theory would go on to incorporate new ideas and theories as they surfaced, often relegating contrary ideas to the status of true but unimportant.[10] Let's review several concepts of Neo-Darwinism.

The gene pool

Neo-Darwinism applied ideas from population statistics and introduced the abstract concept of the "gene pool". Instead of counting, for example, humans and their subpopulations, biologists theorized about populations of genes and how these changed across time. The gene pool represented a moving census of all the genes (including their subvarieties) in a particular population of a given species.

7 In fact, Neo-Darwinism it is a more versatile term, as proponents can be denoted Neo-Darwinians or Neo-Darwinists, avoiding the clumsier, and potentially confusing, "Modern Synthesizers", or the lengthier "proponents of the Modern Synthesis".

8 Sanford, 2014, p. 58.

9 Gale, 1990, pp. 1–2.

10 Stoltzfus, 2017, pp. 4–6.

A beneficial gene variant was expected to increase in frequency in the gene pool, whereas a harmful gene variant was expected to be eliminated. Natural selection acting on random mutations was believed to be the main mechanism of change.

Where the population of organisms was large, such as in plants or insects, the gene pool was correspondingly large and diverse. The reverse was also true; a small population of organisms would comprise a smaller, more restricted gene pool.

In the following sections, we will discuss causal mechanisms of genetic variation and change. These are accepted by both creationists and evolutionists.

Natural selection

Natural selection refers to a reproductive advantage (positive "reproductive fitness") of a subpopulation in a particular environment which causes that subpopulation to expand its representation in the gene pool. Reproductive fitness is improved by genetic advantages which enhance brood size and survivability. Conversely, genetic disadvantages such as sterility, genetic disease or shortened reproductive lifespan in a particular environment could cause negative reproductive fitness. This leads to reduced representation of that variety in the gene pool.

Natural selection is the only known mechanism which can cause adaptive, directional change. By way of example, consider bacteria in a petri dish. Antibiotics are added to the dish, causing most bacteria to die, thereby removing their genes from the pool. A small number of bacteria may be naturally resistant to the effect of the antibiotic and go on to increase in number. Selection has caused the gene pool (and the bacteria) to become more adapted to an environment containing antibiotics. This leaves the resistant bacteria as demographic winners while the antibiotic remains present. Resistant bacteria are often referred to as "superbugs", but all this means is that they are resistant to antibiotics. In fact, they are often less vigorous than the original bacterial population. In some cases, antibiotic selected varieties are relatively information-deficient, due to gene inactivation via mutation.[11] Specialisation is often achieved at the cost of robustness.[12]

A key point to note is that selection (whether natural or artificial) *reduces* information in the gene pool. Those genes that are selected remain in the gene pool, whereas those that are not selected could become rarer or disappear altogether. By itself, natural selection cannot increase information in the gene pool.

---

11 Wieland, 1994.

12 A similar effect is noted in dogs—mongrels can be more robust than some pure-bred types.

Genetic drift: random "selection"

Darwin knew that natural selection, and a subtype he termed sexual selection, were not the only mechanisms operating in nature. However, being unaware of genes, he could not accurately describe these other mechanisms. Genetic drift, which essentially refers to random, or non-selective removal of genes from the gene pool, was the name ascribed by Sewell Wright to this phenomenon in the 1930s, just prior to the formulation of Neo-Darwinism.

To understand genetic drift, we'll cover an example. Consider the case of

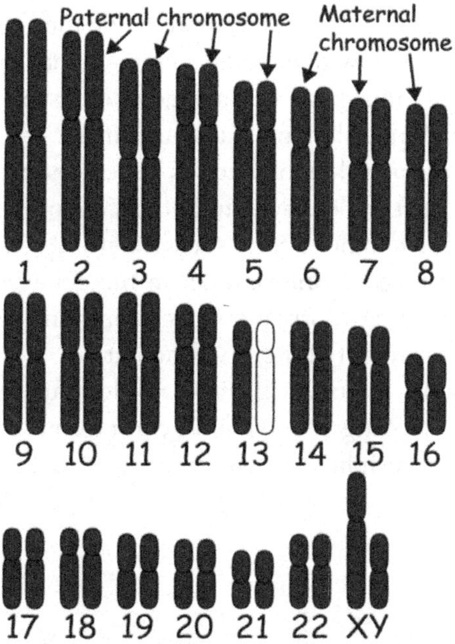

Figure 12.1: Schematic diagram of human male genome comprising 23 pairs of chromosomes

two humans producing offspring. One of the humans has one chromosome[13] with a new beneficial mutation. In the figure 12.1, the sex chromosomes (XY) are shown bottom right. The right chromosome of each pair is paternal (as indicated at top); the left chromosome is maternal. In this example, a random beneficial mutation happens to have occurred in paternal chromosome 13 (indicated in white).

---

13 In organisms that reproduce sexually, each piece of DNA is compacted into a tightly coiled structure called a chromosome. Humans have 46 pieces of DNA in most cells (46 chromosomes). The 46 chromosomes can be sorted into 23 different types, so that each cell has a pair of each type of chromosome. One pair, composed of either two X chromosomes or an X and a Y chromosome, determines the biological sex of a person. Each of the remaining 22 pairs contains different information and scientists number them 1 through 22.

If there is only one offspring, the chance that it will not inherit the beneficial mutation is 50%. For two offspring, the chance that neither will inherit the mutation drops to 25%; in other words, the chance of the beneficial mutation surviving has increased. If there were three offspring, the chance of the mutation not being passed on decreases to 12.5%; for four offspring, this reduces to 6.25%. In short, the more offspring, the greater the chances that the beneficial mutation is inherited by one or more members of the next generation. But it's also important to note that as the number of offspring increases, the power of random effects to remove that chromosome diminish. For instance, the difference between one offspring and two was 25%, but between three and four, it has decreased to 6.25%. Clearly, though, the random effects never fully disappear, irrespective of how many offspring there may be.

With genetic drift, this kind of consideration is applied to the whole population. For small populations, random effects can remove beneficial mutations. In very large populations, such as insects and plants, random effects are less pronounced, that is, beneficial mutations are more likely to survive to the next generation due to other effects such as natural selection.

As genetic drift proceeds, information in the gene pool is reduced, but, unlike natural selection, this information loss is random; it is determined by the effects of chance, independent of the environment.

We need to consider a more realistic example to understand genetic drift more fully. Instead of looking at one pair of chromosomes, we'll now consider human reproduction, which involves 46 chromosomes.

When a sex cell (egg or sperm) forms in a human, it contains only 23 chromosomes, not 46. It has only received one chromosome from each of the 23

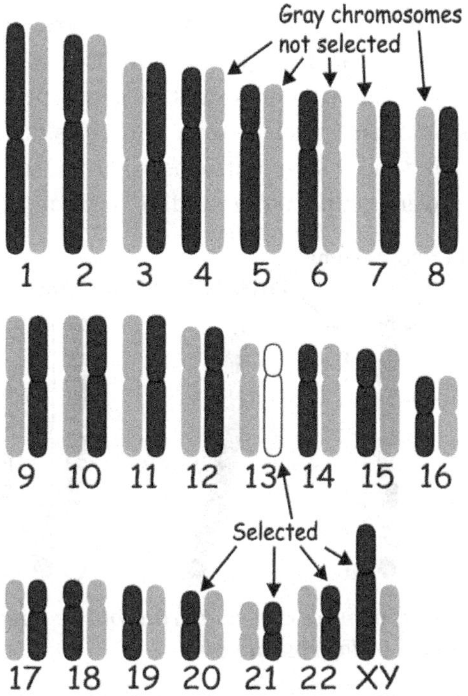

Figure 12.2: Random assortment removes half of the genetic material of a particular sex cell

pairs of chromosomes from the father or mother. Exactly which chromosome from each pair ends up in the sex cell is random, due to the *independent assortment of chromosomes*. As shown in Figure 12.2, half the chromosomes (colored gray) are randomly removed when the sex cell forms. In this example, the beneficial mutation in chromosome 13 happens to be retained by chance.

Either the left or right half of each chromosome pair may be selected for each sex cell. Somewhat like flipping a coin, this is a random process. When one of each of these sex cells combine to make a new human being, the particular 46 chromosomes in each of that human's cells represent one collection out of a set of about 70 trillion different possibilities that he or she could have received from the parents.

Children do not inherit more than half of their father's or mother's genes. For a family of two to four children, it is fairly certain that some of the parent's genes will not be transmitted to their children. In a large community (say the size of a city), that will not likely lead to the loss of any genes, because those genes not transmitted in one family are statistically likely to be transmitted via a different family. But in a small population (say, an isolated town), loss of genes through independent assortment of chromosomes is especially likely. If animals or humans go through a genetic bottleneck (a sharp but temporary drop in population, for example, due to a famine), there will be a significant loss of genes from the gene pool at that time.

Whereas independent assortment results in genetic drift at the level of whole chromosomes, another effect, known as "crossing over", effectively results in relatively random assortment *within* chromosomes, as shown in Figure 12.3. Crossing over occurs before random assortment takes place. The

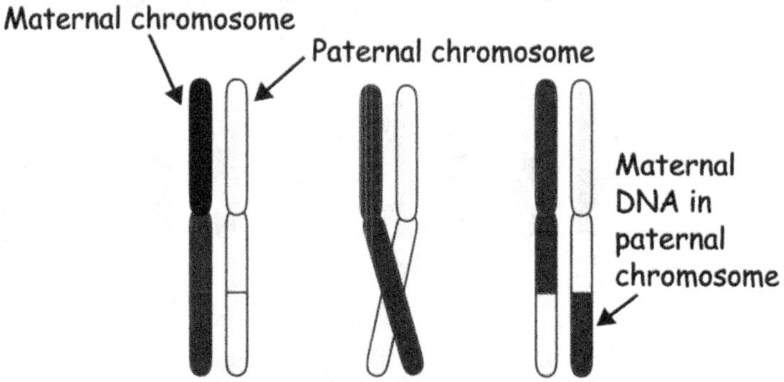

Figure 12.3: Pairs of chromosomes can swap random segments of maternal and paternal DNA via crossing over

initial arrangement (shown at left) begins to change when swapping has commenced (center). This leads to the final arrangement (right) which has some maternal DNA missing from the maternal chromosome as it was swapped into the paternal chromosome (and vice versa).

Thus, when a sex cell forms, each pair of chromosomes can exchange equivalent subsections of DNA between themselves—potentially containing one or more genes—prior to a single chromosome from that pair ending up in the sex cell. Like independent assortment, crossing over appears to be relatively random. A gene that is swapped out by crossing over may be lost to the gene pool (if it is a rare gene variant), even if the chromosome it originated in ends up in a sex cell.

Gene flow

The next causal mechanism affecting the gene pool is gene flow, which occurs when an organism from one isolated population migrates to, and reproduces in another population. An example might be genetically modified pollen blowing into a neighboring field and fertilizing some plants there. Genetic information has been transferred between populations, but no net increase of information in the combined gene pool has occurred.

Summary thus far

None of the above causal mechanisms add information to the gene pool. In the case of natural selection, we saw that it removed information, even though it was directed toward improved adaptation. In the non-directed mechanisms—genetic drift resulting from independent assortment of chromosomes and crossing over—there is also loss of information. The final mechanism we considered, gene flow, does not create new information but shifts existing information between local gene pools. This does not increase information when viewed at the level of a global gene pool.

Despite none of these mechanisms adding information to the gene pool, they are sufficient to explain the "Darwinian" processes which were proposed in *Origin*, such as variation, adaptation, and speciation. We will now turn our attention to the process Darwin emphasized in the title of *Origin*: speciation.

Speciation

Variation, adaptation and speciation are possible using only the mechanisms we have described—natural selection, genetic drift and gene flow. The typical requirements under which speciation occurs are a combination of isolated breeding population, altered environmental conditions, and time. This means that speciation does not require an increase in genetic information. Loss of information from the gene pool by natural selection could lead to the development of new species. Similarly, loss of information from other causal mechanisms, such as genetic drift, or a population bottleneck, could lead to speciation.

Even reassortment of existing genetic information has conceivably led to speciation in the past, particularly in subpopulations which have become isolated geographically or functionally. American evolutionary biologist Richard Lewontin noted that novel genes are not required for "speciation and divergence of full species" because "of the immense array of genetic variation that exists in populations of sexually reproducing organisms".[14 15]

An explanation for novelty

But these mechanisms are insufficient by themselves for a holistic theory of evolution. They explain speciation purely by rearranging or deleting existing genetic information, and therefore require a starting point of large amounts of pre-existing information to work. For a holistic theory, these mechanisms are inadequate because such a theory needs to explain the origin of all of the

14 Lewontin, 1974, p. 186.

15 See "Birds of a feather don't breed together" (Wieland, 2014) for further information.

information from a starting point of near zero information in the putative ancestral primordial organism. Richard Dawkins has made this point forcefully:

> The feature of living matter that most demands explanation is that it is almost unimaginably complicated in directions that convey a powerful illusion of deliberate design.[16]

There must be a mechanism, or mechanisms, to produce this "unimaginable complexity". Such mechanisms must increase the information content of the gene pool in a way that explains new body plans, new organs, new structures, and new biochemical functions, all of which must have arisen in the course of the supposed evolutionary scenario. It is not possible to progress from a simple single-celled organism to *Homo sapiens* by repeatedly deleting or rearranging information in the gene pool.

Biologists working on Neo-Darwinism needed a naturalistic source of new information. The source of novelty they chose, somewhat surprisingly, is mutations acting on the DNA of the sex cells (for sexually reproducing organisms). We regard this as surprising because mutations are essentially random errors in DNA. These errors creep in either while DNA is being copied, or due to random DNA damage, and for this to be heritable, it must occur in a sex cell. Mutations can either randomly delete information, duplicate existing information, move information, or change something in-place.

Point mutations

We begin our consideration of mutations by looking at so-called point mutations, mutations affecting a single "character" in DNA. These are the smallest mutations that can occur. (Refer to the description of DNA in Chapter 10 for background to this section, and to Table C.1 in Appendix C for a list of codons.)

One type of point mutation is analogous to changing an alphabetic character in a page of text. For example, the three-character codon TAT codes for the amino acid tyrosine. If a mutation changes the last T to C giving TAC, the mutation is likely to be silent, because the new codon happens to be the other synonym of tyrosine. In a similar way, the middle character might mutate from A to C, in which case the codon then codes for serine instead of tyrosine. This could change the function of the protein if it is in a sensitive location. It is likely that inducing random changes into a working gene would tend to garble it, rather than improve it. Nevertheless, a minority of random changes may improve the function of a structural protein or enzyme. Another possibility is for the last character of the tyrosine codon in the DNA to change from T to A or

16 Dawkins, 2004, p. 92.

G, altering the codon to a stop codon. In this case, translation of the DNA into a protein is likely to be truncated at the point of the mutated codon since it now terminates at that point. This is likely to render the protein non-functional because it is shortened.

The remaining types of point mutation are also very likely to be harmful. These are *insertions* and *deletions*, when an extra character is added between two adjacent characters, or where one is deleted. Both these types of mutation lead to a *frameshift* in the codons after the mutation. A brief example will serve to explain this point. Suppose there are four sequential codons **TCG | GAT | GTA** | ATA. (For ease of reading, we've added vertical lines between the codons, but in real DNA no such separator exists). We are mainly interested in the first three codons, comprising nine characters, shown in bold (which code for serine, aspartate then valine, although the exact amino acid names are not important). Suppose the second character in the string, C, is deleted. This has the effect of shifting all the other characters after that point one place to the left. This frameshift results in **TGG | ATG | TAA** | TA (we can ignore the final two characters for the purposes of this example). In this case, the first codon containing the deletion changes to tryptophan, while the second and third codons change to methionine and a stop codon respectively. This single deletion has completely altered the message to the right of the deletion. Translation of the protein will terminate at the stop codon, shortening it and probably rendering it non-functional. The other two amino-acid changes before the stop codon may also have rendered the protein less functional or even completely non-functional. Hence, a point *deletion* is likely to be harmful.

An *insertion* also affects the reading frame, so once again the downstream effect is a change in the sequence of amino acids, and the position of the stop codon is likely to change. The stop codon might occur earlier or later in the coding chain, fundamentally changing not just the sequence of amino acids, but also the protein length. These are likely to reduce or destroy the function of the protein.

What this means is that point mutations are more likely to lead to gene damage than to significant information increases. Insertions and deletions are likely to be harmful and might even lead to the death of the organism if they occur in an important protein or enzyme. Most of the substitutional point mutations, where one character is changed from one to another, are also likely to be harmful, at least to some degree. Some are very harmful. The devastating disease sickle cell anemia involves only one character changed in the protein hemoglobin.

Point mutations, by themselves, are hardly an adequate explanation for the emergence of entirely new genes. What other types of mutation could the Neo-Darwinists invoke?

Large-scale mutations

To explain the generation of large numbers of new genes, Neo-Darwinians suggested that larger scale mutations were important. We have already met chromosomal fusion and polyploidy in Chapter 8; these are mutations occurring at the level of chromosomes and can lead to speciation.

In between these very large-scale chromosomal mutations and the tiny point mutations are other mutations which affect segments of DNA:

- Segmental deletions remove a portion of a chromosome.
- Segmental insertions add a section of DNA to the chromosome. The new material could be from the same chromosome, another chromosome, or even an exogenous (outside) source, such as a virus. This could add copies of existing genes to the chromosome.
- Duplications occur when an extra copy of an existing gene or set of genes is inserted into the DNA.
- Inversions occur when a section of a chromosome breaks off and then reattaches in the reverse orientation. (The area where this occurs often becomes non-functional.)
- Translocations occur when a portion of one chromosome breaks off and attaches to another chromosome.

Many of these processes are known to be associated with serious, or even fatal genetic disorders in people and animals, although some may be neutral or even slightly beneficial. In general, the larger the mutation, the more likely it is to disrupt normal cell function and be harmful. Nevertheless, segmental insertions and duplications provide a mechanism for adding extra copies of existing genes into a chromosome.

By themselves, these large-scale mutations are not an adequate explanation for the emergence of large numbers of novel genes. They show how existing genes can be moved, inserted or duplicated, but not how entirely new genes are created.

The final Neo-Darwinian step

From here, the Neo-Darwinists proposed that further mutations affecting duplicated genes/DNA in combination with natural selection could gradually morph the DNA blueprint for one enzyme (or structural protein) into an entirely different type of enzyme (or structural protein).

The explanation for how this might occur was similar to one Darwin used in *Origin* to explain the emergence of a complex structure such as the human eye. Darwin proposed that starting with a primordial eye, natural selection could guide development provided that the change at each step, however small, was a reproductive advantage for the animal. A long sequence of positive mutational changes was required to progress stepwise from the primordial eye to the human eye. Each step had to be susceptible to natural selection, meaning that each step had to be a beneficial mutation. (If it was harmful, it would result in information loss, not gain.)

Similarly, at a DNA level, for one gene to morph into another, this had to be guided by natural selection and it would require a stepwise set of point mutations (or equivalent large-scale mutations), each of which had a survival advantage. There could not be any gaps in the steps. The requirement is effectively "mutational steppingstones" between two different but functional proteins.[17] The nature of mutations is therefore critically important to Neo-Darwinism.

Most mutations are harmful

Experimental evidence available to the Neo-Darwinists suggested that the majority of mutations were harmful, but most of the rest were *assumed* to be beneficial. These are shown indicatively in Figure 12.4. The environment, via

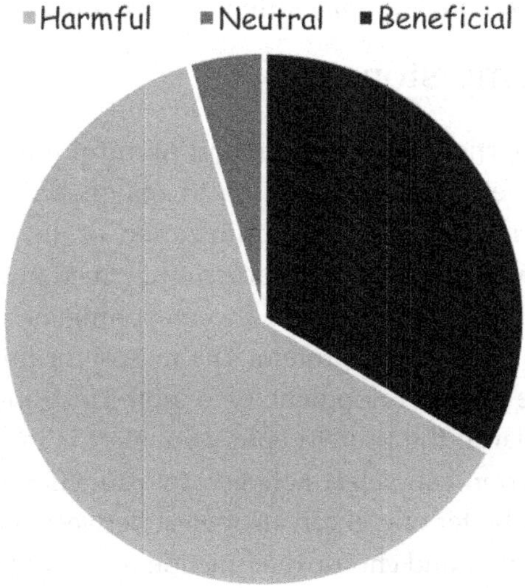

Figure 12.4: Classification of mutations in Neo-Darwinism

natural selection, can influence whether, and to what extent, a particular mutation will be harmful or beneficial, changing the relative frequencies of these in either direction.

Only a small proportion of mutations were thought to be neither beneficial nor harmful—that is, neutral (mid-gray). Neutral mutations are invisible to natural selection, although like all mutations they can be affected by genetic drift. But the small portion of neutral mutations implied that genetic drift was

17 Axe, 2016, chapter 7.

not quantitatively important. Natural selection was believed to rule evolution with a powerful hand, a key belief of Darwin himself.

It is known that some apparently harmful mutations can persist in the gene pool at an unexpectedly high rate, for example, sickle cell anemia. This calls into question how powerful natural selection is, but Neo-Darwinians proposed "balancing selection" to explain this. The example of sickle cell anemia, mentioned previously, is instructive here.

Sickle cell anemia is very serious, even fatal, when the mutated gene is inherited from both parents. But like recessive mutations generally, having only one copy of the gene does not cause the disease, so only in cases where both copies are present, should it be visible to natural selection. This is why such harmful mutations persist, rather than being eliminated by selection. But sickle cell anemia persists at a significantly higher frequency than one would expect. This is because the carrier state is actually favored by selection in parts of the world where malaria is common; carriers are less likely to die in early childhood (when immunity develops) from the cerebral form of falciparum malaria. This selection *for* the gene (in carriers) is said to 'balance' selection *against* it (due to sickness and death) in sufferers with both genes.

Mutational stepping-stones

Assuming the proportions of beneficial and harmful mutations shown in Figure 12.4, it would certainly be plausible to obtain a short sequence of mutational improvements, such as the optimization of an existing enzyme. Indeed, experimental evidence for such processes has been observed.

However, the real question involves the morphing of one functioning protein into a different functioning protein. The reason for this is that natural selection cannot guide the development of a non-functional protein gene towards functionality. Until the protein is functional, it is invisible to natural selection. Natural selection can select between two functioning protein gene variants and choose the better one. It can also select between a functional and a non-functional protein gene and choose the functional one. But it cannot choose between two non-functional protein genes. Until the protein is functional, we only have random mutations and genetic drift to explain its development—neither of which have any direction. Any increase in complexity in one step could all be undone in the next. And as we saw in Chapter 10, relying on randomness to assemble even minimal information is hopeless; harmful mutations are more likely than beneficial ones. Even a non-functional protein gene close to a functional state is more likely to degrade than to reach functionality.

A series of mutational stepping-stones, therefore, is needed between two functional proteins for natural selection to work. If the gene under transition falls off this path and becomes non-functional at any point, natural selection can no longer work on it.

Furthermore, to transition from one functional protein to another would intuitively require a much longer sequence of beneficial mutations than would be required to fine-tune a protein. But because the beneficial mutations are less common than harmful ones, proceeding all the way from one protein to another without falling off the mutational stepping-stones seems unlikely.

In view of the difficulties, not all biologists were convinced that random mutations were suitable for the task of increasing complexity in organisms. They struggled to see how random errors could add significant amounts of new information. Sir James Gray, Professor of Zoology at Cambridge University, and an influential UK evolutionist in the first half of the twentieth century, expressed doubts. "A few readers, perhaps rather pagan in their outlook, may think it a little strange that, if the case is quite so strong as they are asked to believe, it should still be necessary to argue the merits of natural selection with almost evangelistic vigour".[18] Gray went on to admit that "orthodox theory", by which he meant Neo-Darwinism, was unlikely, stating that "[n]o amount of argument, or clever epigram, can disguise the inherent improbability of orthodox theory ...".[19]

Extrapolation

In our discussion of Neo-Darwinism to this point, we see that it contains a causal model that explains variation and speciation of pre-existing complex organisms. If we consider an abbreviated morphological taxonomy shown in Figure 12.5, we note that groups at the bottom of the diagram (shaded dark gray) such as species or varieties, contain organisms which are structurally similar. Variation at the species or varietal levels can be relatively minor and is well explained by mechanisms such as natural selection, genetic drift, mutations and so forth. Neo-Darwinism also explains how a sequence of mutations under the influence of natural selection could tune and optimize the activity of an already functioning protein (for example, an enzyme).

As one ascends to broader categories, such as genus and family (shaded mid gray), the applicability of these mechanisms becomes slightly less certain because some extrapolation is involved.

18 Gray, 1954.
19 Ibid.

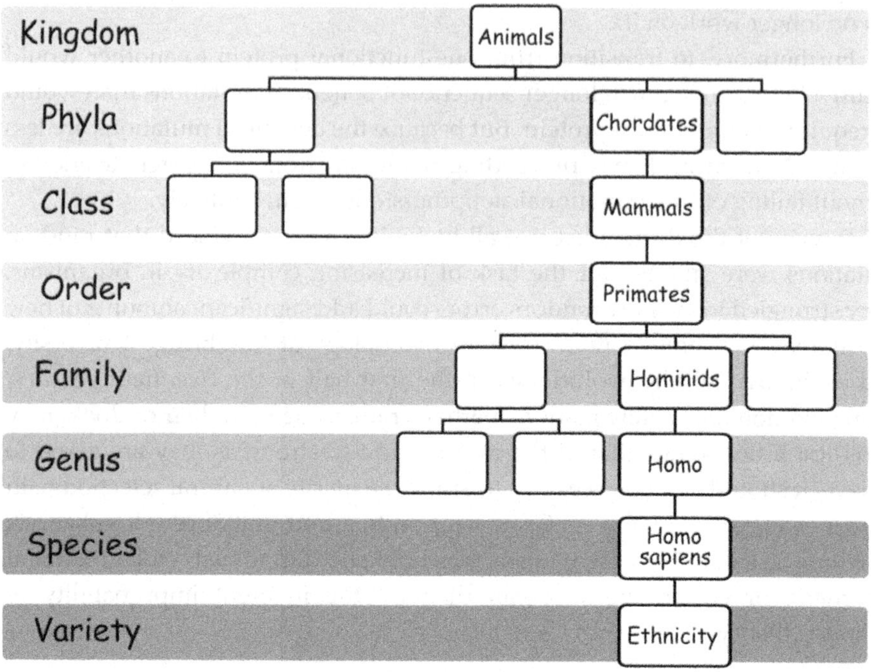

Figure 12.5: Morphological taxonomy

But at the highest and broadest categories in the morphological hierarchy (shaded pale gray), the degree of extrapolation of evolutionary mechanisms is much greater. At these higher levels, the differences between groups on the same level are much larger. Indeed, organs and the body plan can be different—consider the shape and structure of a starfish versus a fish versus a man.

At these highest and broadest categories, it is clear that multiple new genes will be required to permit an increase in complexity and fundamental change in body plan. Recall from Chapter 10 that Koonin suggested a simple primordial organism would require at least 150 genes; current estimates suggest that humans have between 19000 and 21000 genes. The problem for Neo-Darwinism is that it seems very unlikely that it can be extrapolated from where it is known to work—the level of species and varieties—to explain the naturalistic evolution of man all the way from a primordial organism. This requires an explanation for the generation of thousands, if not tens of thousands, of functioning genes.

This question of extrapolation came to a head in a macroevolution conference held at Chicago's Field Museum of Natural History in 1980. It was the first time in more than twenty-five years that geneticists and paleontologists had met for serious discussions[20] and was billed as a "historic conference".[21]

A central question under consideration was whether small changes could be extrapolated to explain large changes, or in the scientific terminology of the period, "whether the mechanisms underlying microevolution could be extrapolated to explain the phenomenon of macroevolution".[22] Roger Lewin, a prize-winning science writer, reported that at "the risk of doing violence to the positions of some of the people at the meeting, the answer can be given as a clear, No." Further, prominent American Neo-Darwinian scientist, Francisco Ayala, said, "We would not have predicted stasis from population genetics, but I am now convinced from what the paleontologists say that small changes do not accumulate".[23]

What this meant was that while Neo-Darwinism could model changes at the level of genera and species, most scientists at the Chicago conference accepted that those mechanisms could not be blindly extrapolated to the highest levels of a putative morphological hierarchy.

At the macroevolution conference, prominent paleontologists such as Stephen Jay Gould were clear: "Certainly the [fossil] record is poor, but the jerkiness you see is not the result of gaps, it is the consequence of the jerky mode of evolutionary change".[24] Stuart Kaufman of the University of Pennsylvania stated that in fruit flies (*Drosophila*), "one did not see intermediate changes between major mutants, implying single gene switches".[25] For some Neo-Darwinians, the idea of jerky change was heresy. The paleontologists were accused by some of monotheism, a charge which Gould's colleague, Niles Eldredge, sought to counter[26] by seeking some sort of accommodation with the ideologues.

Extrapolation of small-scale mechanisms of variation to explain large-scale change over long periods of time was, and is, the heart of the secular explanation for biological diversity. If large-scale changes were to emerge suddenly without precursors, this would imply that small-scale changes have not added up to large-scale change, as a holistic theory of evolution demands. As we saw in the previous chapter, a well-known example of this problem is the so-called Cambrian explosion.

---

20 Lewin, 1980.
21 Ibid.
22 Ibid.
23 Ibid.
24 Ibid.
25 Ibid.
26 Ibid.

From maxi yacht to foiling monohull: without a rudder

We step back briefly from the 1980s to the late 1960s, to consider the emergence of another challenge to Neo-Darwinism. In Neo-Darwinism, natural selection was believed to be a dominant factor affecting mutations. Nevertheless, it was understood that if a mutation was neutral, meaning that it was neither beneficial nor harmful in a selective sense, then natural selection would not act on that mutation. Instead, random effects, referred to as genetic drift, would be the major mechanism affecting the frequency of that mutation in the gene pool. In Neo-Darwinism, genetic drift was deemed a minor player in the process of evolution. But that belief was about to be challenged.

In 1968, Japanese scientist Motoo Kimura introduced the Neutral Theory of evolution, which other scientists further elaborated in articles published in 1969. This theory, which confounded proponents of Neo-Darwinism in the ensuing decades, was mathematical in nature and proposed that while many mutations in DNA were harmful, the bulk of the remainder were effectively neutral, especially in population sizes consistent with the sizes of bird, reptile, or mammalian populations, and the human population.

Colin Patterson noted that Neutral Theory had at last introduced apparent mathematical rigor into evolutionary science:

> Neutral Theory is therefore more like a theory in physics or chemistry than other evolutionary theories; it makes bold testable predictions, and can be put in a law-like form … .[27]

Neutral Theory seemed to have helped to elevate evolution from soft science to something more akin to a hard science, like physics or chemistry. It enabled scientists to make predictions of evolutionary change at a molecular level. If Darwin's original theory was like the Cutty Sark, and Neo-Darwinism was like a maxi yacht, Kimura's Neutral Theory (as an origins theory) was like a foiling monohull. All three vessel types remain in the paradigm of "sailing vessels", just as Kimuran Neutral Theory remained within the overall paradigm of the ancient Greeks.

Neutral Theory meant that in populations of the order of magnitude of those such as mammals, birds and reptiles, natural selection scarcely occurs at a molecular level. Instead, mutations are largely retained according to random chance. Mechanisms of chromosomal assortment and crossing over—manifested in virtually every sexual conception—are what mostly affects the contribution of mutations and variations, rather than natural selection. Even mutations which are weakly beneficial or harmful *act as if neutral* in smallish populations. In these populations, genetic drift is the key causal factor in

---

27 Patterson, 1999, p. 119.

evolution at the level of DNA; the "noise" of random variations overwhelms the "signal" of natural selection, virtually obliterating it.

The diagram of mutations is fundamentally changed under the Neutral paradigm.

Although the majority status of harmful mutations remains, the proportion of beneficial mutations falls dramatically under Neutral Theory. Instead, neutral change becomes dominant over beneficial change. If beneficial mutations are much less likely, it follows that any evolutionary increase in complexity is also less likely. Indeed, Neutral Theory was highly damaging to the Neo-Darwinist conception: natural selection, the only mechanism known to

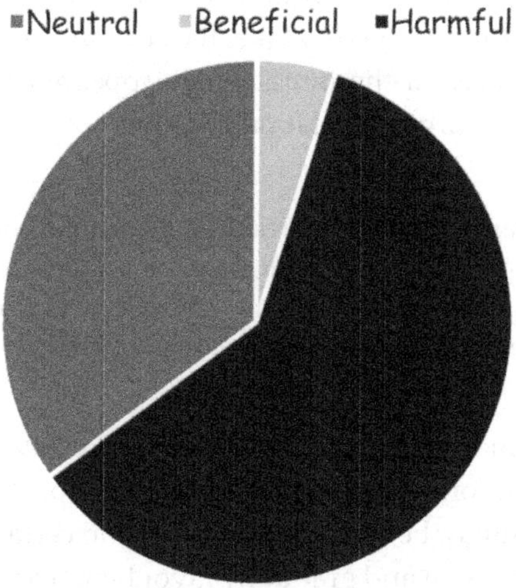

Figure 12.6: Proportions of types of mutation in Neutral theory

cause directed change at a genetic level, was heavily downgraded in importance.

Even before it was published, Neutral Theory was self-evidently divisive. Allegedly, Motoo Kimura could not bring himself to write that natural selection must be involved in adaptive evolution and asked an American friend to write it for him.[28] Even if Kimura doffed his hat to the prevailing scientific orthodoxy, scientists quickly appreciated the significance of Neutral Theory to the origins debate. It landed like a cat among the pigeons for proponents of Neo-Darwinism. Biologist and geneticist Richard Lewontin (whom we met in Chapter 4) succinctly outlined the problem:

28 Dawkins footnoted this anecdote in the essay "Darwin Triumphant", in *A Devil's Chaplain* (Dawkins, 2003).

> If the empirical fact should be that most of the genetic change in species formation is indeed of this non-Darwinian sort, then where is the revolution that Darwin made?[29]

Fred Hoyle and Chandra Wickramasinghe also commented on the difficulties posed by Neutral Theory and noted that the size of populations was an important determinant as to whether Neo-Darwinian evolution would actually work. This argument held that, in relatively small populations, such as humans, rare mutations might only occur, for example, once per century; if an individual with a rare mutation failed to reproduce, perhaps by dying at a young age, then their mutation would disappear from the gene pool.[30] On the other hand, in large populations, such as bacteria or plants, an equivalently rare mutation might occur, for argument's sake, tens or hundreds of times per year, and even if some instances of the mutation disappeared from the gene pool, others might persist. The large population simply affords more chances for the mutation to "stick":

> Although the relevant mathematical results only became known in recent years, they are not particularly difficult to derive, and it is something of a mystery to understand why they were not obtained long ago. Possibly the reason was that the results are not particularly helpful to the Darwinian theory.[31]

Hoyle and Wickramasinghue observed that the key factor is the number of individuals in a species' or subspecies' breeding group. When this number is sufficiently large, as it might be in bacteria, plants, and certain insects, evolution may follow a Darwinian path and gradually favor beneficial changes. However, when the breeding population is small, the trend shifts away from improvement and instead leans toward genetic decline. This makes it difficult to explain how more complex organisms such as reptiles, birds, and mammals could have evolved through Darwinian processes.[32]

Colin Patterson expressed a similar concern, noting that if the major mechanism of modification is neutral change, then evolution occurs randomly rather than being directed by natural selection. "Neutral change in DNA ... [does] nothing to explain the existence of whales and elephants, fleas and butterflies or ferns and oak trees".[33]

---

29 Lewontin, 1974, p. 4.

30 Hoyle cited the example of fur in humans, which is controlled by a single gene which has become inactive. The chances of this gene becoming activated are about one in a billion, which equated, when Hoyle was writing, to one instance in Europe per century. A person who exhibited such a mutation was Adrian Jeftichjew, the "Russian dogman" who was displayed by the PT Barnum circus in the late nineteenth century (Hoyle, 1999, p. 101).

31 Hoyle & Wickramasinghe, 1981, p. 7.

32 Ibid.

33 Patterson, 1999, p. 119.

Theodosius Dobzhansky, one of the originators of Neo-Darwinism, objected to Neutral Theory on metaphysical grounds—it undermined his worldview:

> ... Dr. Kimura and his followers claim evolution to be due to changes which are neither useful nor harmful to their possessors. They are simply neutral and are established merely by chance. If that were so, evolution would have hardly any meaning, and would not be going anywhere in particular. ... This is not simply a quibble among specialists. To a man looking for the meaning of his existence, evolution by natural selection makes sense.[34]

Mathematical modelling in Neutral Theory suggested that natural selection could be significant in environments where the population is very large and competition within the population remained intense.[35] Only under these conditions could the signal of natural selection shine through the fog of random effects strongly enough to be discernible.

As far as causal model science went (that is, explaining changes in the gene pool at a molecular level), Neutral Theory was no threat; it was an advance. Neutral Theory corrected the overemphasis on natural selection in Neo-Darwinism. But Neutral Theory represented an enormous challenge to the orthodox position of Neo-Darwinism as a holistic origins account. It was no longer clear that Darwinian natural selection was the predominant factor in evolution. As a holistic origins account, Neo-Darwinism could not easily incorporate Kimura's Neutral Theory because the key mechanisms of each theory were antithetical to each other, and for Neo-Darwinism to make sense, natural selection must be the key mechanism.

The secular humanist consensus was beginning to shatter. If natural selection is orders of magnitude less important than genetic drift in mammalian populations, then it is impossible for humans to emerge by evolutionary descent.

These seismic developments prompted the new field of molecular evolution to emerge, which moved the debate along further.

Picking up the pieces

After the bombshell of Motoo Kimura's Neutral Theory, proponents of Neo-Darwinism watched on as their theory crumbled around them. They needed to respond— quickly and effectively. Richard Lewontin, the brilliant Harvard educated biologist who was a key supporter of Neo-Darwinism, attempted that in his 1974 book, *The Genetic Basis of Evolutionary Change.*

---

34 Dobzhansky, 2001.

35 By large population levels we mean those found in bacterial, plant or insect populations (Hoyle & Wickramasinghe, 1981, p. 7), which are far greater than those of humans, animals and birds.

Lewontin discussed several ways for those persuaded by Neo-Darwinism to respond to the problems raised by Neutral Theory.

According to Lewontin, an important part of the response was to clarify what science is, and what it is not. He emphasized, that, contrary to "textbook" understandings of what science is, practical experience demonstrated that science is riven by prejudice and conflict related to ideological biases and worldviews. These biases were especially noteworthy in the gray areas:

> But this textbook myth has no congruence with reality. Long before there is any direct evidence, scientific workers have brought to the issue deep-seated prejudices; the more important the issue and the more ambiguous the evidence, the more important are the prejudices, and the greater the likelihood that two diametrically opposed and irreconcilable schools will appear. ...[36]

Lewontin then mentioned that even if evidence appears to settle the question, the "struggle" can be continued by redefining the terms of the debate.

Lewontin's starting point in defending Neo-Darwinism was to position Neutral Theory squarely in a scientific gray zone. Here, the evidence base was disputable, and there were legitimate doubts as to whether a resolution might be achievable. From there, Lewontin identified several approaches which might be able to counter the Neutral threat. Other scientists followed his lead. We summarize several of the broad types of approaches next.

The first approach pointed out that the Neutral Theory evolutionists were not antagonistic to proposed mechanisms of Neo-Darwinism and did not see their theory as a refutation of that system. However, the anecdote about Motoo Kimura and the objectionable sentence casts some doubt on that proposition.

Another approach was to suggest that Neutral Theory is true but unimportant. If it is unimportant then natural selection can continue to be emphasized as critically important. However, the selection of quotations above gives the lie to that approach.

A third approach stated that Neutral Theory applies at the DNA level, but natural selection applies at the phenotypic level (the level of observable traits). Richard Dawkins sought to define the problem away in this manner:

> From a Darwinian point of view, neutral mutations are not mutations at all. But from a molecular point of view they are extremely useful mutations because their fixed rate makes the clock reliable.[37]

He then goes on to offer soft science, that is, mere words, as a counter to the hard science of Neutral Theory:

---

36 Lewontin, 1974, p. 29.

37 Dawkins, 2004, p. 73.

> Darwinian selection remains the only explanation for adaptive evolution and it is arguable (I would argue) that most if not all of the evolutionary changes we actually see in the macroscopic world (as opposed to those concealed among the molecules) are adaptive and Darwinian.[38]

The final approach we consider here is the argument that natural selection can be amplified through linkages. This means that genes in close proximity in the DNA are also influenced by selective effects if a gene near them is selected.[39]

But we return to what Lewontin said—this type of issue can be argued about ad nauseum, without resolution, because positions are potentially determined more by ideology and worldviews than evidence and science.

Colin Patterson noted that most scientists accepted aspects of Neo-Darwinism and Neutral Theory and combined both as causal model science. But it was scientifically questionable to combine both as part of a holistic origins account:

> When these two theories are combined as a general explanation of evolutionary change, that general theory is no longer easily testable.[40]

Any failures to find evidence for natural selection can be ascribed to Neutral Theory, and vice versa. Holistic theories of evolution were once again heading towards unfalsifiability through the 1970s. The post-Kimuran era was a turning point for Neo-Darwinism.

Headwinds affecting Neutral Theory

Although Neo-Darwinism has fared badly with the passage of time, Neutral Theory has also come under increasing strain. Earlier we noted that prominent evolutionists rejected the importance of Neutral Theory from the outset. Neutral Theory also struggles as part of a holistic account because it is inconceivable that any model largely dependent on random processes could generate the human genome—the probability is too low for that to occur, much lower than the probability of abiogenesis (which is practically impossible). Further problems have emerged over time.

From its inception, proponents of Neutral Theory assumed that most DNA in organisms is junk—random byproducts of evolutionary history with no function. This view appeared to be validated when it emerged that less than three percent of the human genome codes for proteins.[41] Large regions of apparently random, useless DNA were assumed by evolutionists to represent "evolutionary scratch pads" enabling organisms to randomly change sequences around continuously, occasionally hitting the jackpot—a new gene.

---

38 Ibid., pp. 73–74.

39 Lewontin, 1974, p. 312.

40 Patterson, 1999, p. 45.

41 Tomkins & Bergman, 2019.

Junk DNA was the label given to much of the human, and other genomes. This all fit very nicely with Neutral Theory. Junk DNA is also an important part of evolutionary theory from a mathematical perspective—without it, the probabilities of evolution occurring are less likely.[42]

But as molecular biology continued to advance, the label "junk DNA" was increasingly called into question. It turns out that almost the entire genome is transcribed into RNA, not just the coding regions. Part of the region formerly known as junk DNA was found to comprise genes for long noncoding RNAs, which are at least twice as common as protein coding genes.[43] Many of the long noncoding RNAs have complex regulation and are expressed depending on the cell type and even the state of the cell.[44] It is clear that those long noncoding RNAs which have been investigated in detail *in vitro* (Latin: in glass—i.e., in dishes in the laboratory) can have important functions, although some of the details remain unclear. Some of the known functions include "gene regulation, chromatin modification, translational regulation, structural and catalytic components integrated with proteins ... [and] intercellular signaling".[45]

Another component of "junk DNA" is retrotransposons, which were believed to represent leftovers from ancient viral infections where random parts of viral DNA had become inserted into the DNA of the host organism.[46] Research has suggested that retrotransposons are not randomly distributed in DNA but are often found adjacent to protein-coding regions, and are "multifaceted regulators of the functional output of the mammalian transcriptome".[47] Thus, they are not junk DNA.

These developments in molecular biology have shed more light on DNA codons and non-coding regions, prompting a reassessment of the term "junk DNA". This impacts negatively on Neutral Theory as a candidate for a holistic evolutionary mechanism.

Timing and textbooks

Niles Eldredge wrote about the difficulty scientists experienced in critiquing Neo-Darwinism, especially if they were not from the discipline of population genetics. Prior to the 1980s, they were simply frozen out of the discussion at the "'High Table' of evolutionary discourse".[48] An "inner group of geneticists" had prevailed over evolutionary biology from the 1950s.[49] Neo-

42 Carter, 2009, p. 12.
43 Tomkins & Bergman, 2019.
44 Ibid.
45 Ibid.
46 Carter, 2009.
47 Ibid., p. 13.
48 Eldredge, 1996, p. 1.
49 Ibid.

Darwinism was promoted as settled science. Darwinism, or more correctly, Neo-Darwinism, had reached an intellectual peak in the 1950s and early 1960s.

Simultaneous with the zenith of Neo-Darwinism were two other direct and powerful supports for the naturalistic worldview. In cosmology, the eternal universe was still a possibility; the steady state theory was only overturned by the Big Bang in the mid-1960s. The existence problem, therefore, had not yet become the serious issue for the naturalists that it would several years hence. Secondly, logical positivism, which held that only scientifically derived knowledge was of value, had not yet been decisively refuted. That would come in 1962, via Thomas Kuhn's book *The Structure of Scientific Revolutions*.

Perhaps a third factor gave additional, though indirect, support to the naturalistic worldview as late as the early 1960s. The church at large had not been able to mount a winning argument against naturalism as Neo-Darwinism reached its peak. The resurgence of creationism only began in earnest in 1961 with the publication of *The Genesis Flood*. As a result of all these factors, secular humanism was riding high.

This period reflects the kind of scheme which people imbibe from schooling today—origins science is hard settled science, and people who believe in a straightforward reading of Genesis are either ignorant or crazy. But such a view is outdated; its plausibility peaked with Neo-Darwinism in the early 1960s.

By the late 1960s, the situation for naturalism had changed sharply for the worse. In cosmology, the steady state model was out; the Big Bang, with its finite past, was in, and the existence problem was back with a vengeance for the naturalists. In biological origins, prominent paleontologists had shown that the fossil evidence did not support slow, gradual change, which both Darwin and Neo-Darwinism had predicted. Kimura's Neutral Theory, which was published in 1968, undermined the mechanism proposed by Darwin and the Neo-Darwinists.

But, in an ironic twist of history, the Epperson v. Arkansas decision of 1968—the same year Neutral Theory was published—codified the belief that a straightforward reading of Genesis is wrong and arguably established secular humanism as the state religion in the U.S. Similar changes occurred throughout the West.[50]

Human evolution was therefore actively propelled into schools, and therefore the mainstream, before the scientific establishment had time to digest Neutral Theory and its implications. Although Neutral Theory is mentioned in some contemporary textbooks, evolutionary education in the school system is largely conducted as if the intellectual milieu were frozen in the heyday of Neo-Darwinism. School students are taught the dogma that Darwinism is in

50 Secular humanists typically maintain that the Epperson decision did not establish a religion. Creationists consider that Epperson supports the secular humanist religious perspective and harms Christianity.

ascendancy, whereas in fact, the foiling sailboat of Neutral Theory arrived in 1968—without a rudder. Evolutionary mechanisms were becoming clearer, but holistic theories of common evolutionary descent were becoming more difficult to explain. It wasn't good for the evolutionists.

The upshot was that radical new developments of evolutionary theory challenging the entrenched consensus increasingly came from outside the traditional discipline of evolutionary biology—this included intellectuals such as Sir Fred Hoyle and the information scientist Lee Spetner. We will review these arguments, and others, in the next chapter.

The Third Way

With Neo-Darwinism under increasing strain, and creationism still officially excluded by the scientific establishment, some evolutionists have sought "The Third Way". The Third Way rejects biblical creationism, and attempts to provide a holistic evolutionary account, also known as the Extended Evolutionary Synthesis. One group of scientists and researchers have created a website to promote these new developments.[51] They criticize Neo-Darwinism because it "ignores much contemporary molecular evidence and invokes a set of unsupported assumptions about the accidental nature of hereditary variation".[52]

Notably, some advocates of the Third Way have stated that the "DNA record does not support the assertion that small random mutations are the main source of new and useful variations".[53] This statement appears to represent a rejection of the molecular mechanisms underlying both Neo-Darwinism and Neutral Theory. The Third Way proponents point out that many processes of variation are well regulated cellular and biological processes, not random processes.

Other supporters of the Extended Synthesis propose mechanisms such as niche construction as important. Niche construction pays attention to the fact that not only does the environment impact on organisms via natural selection, but organisms also impact on the environment and other organisms, signifying a more complex, dynamic, and reciprocal process.

While Third Way proponents reject Neo-Darwinism, they have not established alternative molecular mechanisms.

Conclusion

In our view all these more recent developments foreshadow the demise of Neo-Darwinism and Neutral Theory as supporting a holistic origins theory

51 The Third Way, _.

52 Ibid.

53 Ibid.

based on evolution. They represent a further fracturing of the secular origins science consensus.

None of the developments propose any replacement at the molecular level for the mechanism of evolution provided by Neo-Darwinism[54] or by Neutral Theory; at least, not one compatible with a materialistic or atheistic worldview. Specifically, all these theories fail to adequately model evolution of the sort believed by secular scientists to have occurred during the Cambrian explosion.

Neo-Darwinism and Neutral Theory barely extend beyond explaining speciation, which is an inadequate support for the secular humanist worldview, while being fully compatible with the creationist perspective.

References

Axe, D. (2016). *Undeniable: How Biology Confirms Our Intuition That Life Is Designed*. (ePub edn) London: HarperCollins Publishers Ltd.

Carter, R. W. (2009). The slow, painful death of junk DNA. *Journal of Creation*, 23(3), 12–13.

Dawkins, R. (2004). *A Devil's Chaplain: Selected essays by Richard Dawkins*. London: Weidenfeld & Nicolson.

Dennert, E. (1904). *At the Deathbed of Darwinism: A Series of Papers* (Project Gutenberg, 2007 edn). Burlington: German Literary Board.

Dobzhansky, T. (2001). BBC's The Life Game. Retrieved June 7, 2023, from Perspectives on Molecular Evolution: wayback.archive-it.org/9060/20230418130943/https://authors.library.caltech.edu/5456/1/hrst.mit.edu//hrs/evolution/public/collections/lifegame.html

Eldredge, N. (1996). *Reinventing Darwin—The Great Evolutionary Debate*. London: Phoenix, Orion Books Ltd.

Gale, J. S. (1990). *Theoretical Population Genetics*. London: Unwin Hyman.

Gray, J. (1954, February 6). The case for natural selection: Evolution in Action. *Nature*, 173(4397), 227.

Hoyle, F. (1999). *Mathematics of Evolution*. Memphis: Acorn Enterprises LLC.

Hoyle, F., & Wickramasinghe, C. (1981). *Evolution from Space—A Theory of Cosmic Creationism*. New York: Simon and Schuster, Inc.

Lewin, R. (1980, November 21). Evolutionary Theory Under Fire: An historic conference in Chicago challenges the four-decade long dominance of Neo-Darwinism. *Science* 210(4472), 883–887.

Lewontin, R. C. (1974). *The Genetic Basis of Evolutionary Change*. New York: Columbia University Press.

Patterson, C. (1999). *Evolution* (2nd edn). New York: Comstock Publishing Associates.

Scott, E. C., & Branch, G. (2009). Don't Call it "Darwinism". *Evolution: Education and Outreach*, 2, 90–94. doi: 10.1007/s12052-008-0111-2

Sanford, J. C. (2014). *Genetic Entropy*. (4th edn). Canandaigua: FMS Publications.

Stoltzfus, A. (2017). Why we don't want another "Synthesis". *Biology Direct*, 12. doi: 10.1186/s13062-017-0194-1

The Third Way. (_). Evolution in the era of genomics and epigenomics. Retrieved May 25, 2024, from thethirdwayofevolution.com

Tomkins, J. P., & Bergman, J. (2019, April 5). Neutral Model, genetic drift and the Third Way—a synopsis of the self-inflicted demise of the evolutionary paradigm. Retrieved May 25, 2024, from creation.com/evolutionary-mechanisms

Wieland, C. (2014, August 20). Birds of a feather don't breed together. Retrieved 5 April, 2024, from creation.com/birds-of-a-feather-dont-breed-together

54 Tomkins & Bergman, 2019.

13

The dead-end of common evolutionary descent

We saw in the previous chapter that the original gene pool concept of Neo-Darwinism relied on flawed assumptions. For example, Neo-Darwinians wrongly assumed that most mutations were either selectively beneficial or harmful, rather than neutral. As these flawed assumptions were challenged and corrected, evolution as a holistic origins theory progressively looked less likely. Kimura's Neutral Theory was important in this regard because it brought the challenge into the limelight. By the 1980s, many biologists agreed that the mechanisms of microevolution could not simply be extrapolated to explain large-scale change. Such an extrapolation was contradicted by the fossil record.

In this chapter we'll survey four well-credentialed authors who have published arguments undercutting evolution as a holistic explanation for the diversity of life—Lee Spetner (1997), Stephen C. Meyer (2012), Douglas Axe (2016) and Sir Fred Hoyle (published 1987, 1999)—all of whom have written in the post-Kimuran period. These thinkers discuss the inadequacy of molecular mechanisms of variation to explain the large-scale informational gain required for a holistic theory of evolution.

Some of the material we cover in this chapter is complex. Readers who do not have a scientific background may prefer to skip to the conclusion at the end of each section, or to discuss the material with a scientifically-inclined friend.

Spetner: the information theorist

Lee Spetner, an Israeli American scientist who became skeptical of evolution, outlined the inadequacy of Neo-Darwinian processes in his 1997 book, *Not by chance! Shattering the modern theory of evolution.*

Early in his career, Spetner taught information and communication theory at Johns Hopkins University[1] and was familiar with the highly regarded information theory published by Claude Shannon (1948).[2] Spetner studied the role of random mutations in Neo-Darwinism with a view to determining

1 Wieland, 1998.
2 Shannon, 1948.

whether they could add genetic information. His rationale was that for a primordial single-celled organism to evolve into complex multi-cellular organism such as a human being, a great deal of new genetic information must be added along the way.

Spetner was not interested in speciation resulting from gene inactivation (caused by mutation) or information loss by genetic drift—phenomena which are well accepted on all sides of the origins debate. Instead, he focused on speciation occurring via genetic information gain. Such "upwards evolution" towards increased complexity is the signature process of holistic evolutionary origins accounts.

Spetner rejected the idea that evolution could repeatedly add information by large-scale mutations. This is similar to the problem of abiogenesis—and, as we saw in Chapter 10, arriving at significant information jumps by purely random means is exceptionally unlikely. By contrast, the odds of locking in small gains by natural selection were far higher. Point mutations, according to Spetner, have more power to craft specific sequences than larger segmental mutations. For example, converting a specific sequence of five nucleobases in the DNA from GGGGG to AAAAA cannot be done by segmental inversions alone (which are a type of large-scale mutation). Deletions and insertions can do this, but point mutations are potentially quicker and more precise.[3] From these types of considerations, Spetner concluded that for upwards evolution, a sequence of point mutations was required, each adding some information.[4]

Spetner's analysis centered on a case study of horse evolution. To create a new horse species, Spetner estimated that at least 500 sequential point mutations were required.

First, Spetner wanted to estimate the number of random mutations that would have to occur in a population, on average, for a single point mutation to "stick" in a population long term. He used conservative assumptions for his calculations and explained the steps.

In brief, Spetner approximated "selective fitness" by the selective value, which he defined as the average fraction by which the progeny of the variety exceeds that of the rest of the population. Typical selective values are 0.1% or 1%, the latter meaning that in one generation, the population of the variety in question will increase by 1% over other varieties under natural selection, due

---

3 Spetner, 1999, p. 46.

4 In humans, each chromosome is analogous to a USB memory stick containing a certain amount of information in the DNA. The 23 USB memory sticks in a human sex cell contain about 750 megabytes of digital information. If that amount of information were transcribed onto paper, it would fill more than 1,500 typical paperback books, which in turn would take around 20 meters of shelving to accommodate. In other words, evolution needs to explain the upwards gain in information—by natural selection acting as a ratchet to lock in favourable random changes—from the order of magnitude represented by hundreds of thousands of bytes in a simple primordial organism, to near the order of magnitude represented by 750 megabytes in a human (assuming that in bacteria and humans a significant proportion of the DNA contains biologically relevant information, a proposition which may not be confirmed for some time yet). That's some ratchet.

to some selective advantage. If the selective value is positive, the genetic variety benefits by natural selection. Conversely, if the selective value is negative, the variety loses out.[5]

Spetner argued mathematically that in a large population including a variant with a selective value of 0.1%, the chance of the mutation surviving to become fixed in the population (present long term) is only 0.2%. Otherwise stated, there is a 99.8% chance it will not survive, due to random effects.[6] Having 500 such mutations would increase the chance of one surviving in the long term to about 60%. To get to a 99% chance that a beneficial mutation would survive long term, a population would need about 2500 such mutations.[7] This contradicts Darwin's assertion that even the slightest beneficial variation will be preserved.

Spetner argued that to increase genetic information, three conditions must apply to a mutation:[8]

1. It must have a positive selective value.
2. It must add some information to the genome.[9]
3. It must be dominant—recessive genes are too weak for information generating evolution.[10]

For this to be probable, Spetner calculated there would need to be slightly more than one million of these types of mutations available at each generation.[11]

Neo-Darwinism as two postulates

Spetner arrived at two postulates which must be satisfied for Neo-Darwinism to work in the case of horse evolution:[12]

1. A single nucleotide change in a genome must be able to add a small amount of information and have a selective value more than zero.
2. At every small step of evolution (i.e. a generation), there would need to be about one million mutations which satisfy point (1).

If these two postulates were found to be plausibly satisfied, Spetner would concede that a naturalistic evolutionary account for large mammals is strengthened.

5 Spetner, 1999, p. 54.
6 Ibid., p. 55.
7 Ibid.
8 Ibid., p. 103.
9 Ibid., p. 96.
10 Ibid., p. 101.
11 Ibid., p. 104.
12 Ibid.

Spetner's analysis

Spetner argued that if the two postulates were satisfied, mutations which add information should be so numerous as to be common. He decided to investigate changes in information where specific cases of enzyme change/variation were described in the academic literature.

Readers will recall that enzymes act like nanotechnology machines which chemically modify substrate molecules. Some enzymes are highly specific as to which substrate molecules they accept, while others are less discriminating and can work on a wider range of molecules. Spetner argued that an enzyme which was more specific in which molecule it acted upon had increased information compared with another similar enzyme with less selectivity. For example, if an enzyme changed in such a way that the "evolved" version could act on a wider variety of molecules than formerly, that would represent a reduction of information at the active site.

Spetner provided examples to illustrate the association of specificity and information. For instance, suppose there is a "word-enzyme" that cuts any word matching a template into two pieces.[13] In the first case, the template controlling the active site is a specific sequence of six characters: "ghtsha". This sequence only matches the word *nightshade* (in English). If the information in the active site was reduced to five characters ("ghtsh"), the enzyme can now also match *nightshirt* and *lightship*. The match sequence now carries less information than the first, and it matches more words because it is less specific. If the template at the active site is reduced to four characters ("ghts"), the enzyme would match about 100 different words—*lights, rights* and so on. Ergo, a less specific active site is less specific because it has less information.

Spetner needed to establish how common information increasing mutations are—mutations which increase specificity at the active site of an enzyme. To do this, he analyzed many examples of known evolutionary change. One of these was the development of antibiotic resistance in certain bacteria where this was via mutations (rather than pre-existing resistance genes favored by selection or transferred as plasmids from another bacterial species). Spetner found that the mutational change in each case represented a loss of enzymatic specificity, and hence a loss of information. Spetner concluded: "None of them adds information. All that I know of, actually lose information".[14]

After decades of these investigations, Spetner concluded that since information generating mutations were rare, postulate (2) was negated. Therefore, the holistic evolutionary account of increasing complexity over time does not apply to large mammals.

---

13 Ibid., p. 137.

14 Ibid., p. 107.

Spetner also advanced a second, more complex argument casting doubt on Neo-Darwinism as a holistic origins account.[15] This concerned a specific type of convergent evolution,[16] namely, convergent evolution of nearly identical proteins. One example Spetner gave were genes controlling eye development, found in both insects and humans, which were 94% similar. Spetner explained the difficulty for this to develop via convergent evolution by referencing his horse analysis. A horse has approximately ten thousand genes. If there are a million possible nucleotide changes available in the genome at any point in an evolutionary chain of mutations, there would be an average of 100 possible favorable mutations per gene (which is one million divided by ten thousand). This implies that any given gene has tremendous flexibility in the way that it could evolve. As a consequence, it is very challenging to explain how convergent evolution is possible if mutations are random. A huge number of different possible evolutionary paths is implied by the one million possible beneficial mutations. The probability of two different types of organisms descending the same path, or even a similar path, is infinitesimally small.[17] Once again, convergent evolution of similar genes in unrelated types of organisms is very difficult to explain with Neo-Darwinism.

Conclusion

Spetner concluded that neither Neo-Darwinism nor any evolutionary scheme dependent on random mutations for novelty works for large mammals.

We have merely sketched parts of Spetner's careful analysis, and interested readers are referred to his book for details. Even if secular evolutionists were to quibble with aspects of his analysis or some of his figures, his scheme underlines the sheer inadequacy of Neo-Darwinism, or any similar scheme of naturalistic common evolutionary descent.

Stephen C. Meyer

Epigenetics: the study of non-genetic biological information

We now briefly sketch one of the key arguments in the New York Times bestseller, *Darwin's Doubt: The explosive origin of animal life and the case for intelligent design* (2013) by Dr Stephen C. Meyer, director of the Discovery Institute (see Appendix D for more information about the Discovery Institute

15 Ibid., pp. 107–113.

16 Convergent evolution is the process by which unrelated organisms independently evolve similar traits or adaptations due to similar environmental pressures or ecological niches. For example, birds, bats and insects are unrelated, yet all have wings. Evolutionists regard this as an example of convergent evolution.

17 By way of analogy, consider two similar maps which each have ten separate routes to ten destinations inked in, where two of the routes are the same (or parallel) on both maps. The odds of taking the same (or parallel) route if one is randomly selected on each map is about 1 in 50. Now imagine that the number of separate routes on each map has increased to a million, with a handful overlapping. The odds of choosing the same (or parallel) route is much less than in the first scenario. Choosing the same or parallel path is analogous to convergent evolution.

and the Intelligent Design movement that it represents). Meyer focused his attention on a survey of scientific research relevant to explanations for the Cambrian explosion.

The key argument of Meyer's we will consider here concerns epigenetics. Epigenetics refers to non-genetic information—information not found in DNA—which impacts on the anatomical structure of developing organisms. Some types of epigenetic information are crucial for embryo development, yet this type of information is entirely unexplained by either Neo-Darwinism[18] or Neutral Theory. Evolutionists have assumed that mutations in genes are ultimately sufficient to build new body plans.[19] But increasing numbers of biologists do not believe that DNA alone dictates every biological development in an organism.

Meyer uses an analogy of a building under construction to illustrate the concept.[20] At a building site, several materials might be present: corrugated roofing iron, timber studs, nails, prefabricated windows, bricks, and so on. The structural plan of the building is not determined solely by these materials, but rather by the building blueprints as interpreted by the workers, which detail how these elements are to be combined.

In the same way, the building materials of an organism—proteins, sugars and lipids—do not directly determine the structure of the cell, tissue, or body plan. Other information serves to determine this. What DNA does is provide blueprints for proteins and structural RNAs, etc., and contains control regions which help to ensure that the correct quantity of proteins, and other structural RNA molecules, are made at the right time. So, what are the types of epigenetic information which have been described?

First, cells contain microtubule arrays which in aggregate comprise a filamentous skeleton. Specific proteins are delivered to particular locations within a cell by a delivery mechanism reliant on the cytoskeleton. Without this transport mechanism, the cell or tissue will not develop normally. Microtubules are made from a protein called tubulin, which functions like a building block. The arrangement of the microtubules is not determined by tubulin, in the same way that the shape of a building is not determined by the bricks from which it is constructed, but by their arrangement.[21]

Second, the centrosome is a structure in the cell which supports and regulates the cytoskeleton. During cell division, the centrosome duplicates in a process which does not directly involve DNA. One centrosome ends up in each daughter cell, and each centrosome ensures that a set of chromosomes move into a location which will become a part of each daughter cell. Although the

18 Meyer, 2013, p. 273.
19 Ibid., p. 275.
20 Ibid., p. 277.
21 Ibid., pp. 277–278.

centrosome is composed of proteins, its structure and duplication are not fully specified by genetic information.[22]

Another example Meyer cites concerns the fruit fly species *Drosophila melanogaster*. During egg cell development, special RNAs which code for two separate types of protein are transported to certain attachment locations on the inner cell membrane, one concentrated at the forward pole of the egg and the other concentrated towards the rear pole. Once fertilized, the RNAs are translated into protein. The protein at the forward pole activates genes necessary for the development of head and thoracic structures, while the specific protein at the rear pole suppresses those same genes. Essentially, the arrangement of the RNAs prior to fertilization serves as "spatial coordinates … for embryological development".[23] It is thus the spatial positioning and concentration of these RNAs which influences morphological development of the fruit fly, not just the genetic information contained within the RNA.[24]

Conclusion

Meyer's argument about epigenetic information can be summarized:[25]

1. Epigenetic information provides important information which impacts on the formation of the body plan of multicellular organisms yet is not encoded in the DNA.
2. Neo-Darwinism (and Neutral Theory) states that novelty in living organisms is produced by random mutations in DNA.
3. Hence, neither Neo-Darwinism nor Neutral Theory explains the origin of the epigenetic information required to generate new body plans. Therefore, it cannot explain large-scale information increases required to account for the Cambrian explosion.

We conclude here our brief overview of some of Meyer's arguments against materialistic evolutionary schemes based on novelty from random mutations. Readers are encouraged to refer to Meyer's accessibly written book for greater detail on his arguments and rebuttal of criticisms. However, we caution readers that Meyer, a proponent of Intelligent Design, relies on many secular humanist presuppositions. These lead him to accept a chronological age of the earth of billions of years.

22 Ibid., p. 278.
23 Ibid., p. 280.
24 Ibid., p. 279.
25 Ibid., p. 281.

Douglas Axe

Can mutations generate large scale change?

Sometime after completing his PhD in chemical engineering, Douglas Axe[26] joined the Centre for Protein Engineering in Cambridge to perform post-doctoral research.[27] As an engineer, Axe was skeptical that random mutations combined with natural selection would be sufficiently powerful to "build new organisms"[28] of the type suggested by secular scientific interpretations of the Cambrian explosion.

Axe reasoned that the origin of protein folds could serve as a testbed for the mechanism of Neo-Darwinism.[29] Proteins consisting of single chains have three layers of structure. The first is the sequence of amino acids defining the protein. Second are localized three-dimensional patterns such as alpha-helices and beta-pleated sheets. The third is the overall three-dimensional protein shape which defines the protein's tertiary structure and is determined by the process of protein folding.[30] [31]

Organisms with disparate body plans are found in the Cambrian layer. While some classes of protein might share structural similarities at the tertiary level, the existence of dramatic differences in body plans implies that dramatic differences in protein tertiary structure must also be present.[32] Meyer gives an example of the enzyme lysyl oxidase, required to support the "stout body structures" of some organisms seen in the Cambrian layer, which are not found in simpler organisms.[33] Axe also reasoned that novel tertiary protein structures represented the smallest change in structure possible within an organism.

At the time Axe's research was underway, it was commonly assumed that the formation of functional folded protein structures was straightforward; the amino acids facing outwards were assumed to be relatively interchangeable. However, Axe (and others) showed that changes to amino acids on the external surface of proteins tended to cause much more function loss than previously thought.[34] This meant that as a protein changed, with progressively more amino acid substitutions, it would be increasingly likely that negative natural selection would remove it from the gene pool. Even substitutions involving amino acids with similar properties often resulted in instability of protein folding and loss of function. The number of substitutions needed to change tertiary structure

---

26 Douglas Axe is cited by Meyer in *Darwin's Doubt*.
27 Meyer, 2013, p. 187.
28 Ibid., p. 186.
29 Ibid., p. 189.
30 Ibid.
31 Where a protein has multiple chains combined into a protein complex, it is referred to as quaternary structure.
32 Meyer, 2013, p. 191.
33 Ibid.
34 Ibid., p. 193.

almost always condemned the existing protein to malfunction before the structure was definitively altered. An additional factor is that cells also contain enzymes which destroy unfolded proteins.[35]

Axe concluded that there was no clear path from an existing protein structure to a new one, because it would pass through a period of non-functionality. This in turn would doom the organism[36] via negative natural selection.

Axe and his colleagues examined specific enzymes (named Kbl and BioF respectively) to determine if the function of one could be evolved to the other. They examined all the mutations that could assist this transformation and concluded that it was not possible via evolutionary mechanisms. Another colleague of Axe examined millions of other random mutations that Axe had not tested and obtained the same result.[37] Axe concluded that this path to the formation of a novel tertiary structure was impassable. Some scientists accepted this result, but at least one evolutionist argued that many evolutionary changes were impassable because life has evolved to a form which now resists evolutionary change.[38]

The other possibility Axe considered was whether a gene for a protein with tertiary structure could develop within a nonfunctional part of DNA (by random change/genetic drift). This implies a random search of the "probability space" for a working protein recipe. It is possible to calculate the number of different amino acid sequences that can form a protein 150 amino acids long. The calculation is $20^{150} = 1.4 \times 10^{195}$. Using the beta-lactamase enzyme as a case study, Axe showed that the probability that any random sequence could form a protein with tertiary structure that was functional worked out to about 10^{-74}.[39] [40] A previous calculation of the total number of gene sequences that can have been searched during purported evolutionary history, by British biochemist and Intelligent Design proponent Michael Denton, is about 10^{40}, with single celled organisms such as bacteria making an outsized contribution to that total, hardly relevant to the Cambrian explosion.[41] This leaves us with the probability of evolutionary processes finding functional proteins with tertiary structure as 10^{-34}.

Even if this probability does not breach the level of impossibility proposed by Lee Spetner (at 10^{-45}, see Chapter 10) it remains sufficiently unlikely that Axe concluded "that a reasonable person should reject it".[42] For a start, the time

35 Ibid., p. 196.
36 Ibid., p. 193.
37 Axe, 2016, Chapter 6.
38 Ibid.
39 Meyer, 2013, p. 200.
40 Axe, 2016, Chapter 5.
41 Meyer, 2013, pp. 203, 205.
42 Ibid., p. 203.

available for evolution to search the probability space during the Cambrian explosion was vastly shorter than used in the probability calculation, meaning that 10^{-34} was a dramatic overestimate of the chances of this occurring. Secondly, the calculation also included bacteria as a major source of search potential, which are largely irrelevant to the Cambrian explosion.[43] This meant, again, that 10^{-34} was a gross overestimate of the chance of a functional protein with tertiary structure being found within evolutionary timescales.

With both paths probabilistically impassable, Axe concluded that Neo-Darwinism was an inadequate explanation for body plan novelty found in the Cambrian layer.[44]

A final investigation of Axe's we consider here concerned protein evolution. In the first of two experiments he conducted, Axe exposed bacteria to mutational influences. The bacteria contained a natural bacterial enzyme which inactivates penicillin. Axe isolated bacteria where the gene encoding the enzyme had mutated to the point where the enzyme was barely functional. In the laboratory he induced further mutations and exposed bacteria to selection by progressively increasing the doses of penicillin. After six rounds, the gene and its enzyme had become highly active, even more than the original intact enzyme. Axe concluded that a combination of mutations and selection could improve the function of an enzyme, a process he termed "fiddling", evoking a mechanic tweaking a barely running engine to improve its function.

In the second experiment, Axe damaged the enzyme and its gene to the point it no longer worked enzymatically. But Axe was able to show that the damaged "enzyme" still inactivated penicillin, albeit weakly, by a non-enzymatic pathway. He found that the same rounds of treatment did not repair this non-enzyme protein. Unless a working enzyme was provided, mutations combined with selection did not work.[45] This was an interesting illustration of the famous maxim coined by geneticist Hugo De Vries[46] a few decades after Darwin: "natural selection may explain the *survival* of the fittest, but it cannot explain the *arrival* of the fittest".[47] Natural selection could optimize an existing design, but not create a new one, because a non-functional gene is either neutral or harmful.

Conclusion

Axe's calculation of the probability of "finding" functional proteins by a process of random mutations and selection cast doubt on whether Neo-Darwinism can support a holistic evolutionary origins account. He also cast

43 Ibid., p. 205.

44 Ibid., p. 207.

45 Axe, 2016, Chapter 7.

46 De Vries (1848–1935) was an early geneticist who coined the term "mutation".

47 Wagner, 2014, Chapter 1.

doubt on whether a functional protein can evolve stepwise into a different functional protein without suffering from misfolding at one or more of the steps, potentially dooming the organism.

Hoyle's doubt: a mathematical analysis of evolution

Sir Fred Hoyle was a giant of British academia. He is mostly remembered for his work in cosmology, but he had the kind of intellect that allowed him to take on evolutionary biology and speak with authority.

In his book, *Mathematics of Evolution* (1987, 1999), Hoyle presented a critical analysis of the mechanism of Neo-Darwinism. As a student at the University of Cambridge, Hoyle had topped his mathematics class, so he was well suited to this task. Hoyle's work is filled with pages of complex equations but also contains much readable prose.

Hoyle accepted evolution as causal model science in relation to variation and speciation but found the consensus theory unconvincing as a holistic origins explanation. He doubted the theory could be successfully extended to large-scale change, specifically critiquing the concept of evolution leading to increased complexity of organisms such as mammals, which is an essential component of evolution as origins theory.

Majority of mutations are harmful

Under Neo-Darwinism, mutations can be represented schematically as shown in the left panel of Figure 13.1 (presented again for convenience). As shown, most mutations are harmful (large slice).

Neutral changes were considered relatively insignificant (thin dark gray slice), leaving harmful and beneficial (light gray slice) mutations as the major types of novelty. In this scheme, natural selection acting on positive or negative mutations was the key driver of evolution at a molecular level.

But as per the previous chapter, this all changed in 1968 with Neutral Theory, as shown schematically in the right panel of Figure 13.1. While dominant harmful mutations were eliminated by natural selection, the large number of neutral or recessive (where the trait is only expressed if the offspring inherits a copy from both parents) changes were invisible to this mechanism. A mutation which is neither good nor bad (hence neutral) cannot be selected for or against. Instead, genetic drift (random changes in the genetic makeup of a population) was the key mechanism. Under Neutral Theory, beneficial mutations, which are needed for large scale change in the direction of increased complexity, were hypothesized to be relatively uncommon.

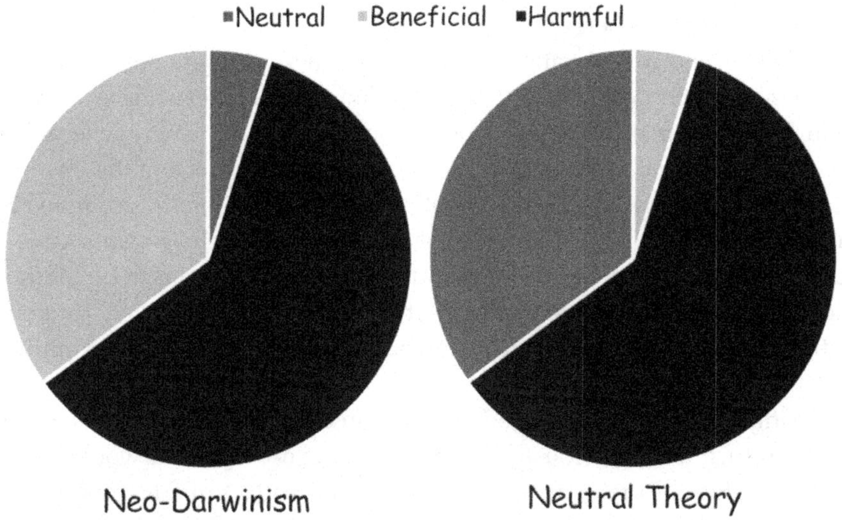

Figure 13.1: Mutation types as envisaged under Neo-Darwinism and Neutral theory

Hoyle explained that the fact that mutations are mostly harmful makes sense, considering that mutations represent random DNA copying errors or damage. Consider a page of text out of a book. One would hardly expect that the prose would improve in quality or meaning by adding random changes. A loss of meaning or correctness would be far more likely. Similarly, if a mammal has one or more mutations, most would be harmful, and only rarely would one be weakly beneficial.[48] Hoyle pointed out that inducing random changes in a complex structure such as a functioning protein must risk breaking one of many critical areas, whereas there could only be a limited number of possible beneficial tweaks.[49]

48 Hoyle, 1999, p. 9.

49 Consider how many computer game developers employ chimpanzees at keyboards to randomly edit the programming code of their latest game. Would chimps improve the code or introduce bugs?

Beneficial mutations in relation to generating new information

Hoyle asked what sorts of mutations could be helpful in generating new genes and identified several considerations (his discussion includes both sexual and asexual reproduction, but we will limit our discussion to the former). First, Neutral Theory tells us that a single beneficial mutation occurring in the gene pool is likely to disappear due to random effects. For such a mutation to survive in the gene pool to the point where it becomes permanent, it will have to have independent recurrences multiple times. The most likely mutations to recur must be simple—for example, point mutations. While it's possible that a complex, beneficial mutation may occur, it is extremely unlikely, and such a mutation is unlikely to persist in the gene pool.

Hoyle's second consideration is linked to the first. A single point mutation might be beneficial because it optimizes the function of an existing enzyme or protein, but it cannot create a new gene by itself. To generate a new gene, a long sequence of beneficial point mutations would be needed. Each of these would need to lead to slightly increased adaptation so that they could be promoted by natural selection. However, Hoyle considered this unlikely, because once a supply of possible beneficial point mutations is exhausted, adaptive evolution in the direction of increased complexity stops. It is much more likely for point mutations to optimize a gene which is already in striking distance of an optimal state.[50]

If a single point mutation could generate an entirely new gene by acting as a switch to turn on a previously non-functional gene, then that would imply that most of the information was already there, ready to be activated. This simply shifts the problem of generating new genes back one or more steps but does not eliminate it.

Finally, a complex mutation is probabilistically equivalent to multiple simultaneous point mutations. In the scenario of multiple simultaneous point mutations, the probability of some harmful mutation being linked with a beneficial mutation is high. There is a substantial risk that the beneficial mutation will be swept away due to natural selection working against the harmful mutation. Once again, this suggests that any beneficial mutation is likely to be a single point mutation, not a multiplicity of point mutations.

In summary, evolution resulting from beneficial mutations must proceed in small steps in a large population (to enable beneficial mutations to occur and have a fair chance of recurrence). Not only that, but for larger changes, a sequence of beneficial single point mutations must occur, each leading to slightly increased adaptation, so that they can be chosen by natural selection. (A temporal sequence of single point mutations under the influence of natural

50 Hoyle, 1999, p. 137.

selection is probabilistically more likely to occur than all of the mutations occurring simultaneously.)

Small steps can optimize a gene, and different taxa are known to optimize genes in different ways.[51] But to turn the chromosomes of a reptile into those of a mammal, for example, many small steps would be required. A large mutational change in one step does not work, because of the very low probability it would be matched in a member of the opposite sex, which is often required for effective sexual reproduction. Since there must have been many small steps, they should be abundant in the fossil record. However, small steps leading to large changes are never seen in the fossil record, which casts doubt on the possibility of large-scale evolutionary processes extrapolated from causal evolutionary mechanisms.[52]

Hence, Hoyle believed that with respect to mammals, including humans, Neutral Theory suggested that the evolutionary table was tilted towards loss of information and against large-scale gains of information. Beneficial mutations were relatively uncommon, and tended to be overshadowed by harmful ones. A long sequence of beneficial mutations would be required to generate increased complexity, a situation which is extremely unlikely.

How beneficial mutations survive

Hoyle suggested that because only a minority of mutations are beneficial, it is difficult to explain how they can survive.

As we said, in an individual with more than one mutation, any beneficial mutation is likely to be linked with a harmful mutation, and it is likely that the effect of natural selection on the coexisting harmful mutations present in an individual will outweigh the effect of a beneficial mutation, with the good swept away with the bad. How then can beneficial mutations survive long enough to lead to adaptation? How can a beneficial mutation become functionally separated from any coexisting harmful mutations in the individual? Hoyle considered several possible mechanisms which might allow a beneficial mutation to persist.

One possibility was a reverse mutation. Theoretically, if a harmful mutation occurs, it is possible for the harmful change to be subsequently reversed in a future generation by a mutation occurring at the same site, converting the change back to the original state. However, this is incredibly unlikely, so Hoyle concluded that reverse mutations cannot be a key mechanism to decouple a beneficial mutation from other contemporaneous harmful mutations.[53]

Since reverse mutation was not the answer, but beneficial mutations had to become functionally separated from harmful mutations to survive, Hoyle

---

51 Ibid., p. 130.
52 Ibid., p. 138.
53 Ibid., p. 10.

proposed that crossing over (refer Chapter 12) was the key mechanism in organisms which reproduce sexually. As a random, common process, it was possible for crossing over to detach a beneficial mutation from contemporaneous harmful mutations existing in the same individual. Hoyle's mathematical modelling suggested that uncoupling of beneficial mutations from harmful mutations could occur within tens of generations as long as the mutations were not situated close together on the same chromosome. If they were close together on the same chromosome, thousands of generations could be required for uncoupling.[54]

So, Hoyle identified both the presence of chromosomes and crossing over as essential components of enabling slightly helpful mutations to make headway against the flood of harmful mutations in organisms which reproduce sexually. But he noted that slightly harmful mutations that are very small could fly under the radar of natural selection and progressively accumulate. With this in mind, Hoyle estimated mathematically that the gene pool would become progressively corrupted sometime around a million to ten million generations, which he identified as the effective genetic lifespan of corporate humanity, or other biologically similar mammals.[55]

Rather than the genetic code improving across time, Hoyle identified natural selection as one of several mechanisms that (incompletely) held back the flood of harmful mutations. In other words, to suggest that evolutionary mechanisms were adequate to explain all increased complexity in organisms on Earth mischaracterized the nature of the problem.

Adaptation by gene inactivation

Was Hoyle proposing that evolution by mutation and natural selection did not occur? Not at all. Indeed, these processes can lead to relatively rapid, large-scale change, but not in the direction of increased complexity. We term this *gene inactivation*, and there are several ways in which it can occur.

Occasionally, a single point mutation might inactivate a whole gene, leading to increased adaptation (greater specialization) in a particular environment. This latter effect represents a reduction in genetic information, not an increase. For example, tuskless elephants enjoy a survival advantage where poachers are present. However, if poaching is banned, tuskless-ness is no longer an advantage. On the contrary, tuskless elephants might be less able to defend themselves and their offspring from predators. And, if the genes for tusks are lost from the gene pool, future generations of elephants will not be able to grow tusks again.

Where gene inactivation leads to improved adaption, a great selective advantage might be enjoyed, as long as the environmental conditions remain

54 Ibid., p. 136.
55 Ibid.

constant. If the environment reverts to the former state, the gene might be able to be reactivated if a reverse mutation occurs. However, if several mutations are responsible for the inactivation of the gene, it becomes effectively dead.[56] It is very unlikely for that gene to be reactivated as multiple reverse mutations are required. In this case the organism may be at increased risk of extinction because it is unable to reinstate corrupted genetic information needed to cope with reversion of the environmental conditions.

So, under conditions where an animal becomes highly adapted to a niche environment by inactivation of genes, natural selection can impair the potential for further adaptation. Where adaptation occurs by losing genes, there is a limit as to how many genes the animal can afford to lose.

Hence, mutations combined with natural selection can lead to rapid adaptation by gene inactivation, but this represents a reduction in genetic information, not an increase.

The problem of Histone H4

Because large scale changes in evolution would have to occur in small steps, Hoyle observed that this causes a problem for genes which are all-or-nothing cases and cannot have evolved by a sequence of small steps. An example of this is Histone H4, a small protein molecule critical to the packaging of DNA within the cell nucleus.[57]

Histone H4 is found in all organisms which have a membrane-bound nucleus, including many single-celled organisms.[58] The sequence of amino acids in Histone H4 is almost identical across all organisms containing it; even minor alterations in the sequence of amino acids tend to be fatal or induce severe disability. Hence, from the perspective of a biologist, the structure of Histone H4 probably needed to emerge in one step. It is critical to the function of the cells that feature it.

Arriving at the current structure of Histone H4 by random point mutations would have involved many harmful or fatal precursors. But to get a sequence of over 100 amino acids correct in one step is impossible. Hoyle reasoned, therefore, that Histone H4, and molecules like it, cast doubt on the adequacy of evolutionary causal mechanisms to explain the origin of some critical proteins needed for life.

The problem of the male Y-chromosome

Hoyle noted that the Y-chromosome—the male chromosome—has the X-chromosome as its homologous pair. The Y-chromosome is substantially smaller than the X-chromosome, and only very small regions are known to

56 Ibid., p. 101.
57 Ibid., p. 103.
58 Bacteria are an exception: they are single-celled organisms, but do not have membrane-bound nuclei.

participate in crossing over with an X-chromosome. Most of the Y-chromosome does not participate in crossing over at all. Hence it is very unlikely for beneficial mutations on the Y-chromosome to be uncoupled from harmful mutations.

This suggests that over time, Y-chromosomes must become progressively damaged in the gene pool. Some evidence supports the contention that human Y-chromosomes are, on average, undergoing "genetic erosion" and have been shrinking. Other evidence points to self-repair mechanisms built into the Y-chromosome which limit damage.[59] This fits with Hoyle's point that the general direction of evolution is towards degradation of the genome, which is countered, perhaps incompletely, by various protective mechanisms. This is entirely opposite to the general trend of evolution postulated by the secular origins account.

Genetic entropy

Aspects of Hoyle's arguments have been amplified and made more accessible to general audiences by John C. Sanford, one of the co-inventors of the "gene gun".[60] Sanford uses the term *genetic entropy* to describe the steady accumulation of harmful mutations in organisms, leading to aging and the certainty of death. Genetic entropy, Sanford argues, also applies to species as a whole. For example, all humans accumulate an extra 100 mutations, on average, per generation. The gene pool degrades over time. Natural selection can hold back the tide of harmful mutations, but it does so incompletely.

Not only are most mutations harmful, but most of those that are "neutral" are also very slightly harmful. They are neutral in their susceptibility to selection, but not to their overall impact. Harmful mutations must therefore accumulate as Hoyle suggested, including in the Y chromosome, because natural selection cannot "see" many of them. Once these mutations become widespread in the population, they cannot be removed by selection. The general direction of change across the gene pool is toward information degradation, not increase.

Conclusion

Hoyle considered that proponents of Neo-Darwinism had wrongly conceptualized how mutations affect populations. He concluded that "the essential problem for [Neo-Darwinism] ... is how to cope with this continuing flood of adverse mutations".[61] By this he meant that the naive view of many evolutionary biologists (including Darwin himself) that most beneficial

59 White, 2013.

60 See Sanford's book *Genetic Entropy* for further details (Sanford, 2014).

61 Hoyle, 1999, p. 9.

hereditary changes must inevitably become prominent in a population, was mistaken.

As a committed secularist, Hoyle did not despair at the immense problems facing Neo-Darwinism. Instead, he proposed a hypothetical evolutionary mechanism, which he termed a *genetic storm,* to account for large scale genetic changes. Hoyle claimed this could lead to extensive genetic reorganization.[62] He concluded that the evidence supports a scheme where genes must have already existed right from the beginning of life on Earth, ready to be activated in response to ecological need.

While Hoyle's overall scheme was naturalistic, it is notable that this latter conclusion of living organisms containing abundant genetic information from the outset is compatible with biblical creation. Equally important, Hoyle had opened a new and fruitful analytic approach for investigating evolution as origins science.

Hoyle's analysis cast serious doubt on the claims of Neo-Darwinism. His scheme suggested that genetic information was much more likely to degrade, at a molecular level, than to increase. Speciation could certainly occur due to loss of information, but large-scale positive changes were extremely unlikely. Evolution as causal model science was intact, but evolution as a holistic origins theory was rendered implausible.

"Evolution" in origins accounts: the real story

The advances in understanding the mechanisms of "evolution" have severely undercut all purely naturalistic and materialistic theories of common evolutionary descent which have been advanced to date. "Evolution", it seems, is not as the secular consensus suggests. The characteristics of "evolution", if we piece together the evidence base and modelling discussed above and in previous chapters, is:

1. As per Hoyle, a large amount of genetic information was present in organisms from the outset of life on earth.
2. Darwin's theory explains speciation and variation, probably up to at least the level of the genus, but not much further.
3. Natural selection is important in holding back the flood of harmful mutations from overtaking the gene pool
4. It is likely to be extremely rare for "evolution" to proceed in a direction which adds significant amounts of information. Certainly, nowhere near common enough to explain common evolutionary descent from a primordial single-cell organism.

62 Ibid., p. 114.

5. Some animals have genetic mechanisms which can be activated to enable adaptation to a range of environments, often by turning off genes.
6. Most "evolution" at a molecular level involves loss of information. Hence, as organisms become more adapted, they may be more specialized, but likely less genetically flexible. Such organisms have less potential for further variation/speciation and thus are more prone to extinction overall. This is especially so if the loss of a gene involves more than a single mutation and environmental changes revert to their previous state. Also, as information is progressively lost, the capacity for further change is, if anything, diminished. Hence species can eventually come to appear "fixed".
7. Natural selection can fine tune genes but can scarcely create them.
8. "Evolution" can occur quickly.

Notably, all these characteristics of "evolution" are entirely compatible with creationism. Most modern creationists would accept these empirical causal mechanisms and accept that they were operative in history and can help explain adaptation and speciation on a large scale. The key debate between secular humanists and creationists is not whether mutation, natural selection, genetic drift or genetic bottlenecks have occurred—they have—but whether these mechanisms can be extrapolated to account for all the biological diversity and complexity found on earth. We reiterate Roger Lewin's summary of the outcome of the 1980 macroevolution conference at the Chicago Field Museum of Natural History:

> The central question of the Chicago conference was whether the mechanisms underlying microevolution can be extrapolated to explain the phenomenon of macroevolution. At the risk of doing violence to the positions of some of the people at the meeting, the answer can be given as a clear, No.[63]

Conclusion

The various mechanisms of common evolutionary descent remain points of contention to this day. This lack of clarity impacts on the validity of naturalistic evolution as an effective origins speculation. Naturalistic schemes of common evolutionary descent are left in an awkward position.

To illustrate the predicament, we cast our minds back to the days of the French naturalist Baron Cuvier (1769–1832). Cuvier's entire life, we should point out, unfolded in what we would now call the pre-Darwinian era. Nevertheless, proto-evolutionary thought was in circulation, exemplified by the unrefined ideas that Erasmus Darwin expressed in his poetry. Cuvier strenuously rejected these evolutionary ideas, and his objections were based on

---

63 Lewin, 1980.

the best evidence available at the time—the fossil record. One of his biographers, writing in 1833, left to posterity a revealing glimpse into the mindset of the pre-Darwinian evolutionists:

> One of the most important questions treated of in this work is that of the alteration in animal forms; whether the forms of lost animals, which differ so much from those which are now living, really indicate species and genera distinct from species and genera now existing, or if time alone has modified the primitive forms, so as to attain the present form. The examination of this question alone would give a satisfactory answer (could they be convinced) to those who believe in the indefinite alteration of forms in organised beings, and who think that, with time and habits, each species might have made an exchange with another, and thus have resulted from one single species ... M. Cuvier seriously refutes it, and destroys it with one objection, that of not finding intermediate modifications between an animal of the former and present world ... .[64]

Think for a moment about what this means. These pre-Darwinian evolutionists held unflinchingly to their belief despite two great difficulties. In the first place, they were unable to point to any compelling mechanism by which organisms might evolve. Second, the evidence of the fossil record provided no support for their hypothesis. But look at what the biographer writes: *could they be convinced*. Despite no compelling mechanism, and no fossil evidence, these evolutionists were wedded to the idea. It was simply part of their worldview.

Now fast forward to the present. Nearly two centuries of careful paleontological investigation has failed to turn up unequivocal circumstantial evidence for common evolutionary descent. On that score, the evidence is largely in the same state that Baron Cuvier dismissed—at best it points to saltatory evolution or punctuated equilibrium, assuming secular scientific presuppositions such as philosophical uniformitarianism informing the interpretation of the fossil record.

But now consider the second point. We saw above that for us, living in the post-Kimuran era, the evolutionary mechanism has become entirely ambiguous. There remains a significant chance that, in Dobzhansky's words, "evolution would have hardly any meaning, and would not be going anywhere in particular". Certainly, Neo-Darwinism, and any scheme based on random mechanisms for genetic novelty, is mechanistically thwarted by the evidence marshalled by Hoyle, Spetner, Meyer and Axe.

So, it turns out that the difficulties faced by theories of common evolutionary descent in the post-Kimuran era are much the same as the difficulties faced in the pre-Darwinian era. And like the pre-Darwinians, the

64 Lee, 1833, pp. 98–99.

post-Kimuran secular scientists are determined to persist with common evolutionary descent. Dobzhansky again: "To a man looking for the meaning of his existence, evolution by natural selection makes sense."

Belief in common evolutionary descent was, and still is, largely a personal choice. It is based, in the final analysis, not on persuasive direct or indirect evidence, nor on compelling empirical causal model science, but on the worldview of the individual.

References

Axe, D. (2016). *Undeniable: How Biology Confirms Our Intuition That Life Is Designed* (eBook). New York: HarperOne.

Hoyle, F. (1999). *Mathematics of Evolution*. Memphis: Acorn Enterprises LLC.

Lee, M. R. (1833). *Memoirs of Baron Cuvier*. London: Longman, Rees, Orme, Brown, Green & Longman.

Lewin, R. (1980, November 21). Evolutionary Theory Under Fire: An historic conference in Chicago challenges the four-decade long dominance of Neo-Darwinism. *Science* 210(4472), 883–887.

Meyer, S. C. (2013). *Darwin's Doubt: The Explosive Origin of Animal Life and the Case for Intelligent Design*. New York: HarperCollins Publishers.

Sanford, J. C. (2014). *Genetic Entropy*. Canandaigua: FMS Publishers.

Shannon, C. (1949). *A Mathematical Theory of Communication*. Illinois: University of Illinois Press.

Spetner, L. M. (1999). *Not By Chance! Shattering the Modern Theory of Evolution*. New York: The Judaica Press, Inc.

Wagner, A. (2014). *The Arrival Of The Fittest: Solving Evolutions Greatest Puzzle*. London: Oneworld Publications.

White, D. (2013, February 15). Extinction of the human male. Retrieved from creation.com/y-chromosome-extinction

Wieland, C. (1998, December 7). Not By Chance! Shattering the modern theory of evolution. Retrieved May 12, 2024, from creation.com/not-by-chance

14
Darwinian aftermath: the rise of neopaganism

Before we conclude our rebuttal of the origins account of secular science, it is worth a brief stocktake of the ground covered thus far. In Part 1, we reviewed five criteria for assessing worldviews and found that proponents of the holistic secular origins account have been unable to conclusively identify an entity with no beginning, leaving the problem of existence unresolved for now. We showed that several prominent physicists, such as Einstein, Wigner and Feynman, have found the secular account unable to account for the coherence of the universe. Secular origins accounts have failed to provide an adequate explanation for non-biological and biological information, except by invoking non-scientific philosophical devices such as the multiverse, quasi-scientific theories such as panspermia (either directed or undirected), or unscientific explanations such as the "miracle of time", and a range of "just-so" stories.

We also noted that not only is consciousness unexplained by current science, but that the origin and diversity of life itself is not adequately explained. We showed that abiogenesis was disproven on probabilistic grounds, and that there is a lack of direct and indirect evidence for Neo-Darwinism as a holistic evolutionary account. There is also a lack of clarity regarding the causal mechanism for common evolutionary descent as part of a holistic naturalistic origins account.

In short, the secular origins account failed across the board. The account is far from the victorious enterprise that it is made out to be.

To complete our overview of the secular side of the origins debate, our remaining tasks are to investigate what underlies the secular origins account and the effects it has on society. A good starting point is to investigate modern atheism. The term *atheist* is negative in that it describes what its adherents *do not* believe. So, what is a positive term that captures what atheists *do* believe?

Polyinfinitism

As we have seen, scientifically minded atheists must—necessarily, even if implicitly—believe in one or more non-personal infinities that underpin their origins account, a belief we term *infinitism*. Comparing the two terms *theism* (as applied to belief in the biblical God) and *infinitism*, we may say that both are beliefs associated with origins accounts, and both postulate one or more empirically untestable infinite entities that underpin that account. These infinities, being inherently outside the scope of empirical science, are of a philosophical or religious nature. We saw in Chapter 9 that the Big Bang model postulates a singularity of infinite density and temperature; a quasi-infinity of 13.8 billion years duration; and an implicit infinity, not clearly identified, to account for existence. In abiogenesis, although panspermia is certainly invoked to raise the possibility of an extremely life-friendly niche found somewhere else in the cosmos, it often effectively (and on occasion overtly) involves an appeal to another infinity—a quasi-infinity of life-compatible planets. Panspermia also conveniently removes the problems of abiogenesis to a non-testable, hypothetical 'elsewhere'. The fine-tuning of the universe may be addressed by a purported infinity of universes—the usual version of the multiverse. And unlike the eternal universe concept, in which the observable universe was simply believed to have had no beginning, these more recently postulated infinities cannot be directly observed at all.

These infinities also mark the process by which the mono-infinitism of the eternal universe, accepted by secular science in the eighteenth and early nineteenth centuries, has become the polyinfinitism of current secular origins science. This process is reminiscent of the development of poly*theism*, in which ancient societies generated a variety of gods to explain various phenomena—mountains, celestial bodies, fertility and so forth. In an analogous way, secular origins science has concocted new infinities. The term polyinfinitism may be equally applicable to both agnostics and atheists, and indeed to scientifically minded secular humanists in general. The earlier eternal universe mono-infinitism of Darwin or Clodd was compatible with atheism or Deism—the latter a worldview consistent with belief in a creator, but one which denies God's miraculous intervention in history. Contemporary polyinfinitism is different—it has no commonality with either monotheism or Christianity whatsoever. From a Christian perspective, scientific polyinfinitism exhibits some resemblance to ancient paganism but with a modern secular twist.

A broad and diverse origins community

Clearly, secular humanists do not object to the notion of the infinite being applied to origin accounts; what most find objectionable is the idea that the infinite is personal, rather than impersonal. Fred Hoyle and Chandra Wickramasinghe noted that the ultimate sin in the eyes of many secular scientists is to posit a theory which might imply cosmic "purpose"—invoking the word "purpose" is worse even than denying Darwinian evolution.[1 2]

However, a minority of secularists do posit a purposive universe as an alternative to polyinfinitism. In this scenario, life has not arisen through strictly random processes, but via a "teleological" or goal directed aspect to matter that "strives", or at least greatly facilitates, the development of living organisms. This does away with the need for several infinities. Life just had to happen.

This idea predates the finite universe of the 1960s. Henri Bergson wrote about it in his book *Creative Evolution* (1907). A more recent exponent was Belgian biochemist and Nobel prize winner Christian de Duve, who argued for "a cosmic imperative" of life.[3] De Duve wrote that he did not believe the universe had a purpose per se, preferring "the indisputable statement that the universe happens to be such that certain events, including the generation of life and mind, were possible, perhaps even probable, if not obligatory".[4] Another recent adherent of this type of view is Thomas Nagel, who as we have already seen, finds the Neo-Darwinian view of life to be untenable and proposes as an alternative the "teleological" emergence of life.

As shown in Figure 14.1, the problems of existence and information can move secularists into several different positions. They might subscribe to polyinfinitism, abandoning strict empiricism in the process, because infinity is not a part of empirical science, as we saw in Part 1.

An alternative is to postulate a purposive universe. Once a purposive universe is entertained, a reasonable question is where purpose originates. Some secular thinkers propose purpose as an emergent property of natural processes. For example, Richard Dawkins argues that genes have purpose and are "selfish" insofar as they always act to maximize their survival. Others suggest the purpose of the universe is both intrinsic and emergent, concluding the universe itself has a divine or sacred aspect. This could lead in the direction of pantheism; Albert Einstein is regarded as having held this view. If purpose is attributed to an intelligent designer who initiated the universe, Deism could become a logical direction. It is possible that philosopher Antony Flew came to accept this position after his switch from atheism to theism. Most of these options arguably require a rejection of the *strictly* secular scientific worldview that many secular scientists profess to hold dear.

---

1 Paul Davies, the science communicator we met in Chapter 10, stated that he is "always careful about using words like purpose or goal because I get jumped on ..." (Closer to Truth, 2020, from 3:29–3:33).

2 Hoyle & Wickramasinghe, 1981, p. 32.

3 Refer to de Duve's 1995 book, *Vital Dust: Life as a cosmic imperative*.

4 De Duve, 2013.

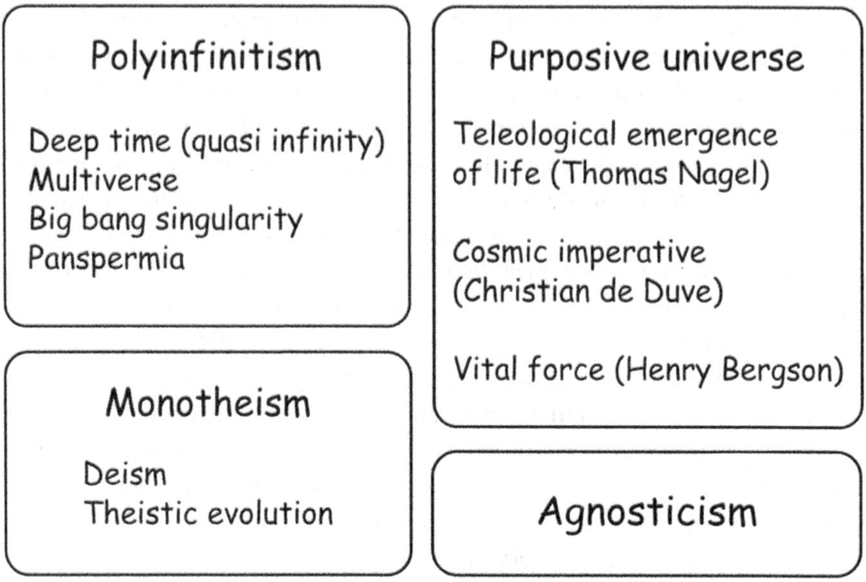

Figure 14.1: Some of the diverse viewpoints encompassed by evolutionism

Agnosticism, or unbelief in relation to philosophical or religious matters, is another option. It is a position which may reject holding to a firm origins account—removing the possibility of certainty. Agnosticism leaves unsettled the identity of the entity with no beginning—although secular humanist agnostics typically express certainty that the entity with no beginning is *not* the God of Genesis.

The upshot is that proponents of the secular account comprise a diverse range of views about ultimate reality, constituting a "broad tent", so to speak.

The religious aspect of evolutionism

Several secular thinkers have admitted that their views constitute a religious perspective. Richard Dawkins, for example, wrote:

> Most religions offer a cosmology and a biology, a theory of life, a theory of origins, and reasons for existence. In doing so, they demonstrate that religion is, in a sense, science; it's just bad science.[5]

If religion is "bad science", then "good science" (from the perspective of Dawkins and his fellow travelers) also represents a religious perspective. Dawkins confirmed this understanding of his position later in his essay:

5 Dawkins, 2002.

> [W]hen it is put to me that … evolutionary theory … is just a religion like any other, I usually deny it with indignation. But I've begun to wonder whether perhaps that's the wrong tactic. Perhaps the right tactic is to accept the charge gratefully and demand equal time for science in religious education classes. And the more I think about it, the more I realize that an excellent case could be made for this.[6]

We agree with Dawkins that the domains of origins science and religion overlap significantly. But there is another similarity. Journalist Fred Reed was surprised by the dogmatic response of evolutionists—the kind of responses commonly ascribed to religious fundamentalists—when he started asking questions:

> I wanted to see how serious proponents of evolutionary biology would respond to awkward questions.
>
> It was like giving a bobcat a prostate exam. I got everything but answers. They told me I was a crank, implied over and over (again) that I was a Creationist, said that I was an enemy of science (someone who asks for evidence is an enemy of science). They said that I was trying to pull down modern biology (if you ask questions about an aspect of biology, you want to pull down biology). They told me I didn't know anything (that's why I was asking questions), and that I was a mere journalist (the validity of a question depends on its source rather than its content).
>
> But they didn't answer the questions. They ducked and dodged and evaded. After thirty years in journalism, I know ducking and dodging when I see it. It was like cross-examining hostile witnesses.
>
> This is the behavior not of scientists, but of advocates, of True Believers. I used to think that science was about asking questions, not about defending things you didn't really know.[7]

Why are some secular scientists so intent on an aggressive defense of evolution as a holistic explanation for the diversity of life, rather than, for example, redirecting their efforts to explain abiogenesis? Writer Gary North provided a compelling explanation for the aggressive championing of evolution. North pointed out that evolutionary descent, as the scheme which does away with the Christian trinitarian God, supplies the sanction for man to determine good and evil.[8] This is the heart of what Darwin described as the "Devil's gospel", concerning which abiogenesis and secular cosmology are mere supporting acts. The devil promised Eve that eating the forbidden fruit would enable her to "be like God, knowing good and evil" (Genesis 3:4) and

6 Ibid.

7 Reed, 2020.

8 North, 2012, pp. 325–331, 337–338.

Eve's children have apparently turned up to collect, recognizing evolutionary descent as the linchpin on which their "god-likeness" depends.

North added that theories of evolutionary descent work in a two-step process. First, God is relegated to the dustbin of history, a figment of superstitious imaginations. Next, *some men*, those who understand the implications of evolution, have an opportunity, indeed a moral imperative, to direct human evolution along a favorable path.[9] Evolution is a doctrine, a religious idea which, in masquerading as science, is tremendously subtle.

Francis Galton (1822–1911), Darwin's half-cousin and eugenics pioneer whom we met in Chapter 7, described the religious/moral duties of the evolutionist:

> To sum up in a few words. The chief result of these Inquiries has been to elicit the religious significance of the doctrine of evolution. It suggests an alteration in our mental attitude, and imposes a new moral duty. The new mental attitude is one of a greater sense of moral freedom, responsibility, and opportunity; the new duty which is supposed to be exercised concurrently with, and not in opposition to the old ones upon which the social fabric depends, is an endeavour to further evolution, especially that of the human race.[10]

That *some* evolutionists regard their doctrine as having religious connotations is unsurprising. American evolutionist Ernst Mayr stated that

> [v]irtually all biologists are religious, in the deeper sense of this word, even though it may be a religion without revelation, as it was called by Julian Huxley. The unknown and maybe unknowable instills in us a sense of humility and awe … .[11]

Mayr went on to propose that if biologists moved away from defining mankind as a multiplicity of individuals, towards defining man as unitary mankind (i.e. emphasizing "the one" instead of "the many"), this would allow an "ethics and ideology" he found congenial, and which he felt would enable the conflict between evolutionary science and more traditional social values to be solved. Mayr's comments underline the fact that prominent evolutionists have found in the theory not only a foundation for ethics, but a basis for views which they themselves characterize as religious.

Richard Dawkins again tipped his hand in an essay titled "Darwin triumphant: Darwinism as a universal truth".[12] In positing the "truth" of Darwinism, Dawkins decisively oversteps the usual scientific categories of hypothesis, theory, and evidence. The word "truth" implies a philosophical, or

---

9 Ibid., pp. 350–351.
10 Galton, 1883, p. 337.
11 Mayr, 1982, p. 81.
12 Dawkins, 2003, p. 78.

religious aspect to Darwinism that its proponents usually like to conceal, preferring instead to discuss the supposed conflict between religion and science. The other word that stands out is "triumphant". While this may be used to describe a scientific theory that displaces a rival hypothesis, the clear implication is that Darwinism has triumphed over religious origins accounts, including the biblical. As Dawkins notes in his essay:

> Adaptations, especially complex adaptations, awake such a powerful hunger that they have traditionally provided one of the main motivations for belief in a supernatural Creator. The problem of adaptation, therefore, really was a big problem, a problem worthy of the big solution that Darwin provided.[13]

Indeed, the word "triumph" is one that the Bible repeatedly applies to Christ and His work (refer 2 Corinthians 2:14 and Colossians 2:15). Dawkins' message, therefore, carries the connotation that Darwin provided a truth that superseded the truth of the Bible and of Christ. This is a religious battle. In line with that, Stephen Jay Gould observed that scientists have been happy to appear as a "new priesthood", a situation from which they "derive benefits".[14]

Non-Christian philosopher of science Michael Ruse has used the phrase "secular religious perspective"[15] to describe evolutionary origins schemes, such as Neo-Darwinism. We consider this an apt description: "religious perspective" is a category at arm's length from organized religion, but the religious aspect of the worldview is explicitly acknowledged. Ruse identifies Darwinism as a bespoke anti-Christian religious perspective; it was Christianity against which it "was formed and fought and grew and matured."[16] Darwinism was acknowledged, from the very start, as having an explicitly religious connotation.

Evolutionary dogma became the established religious perspective throughout the West from the late 1960s, in the wake of the Epperson v. Arkansas decision in the United States Supreme Court. Although non-Christian and anti-Christian sentiment had been rising for centuries, especially among the Western elite, this was the decisive point at which a leading nation of the West officially (legally) changed religion. At this point, evolutionary dogma—that humans have, like all other creatures, descended from lower forms of life—became part of the science curriculum, taught as part of a holistic scientific origins account in schools and universities. Secular humanism had graduated from teaching human evolution from lower forms of life implicitly and covertly in the pre-Epperson period, to teaching evolution explicitly and legally. The foundations underpinning more than a millennium of Christian understanding

13 Ibid., p. 81.
14 Gould, 2012.
15 Ruse, 2017, p. ix.
16 Ibid., p. 282.

were officially rejected as incompatible with science. Secular humanists have since gone on to attain enormous cultural and political power in the West.

Note the significant shift that this portended: prior to the late 1960s, Protestant secular nations such as the U.S. and Australia espoused such ideals as religious neutrality, freedom of religion, and separation of religion and state. From the 1970s, this morphed into the *secular humanist state*, which is notably anti-Christian. This was a subtle, yet highly significant change.

Damaging social effects of evolutionism

While the evolutionary worldview certainly has some of the trappings of religion, its greatest area of divergence from mainstream religions is its non-corporate nature. For example, the "Brights", a group of rebranded atheists that launched in 2003, attempted to establish monthly meetings (known as "meetups") in London. Author Alistair McGrath reported on the attendance at these meetings between 2003 and 2010.[17] Of the 86 months that fall under McGrath's reporting window, 47 meetups are noted, which averages out to about one meeting every other month. The program got off to a reasonable start; the meetups were mostly monthly in the period 2003–2006. But from that point, a serious loss of momentum progressively scuppered the program. Four meetups in 2007 were followed by a couple in 2008 and only one in 2009. No meetings were held in the first ten months of 2010, at which point McGrath's records stop.

We might expect the large, cosmopolitan city of London to contain many thousands of atheists willing to add their support to the cause. But in fact, attendance at these monthly meetups averaged out at just 17, a level at which most churches would face closure. Atheists, apparently, have no reason to meet regularly; their movement appears more suited to large, infrequently held conferences at which famous atheists are invited to speak. Naturally, this lack of regular meetups is their prerogative and right, but at a societal level, a shift from Christianity to atheistic evolutionism results in a serious loss of what is termed *social capital*. This latter concept was brought to the fore by American political scientist Robert D. Putnam in his 2000 book *Bowling Alone: The collapse and revival of American community*.

Social capital refers to the capacity for economic benefit resulting from connectedness in the community. Historically, economists used the term *capital* to refer to assets used in the production of economic value. Roads and bridges are a good example—these facilitate efficient transport of goods throughout the community. Hospitals are another prominent example; these benefit the economy by maintaining or improving the health of the community. Economists then expanded the concept of capital to *human capital*, referring to

17 McGrath, 2011, p. 36.

the ability of a population to add economic value, usually through advanced education and training. (Countries invest in their human capital by rolling out improved educational opportunities for the population.) But more recently, economists have developed the concept of social capital, which refers to the measurable economic benefits available to societies with higher levels of connectedness.

A few examples will serve to illustrate the concept. As the level of trust in a society increases, commerce is facilitated; agreements can be reached via handshake rather than by contracts drawn up by legal experts (at considerable expense). Another benefit of a society with high levels of trust is the reduced expenditure on security. Our father recalls that in his community in 1950s Australia, house doors were rarely locked. These days, houses are not only locked, but many of them bristle with a variety of security features, often installed at considerable expense.

Australian economics professor-turned-politician Andrew Leigh has investigated the issue of social capital[18] and concluded that churches are good for society.[19] Leigh points out that at church, attendees could meet an unemployed person, and, because of that personal connection, become less likely to stigmatize the unemployed generally. Conversely, they could meet a CEO and become more likely to see these types of people as fellow human beings too.

Ironically, Leigh is an atheist, stating that "the science-based critiques of religion have taken me to the point where I do not believe in an external being".[20] Leigh is in some respects like Ronald Numbers; Leigh is from a religious background (including a grandfather who was a Methodist preacher), has been influenced away from the religious beliefs of his forebears by "science-based critiques of religion", yet he retains a healthy respect for religious communities and the positive impact they have on society. "Churchgoers are more likely to give money, including to non-church causes, more likely to donate blood and turn out to vote," Leigh says, adding that "going to church imbues that ethic of helping others and we have lost a little of that".[21]

Leigh's conclusions make sense. Evolutionism is not in line with the realities of a created humanity made for fellowship with God and each other. It can lead to impoverished societies in which the needy are less likely to be helped and fewer people volunteer. The escalating risk of societal fragmentation is a largely unrecognized side-effect of the rise of evolutionism.

---

18 During his PhD studies at Harvard, Leigh worked on the research team of Robert Putnam, whose book *Bowling Alone* discusses social capital in the American context.

19 Eternity News Podcasts, 2019.

20 Herald, 2010.

21 Ibid.

Perhaps as a byproduct of their more disconnected lives, the secular evolutionist cohort tends to exhibit relatively low rates of marriage and partnering. A Pew Research study into marriage rates among various religious groups in the U.S. (2018) found that the cohorts with religious statuses of "atheist", "agnostic" or "nothing in particular" all demonstrated combined marriage/cohabitation rates of less than 50%—among the lowest in the study. At the opposite end of the scale were those identifying as Presbyterians and Anglicans, who were among the most married/partnered cohorts.[22]

But how does an ideology lacking firm scientific support achieve such a rhetorical advantage over biblical Christianity? There are several reasons, of which we will investigate two. The first is in the educational arena.

Government funded schooling

We saw earlier that universal education was established by sixteenth century Protestants so that everyone could read the Bible for themselves. Jesuits set up a similar system in the Roman Catholic sphere. Over time, this educational impulse became institutionalized, such that governments around the world required all their citizens to attend a set minimum of education. In recent times, the United Nations published its Convention on the Rights of the Child, Article 28,[23] which sets out the right of the child to an education.

Thus, the system set up by Protestants at a grassroots level has become universalized. Secular humanists have long shown a cuckoo-like tendency to commandeer programs and institutions established by Christians, and education is no exception. Compulsory education has enabled secular humanists to take control of school curricula via the top-down process of influencing government. Lawsuits that then keep Christian viewpoints out have been very successful to date, especially in the United States.

In the 1880's, the Presbyterian leader A. A. Hodge (1823–1886) predicted that government funded education must lead to widespread atheism. Hodge's rationale was that in government funded education, Protestant parents would complain vigorously if Roman Catholic doctrine were taught, and vice versa. Jews would complain if Christianity were taught, and in like manner Christians would complain if Talmudic Judaism were taught. In the end, Hodge reasoned, *those who believe the least*—atheists and secularists—must eventually remove all other beliefs from government mandated curricula, even if they are a minority player to begin with. Thus, government funded education must become progressively secularized.

The "hidden curriculum" of secular pedagogy teaches that religious beliefs are superfluous to a functional society, and that religious beliefs do not belong

22 PewResearch.org, 2018.

23 United Nations Convention on the Rights of the Child, 2023.

in the public sphere—they should be kept private. Secular education also implicitly suggests that the world around us can be fully understood in purely secular terms, a concept which contradicts the Bible. A secular worldview becomes internalized by some students. Even a very low annual rate of conversion to secular humanism (or other non-Christian worldviews) is sufficient to secularize the state, because, if this rate is maintained over time, the proportion of non-Christians will grow exponentially (assuming most children of secularists adopt their parent's philosophy). Secular education, being heavily subsidized by the taxpayer, is more affordable than private church-affiliated education, and only the truly committed are prepared to find ways to shoulder the extra cost. As time proceeds, more and more voters become secularists and atheists, such that democracy itself develops a secularizing tendency. While Christians are raising money and volunteers for local and overseas missionary work, some secular humanists are proselytizing children in classrooms, on the public purse, during office hours.[24] These secular humanist "missionaries" are at least partly funded through the tax receipts of Christians. Hodge's analysis has proved prescient.

This brings us to the second reason for the rhetorical advantage of the secular origins account.

Crushing of dissent by the atheistic establishment

Secular humanism has unquestionably expanded in influence by crushing dissent in academia. This is related to the scientific consensus that we saw earlier in McElreath's social construct view of science. Once the Darwinian origins account became the default, academia was largely forced to adopt the consensus. As Hoyle and Wickramasinghe noted,

> Most scientists took the correctness of Darwinism to be axiomatic, and they simply argued that any hypothesis needed to make Darwinism work had to be true.[25]

The scientific consensus assumed Darwinism was true, and the "just-so" stories needed for its support were assumed to be true as well. Having achieved that consensus, the establishment has not brooked any dissent.

An example of the take-no-prisoners attitude that spilled out into the public sphere was the expulsion of Michael Reiss, an educator who worked for the

---

24 For example, prominent secular humanist C. F. Potter wrote in 1930 that "...every American school is a school of Humanism. What can the theistic Sunday-schools, meeting for an hour once a week, and teaching only a fraction of the children, do to stem the tide of a five-day program of humanistic teaching?" (Potter, 1930). In 1983, John J. Dunphy wrote a prize-winning essay published in "The Humanist" magazine which stated that "... the battle for humankind's future must be waged and won in the public school classroom by teachers who correctly perceive their role as the proselytizers of a new faith ... These teachers ... will be ministers of another sort, utilizing a classroom instead of a pulpit to convey humanists values in whatever subject they teach ..." (Dunphy, 1983).
25 Hoyle & Wickramasinghe, 1981, p. 15.

Royal Society in the UK. Reiss had the temerity to suggest that the non-scientific nature of creationism should be clarified in science classes, rather than dismissed out of hand.[26] Reiss's views were misrepresented by atheistic scientists, and he resigned under pressure.[27] Even though Reiss was an evolutionist, "several Royal Society fellows [said] his religious views [made] him an inappropriate choice for the post".[28] On the face of it, the situation in which Reiss was placed before he resigned from his position appears to be unethical and discriminatory.

The commercially successful 2008 documentary *Expelled: No Intelligence Allowed* by controversial American lawyer and commentator Ben Stein, investigated the alleged persecution of proponents of Intelligent Design by the scientific establishment. Stein stated in the documentary that "Intelligent design was being suppressed in a systematic and ruthless fashion". This documentary was widely derided and labeled as propaganda by members of the scientific establishment.

Christians have long lamented the coercive control of the secular humanist establishment, which threatens careers of those who disagree with the evolutionary origins account. Jerry Bergman's book *Slaughter of the Dissidents: The Shocking Truth About Killing The Careers Of Darwin Doubters*, Volume 1, quotes the opinion of Thomas Dwight of Harvard: "The tyranny in the matter of evolution is overwhelming to a degree of which the outsider has no idea".[29] Hoyle and Wickramasinghe, who regarded Darwinism as a provably incorrect origins theory, described several encounters they had with biologists who communicated as follows:

> I will admit that your views are consistent with the facts, and that they even have a certain driving logical quality which normal [Darwinian] theory lacks, but I just cannot bring myself to face the upheaval in my thinking that would follow if I agreed with anything you say.[30]

Radiating from the universities and the media, the scientific consensus of common evolutionary descent of humans and all other animals from lower life-forms has filtered down through the layers of society. With constant repetition and with secular humanist origins mantras emanating from multiple sources, a powerful propaganda-like effect is in place.

---

26 Reiss, 2008.
27 Halloway, 2008.
28 Ibid.
29 Bergman, 2012, p. 23.
30 Hoyle & Wickramasinghe, 1981, p. 137.

Satanic discourse

With one of their aims being to defeat the God of the Bible, it is hardly surprising that some evolutionists have historically used provocative metaphors to mock religion. (We recall Darwin's letter to Huxley, in which he described Huxley as *My good & kind agent for the propagation of the Gospel ie the Devil's gospel.*) This language, while not literal, reflects the deep cultural tension between these worldviews.

For Dawkins' part, the collection of essays containing "Darwin triumphant", which we cited above, is a book titled *A Devil's Chaplain: selected essays by Richard Dawkins.*[31 32] Dawkins also wrote the foreword for Dan Barker's *God: the most unpleasant character in all fiction*, a book which drew its title from a sentence in Dawkins' *The God Delusion*. Here is Barker's view of the devil:

> Satan, in the Bible, murdered only ten people. And even this was accomplished only with the direct authority and approval of God. Poor Satan gets all the blame and none of the respect he desires.[33]

Barker points out later in the book that he regards Satan as a fictional character,[34] but it is disturbing to see the blasé attitudes that evolutionists have to the devil.

The above are examples of what Swedish historian Per Faxneld terms *satanic discourse*. According to Faxneld, satanic discourse occurs on a spectrum; occasional mentions by the likes of Darwin and Dawkins are at one end of the spectrum, and do not imply that the people believe that Satan exists. Instead, they mention Satan as a way of showing where they stand regarding Christianity. Further along the spectrum are those who, while still not believing in Satan, nonetheless identify strongly with what the devil stands for. Unlike Darwin and Dawkins, these people engage in satanic discourse frequently and deliberately. Satanism is a part of their identity. It is worth noting that most contemporary satanists self-identify as atheists and do not believe in Satan as a real entity. For example, Peter H. Gilmore, leader of the Church of Satan (and successor to founder Anton LaVey), stated in an interview in 2007:

> We satanists see ourselves as the adversaries to all faith-based approaches to belief. So, we are atheists, we are skeptics, we embrace science as the method for understanding the universe, we look at satanism as being unreligion ... . The aesthetic component of satanism is the symbol of Satan himself, and we see it not as a representation of evil, but as one of

---

31 Dawkins, 2004.

32 The phrase "Devil's chaplain" comes from a letter Charles Darwin sent to J. D. Hooker in 1856, where he wrote "What a book a Devil's chaplain might write on the clumsy, wasteful, blunderingly low & horridly cruel works of nature!" (Darwin Correspondence Project, n.d.)

33 Barker, 2016, p. 7.

34 Ibid., p. 203.

> individualism, liberty and pride, which are all things we value … We satanists see ourselves as a subset of atheism … . Evolution is the way things function … . Darwin is correct. There is plenty of evidence that supports that. There is a certain struggle for survival … . I think that science is utterly compatible with satanism, it's our main approach to understanding the universe around us … .[35]

Evolutionists engaging in satanic discourse do so to mock Christians, while also publicly communicating implacable opposition to organized monotheism, especially Christianity. But these statements are no longer made quietly and inconspicuously, as in Darwin's day—they are shouted from the rooftops. Evolutionists are no longer "afraid of the people"; some prominent spokespeople fearlessly and openly mock Christian belief.

The rise of neopaganism

Richard Dawkins has admitted that one aspect of evolutionary theory has long baffled him—its timing. Dawkins has never understood why humanity did not discover natural selection until relatively recent times. He expressed his astonishment in *A Devil's Chaplain: Selected Essays*:

> I have long found it a complete mystery why it had to wait until the nineteenth century before anyone thought of it. Archimedes' and Newton's achievements seem, on the face of it, far more difficult.[36]

Clearly, Dawkins is missing a view of the big picture. The reason Darwin's theory of natural selection came about when it did is remarkably simple—and highly revealing about evolutionary thought. In fact, Darwinism is not even the only facet of modern life to share this timing.

We again turn to Edward Clodd's *Pioneers of Evolution: From Thales to Huxley* (1897). Clodd's Preface outlines his reason for writing, and also contains a fascinating insight in the the origins of modern origins science:

> This book needs only brief introduction. It attempts to tell the story of the origin of the Evolution idea in Ionia, and, after long arrest, of the revival of that idea in modern times …
>
> Between birth and revival there were the centuries of suspended animation, when the nepenthe [an ancient potion mentioned by Homer in *The Odyssey*] of dogma drugged the reason; the Church teaching, and the laity mechanically accepting, the sufficiency of the Scriptures and of the General Councils to decide on matters which lie outside the domain of both.

---

35 Grothe, 2007.

36 Dawkins, 2004, p. 94.

As we saw earlier, and as Clodd's Preface indicates, evolutionary concepts were thought to have roots in ancient Greece, or Ionia.[37] The philosophers of that time had speculated on the origins of the universe and earthly life forms, and this thinking extended to proto-evolutionary concepts. What Clodd makes clear, however, is that the modern incarnation of evolutionary thought is a revival of these ancient speculations. In Clodd's mind, the ancients had made a start on scientific secular origins, which was later "arrested" by the rise of Christianity and its drug-like effect on reason. But then "modern times" arrived, and with them, "the Renascence [sic] of Evolution", as Clodd later puts it. But what was it that caused these "centuries of suspended aimation" to come to an end, and bring evolution back to the fore?

An answer is found in the "immense fascination for ancient Greece"[38] that arose in the intellectual life of the Western elite in the eighteenth and nineteenth centuries. Prior to this, educated Europeans had largely thought of their civilisation as being "Roman or Christian in origin, with merely peripheral roots in Greece".[39] But from the mid-eighteenth century onwards, the elite sought to cast off any perceived links to Christianity. The literati in particular sought inspiration from non-Christian sources, and identified the pagan society of ancient Greece as a wellspring of "artistic patterns, ethical values, and concepts of human nature that could displace those of Christianity and ossified French classicism".[40]

This new intellectual climate spread across Europe and to the United States. Ideas from ancient Greece manifested themselves in *neoclassicism*, a movement which influenced architecture, art and literature. In the U.S., the Capitol building and the White House (both competed in the early 1800s) are examples of neoclassical architecture; in the U.K., the British Museum and Buckingham Palace are prominent examples. In literature, Edgar Allan Poe's poem *To Helen* (1846) seemed to sum up the fascination for the ancients with the now-famous phrase "the glory that was Greece". In children's literature, Kenneth Grahame's *The Wind in the Willows* (1908) contained a scene in which the main characters worship the Greek god Pan.

The impetus to revitalise ancient Greek culture also spread into the sporting domain. A British doctor, William Penny Brookes (1809–1895), who resided in the Shropshire town of Much Wenlock, was keen to promote the virtues of exercise and sport. But as a *philhellene*[41] (that is, one who admired the ancient Greeks), Brookes sought to achieve his purpose by reviving the

---

37 We saw this in Chapter 5, where we also cited Henry Fairfield Osborn's 1894 book *From the Greeks to Darwin: a history of the development of the evolution idea.*

38 Turner, 1981, p. 1.

39 Ibid., p. 2.

40 Ibid.

41 Beale, 2011, p. 110.

Olympic Games. Accordingly, he started the Wenlock Olympian Games in 1850.

There are several important similarities between the emergence of Darwinian evolution and the modern Olympics. Both resulted from the eighteenth and nineteenth century turn towards ancient Greece as a source of non-Christian inspiration. As Edward Clodd's Preface states, modern evolution was a "revival" of the ancient ideas. Similar language is used for the Olympics; Brookes "urged [the Greeks] to revive their ancient games."[42]

From this common impetus, both Darwinan evolution and the modern Olympics developed into juggernauts through similar processes. Brookes was analogous to Erasmus Darwin; both succeeded in popularising the Olympics and evolutionary thought, respectively, to a limited extent. However, neither formulated the outstanding ideas that would catapault their domain into the mainstream. For that to happen, brilliant successors would be needed; successors whose ideas were enabled by external developments, coupled with an influential set of elite backers.

In the case of evolution, the brilliant successor was Charles Darwin. Darwin's ideas relied on the long ages that Charles Lyell had popularised in 1830. Post-1830, developments to the evolution idea were made by several intellectuals, most famously by Darwin and Wallace. The influential backers that gave the idea critical support came in the form of the X Club, who advocated for Darwinism at the highest levels, and in an organised way.

In the modern Olympic movement, the person analogous to Charles Darwin was the French aristocrat Pierre de Coubertin (1863–1937), who also played a role analogous to the X Club. Coubertin visited the Wenlock Olympian Society in 1890, and was impressed with what he saw. But Coubertin's ingenious development was his idea to internationalise the Olympics—that is, host the games in a succession of different cities around the world. This idea,

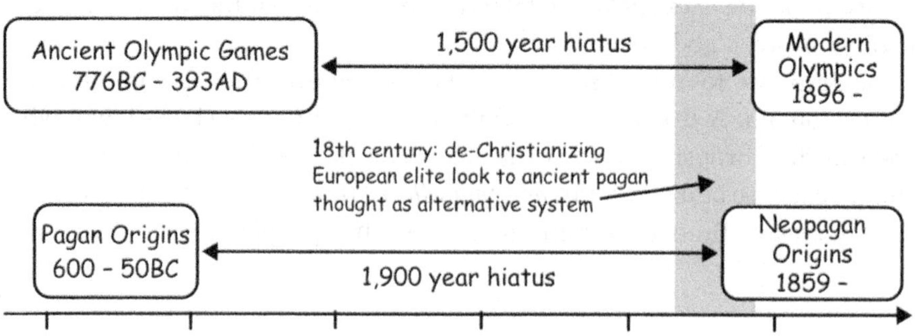

Figure 14.2: Timeline of the emergence of neopagan origins (Darwin) and the Olympics

42 Ibid., p. 10.

coming when it did, depended on external developments that were not in place when Brookes organised the early Wenlock Olympian Games. In transport technology, the steamers of the 1850s were giving way to ocean liners in the 1890s. Fast, comfortable transport, on which the international Olympics would rely, was therefore increasingly available to the masses. In communications, a network of undersea cables was operational by the 1870s; these enabled rapid worldwide communications, crucial to the success of the international Olympics. Coubertin, as a well-connected aristocrat, was able to drum up support and form the International Olympic Committee (IOC) in 1894. The IOC organised the first modern Olympics in Athens in 1896, and the Olympic movement went on to become an institution.

Darwinism and the modern Olympics stand as reminders of the revival of ancient Greek ideas. Both arrived on the world scene at almost exactly the same time (separated by less than 40 years after a hiatus of at least 1,500 years,[43] as shown in Figure 14.2) and for the same reasons.

But the similarities end there. While the modern Olympics and the broader neoclassicist movement were visible evidence that pagan Greek ideas were in vogue with the de-Christianised Western elite, they were not the kind of phenomena to provoke a broader anti-Christian impulse. Darwinism was different. It would eventually drive a presuppositional shift in the masses that would radically alter the religious perspective of the West.

This new religious perspective is the reemergence of paganism itself. Steven D. Smith, Professor of Law at University of San Diego, terms this broad religious perspective *neopaganism*.[44] Smith views paganism and Christianity as deeply incompatible and considers the 2000-year history of the West as a culture war between these two religious systems. This represents a different emphasis from the conventional account of Western history, which posits a progression from paganism to Christianity to secularism, then modernism and now to post-modernism. According to Smith, the West has witnessed a long-term back-and-forth tussle between paganism and Christianity.

Smith notes that the distinguishing mark of the pagan religious perspective, such as was found in ancient Rome prior to the fourth century AD, is that it locates the sacred within this world (a so-called *immanent sacred*) and is generally tolerant towards similar or compatible perspectives. Indeed, history tells us that ancient Greece and Rome incorporated the gods of foreign peoples into their religious systems. This contrasts with transcendent religions,

---

43 The ancient Olympics are thought to have ceased from 393 AD, when the Roman Emperor Theodosius, a Christian, banned pagan activities.

44 The term neopaganism is often applied to pagan religions present in modern times, e.g. Wicca, Druidry, Heathenry and Hellenism. Steven Smith is not using neopaganism is this narrow sense. Instead, Smith uses the term to signify a broad worldview system which contains a variety of differing viewpoints including atheism, secular humanism, secularism, Neo-Darwinism, the New Age movement and pagan religions such as Wicca etc.

especially Christianity, which locates the sacred outside this world; the sacred is a *transcendent Creator* whose glory is reflected in nature.[45]

Rome represented the most successful and devout pagan empire in history. Paganism was "tolerant" to those who were "reasonable" in accepting its overarching framework, values and civic religious observances. Roman paganism was a system, analogous to a "big tent"; religious perspectives existed in a number of forms,[46] including the mythical (a modern version might be New-Age thinking), civic (a contemporary incarnation affirms a range of moral frameworks which conflict with the biblical worldview) and philosophic (modern versions include scientifically-minded Neo-Darwinians who find nature awe inspiring and perhaps sacred). Sometimes paganism was viciously intolerant to those who refused to submit to its attitudes and observances—those that would not demonstrate that they were under the big tent.

Transcendent religions, such as Christianity, posed challenges to the paganism of ancient Rome. Christianity was incompatible with paganism in multiple domains: in defining what was sacred; in relation to authority, obligation, law and justice; notions of the common good; the nature and value of truth; and sex.[47]

After Christianity achieved politico-cultural supremacy in fourth century Rome, it critiqued and reformed aspects of the pagan world: "infanticide, slavery, inequality, neglect of the poor and diseased".[48] But Western Christianity never uprooted paganism completely. Roman Catholicism "Christianised" and preserved many aspects of paganism[49] including classical literature and learning. It retained various pagan names for the days of the week and months[50] and repurposed several pagan festivals. Paganism also persisted as an intellectual orientation, identifiable in the writings of numerous "secular" scholars.[51]

The neopaganism that Smith describes has now reversed the famous but incomplete victory of Christianity in fourth century Rome. In recent decades, secular humanists, "liberals" and "progressives" have begun to abandon their pretence of neutrality, having achieved a string of successes in the culture war.[52]

---

45 The Christian God is also immanent in some respects.

46 Smith, 2018, p. 88.

47 Ibid., p. xii, 108, 111.

48 Ibid., p. 128.

49 Ibid., pp. 195–197.

50 "Days of our week trace their names back to Greco-Roman or Nordic deities (the sun, the moon, Tiw, Wodan, Thor, Frige, and Saturn). Likewise for our months: January is named for the Roman god Janus, March for the god Mars, April for Aphrodite, May for Maia, June for Juno, July and August for the first two of the divinized Caesars" (Smith, 2018, pp. 195–196).

51 Smith, 2018, p. 14.

52 Ibid., pp. ix–x.

[53] This includes conflict over religious symbols; pagan lawmakers ban those of transcendent religion, such as crosses in public places.[54]

In the context of a culture war between Christianity and neopaganism, Darwinism and Neo-Darwinism have been critical, and indeed foundational components of the battle. For a generation or more, the paganized Western elite have attempted to reimpose a pagan conception of origins science. Darwinism and Neo-Darwinism have been decisive in bamboozling the public at large to regard Christianity as outdated and implausible, thereby facilitating the pagan turn in the contemporary West.

The taunts of paganism

Paganism stretches back in history much further than the ancient Greeks and Romans. The Old Testament contains many examples of pagan societies that the Israelites encountered. These Old Testament accounts record that paganism's hostility to Bible believers is a very old phenomenon—almost a constant. In fact, some of the rhetorical devices that ancient pagans used against the Israelites are still being used today.

For example, some atheists like to point out that Christians are already atheists regarding foreign gods and then suggest that Christians should go one step further by dismissing the God of the Bible. Ron Reagan (son of the late U.S. president) put it this way in a speech to the Freedom from Religion Foundation in 2015:

> We're all atheists, of course, but even the religionists—they're atheists. They're atheists about every other God except theirs. None of them believe in Zeus, few of them believe in Vishnu, or Thor … . They're fine being atheists about that; we just go one God farther, we just take it all the way.[55]

Richard Dawkins offered the same kind of argument in *Outgrowing God:*

> Until somebody offers a reason to believe, we are wasting our time bothering to do so. That is the approach we all take to Thor and Apollo and Ra and Marduk and Mithras and the Great Juju up the Mountain. Couldn't we go a little further and think the same way about Yahweh or Allah?[56]

At the core of this argument lies the premise: don't think your God is special. He's the same as all the other "gods"—non-existent.

---

53 Mark Tushnet, an American legal scholar specializing in constitutional law and theory wrote "The culture wars are over; they lost, we won. … For liberals, the question now is how to deal with the losers in the culture wars. … My own judgment is that taking a hard line ("You lost, live with it") is better than trying to accommodate the losers, who—remember—defended, and are defending, positions that liberals regard as having no normative pull at all." (Tushnet, 2016).

54 Smith, 2018, pp. 267–278.

55 Ron Reagan: Not Afraid of Burning in Hell, 2015.

56 Dawkins, 2019, pp. 13–14.

This line of argument is not new. The ancient king of Assyria, Sennacherib (c. 745–681BC) expressed similar sentiments, albeit in very different circumstances. Sennacherib was set to launch a military campaign against Jerusalem, and he was keen that the residents of the city, led by Hezekiah, the king of Judah, simply lay down their arms, rather than trusting in the God of Israel for deliverance. Here's how Sennacherib put it:

> Now therefore don't let Hezekiah deceive you, nor persuade you in this way. Don't believe him, for no god of any nation or kingdom was able to deliver his people out of my hand, and out of the hand of my fathers. How much less will your God deliver you out of my hand?[57]

Sennacherib's thinking was underpinned by the same kind of logic that we saw above: don't think your God is special. He's the same as all the other "gods"—unable to deliver.[58]

The Reversi Dictum

As a final point, we draw attention to the state of cognitive dissonance that is common among those who believe the secular origins account. The following quote from The Third Way website is typical of this mindset:

> The vast majority of people believe that there are only two alternative ways to explain the origins of biological diversity. One way is Creationism that depends upon intervention by a divine Creator. That is clearly unscientific because it brings an arbitrary supernatural force into the evolution process. The commonly accepted alternative is Neo-Darwinism, which is clearly naturalistic science … .[59]

There are several issues with this blurb, but we will focus on the heart of the argument, which holds that belief in a Creator is unscientific, whereas Neo-Darwinism (or other evolutionary schemes) is naturalistic science.

This assertion is false. We have shown that the Christian and secular origins accounts share many features in common. Table 14.1 summarizes the foundational and other features of both accounts. The main distinguishing point between their foundational features is the nature of the underpinning infinity. The Bible posits a personal infinity, whereas infinitists hold to an impersonal infinity or infinities (polyinfinitism). These are, by definition, outside the scope of empirical science, but they are necessary to render certain

57 2 Chronicles 32:15.

58 However, Sennacherib's slight against the God of Judah proved disastrous. God destroyed his army, and Sennacherib himself was later killed by his own sons, at which point another of his sons took over the kingdom (2 Kings 19:35–37). From Israel's perspective, the defeat of Sennacherib was unique. It marked the only instance in Old Testament history where an opposing force was defeated militarily without any involvement from the Israelites. It was as if God, having been personally slighted by Sennacherib and his emissaries, took matters entirely into His own hands.

59 The Third Way, 2024.

phenomena explainable. In incorporating them, secular scientists smuggle God-like qualities of infiniteness or purpose into their origin account, which results in what we term a secular miracle.

The result of this conceptual similarity is that every philosophically based argument levelled at the Christian origins account from the polyinfinitist viewpoint has an equivalent argument that can be levelled from the Christian viewpoint against the polyinfinitist account—except in relation to the nature of the underpinning infinity.[60] We call this the Reversi dictum, named after the popular board game (also known as Othello), in which players flip white and black sided disks.

60 A corollary of this dictum is that debates between polyinfinitists and Christians on origins matters may not be particularly effective in convincing audience members one way or the other.

Table 14.1: A comparison of the foundations and features of the biblical and polyinfinitist origins accounts

	Origins account foundations and features	Biblical Christianity	Polyinfinitism (Secular Humanist)
Features	Incorporates circumstantial narratives (criticized as "fairy tales" or "just-so stories")	Yes: biblical and extra-biblical explanations	Yes: cosmological, emergence-of-life and evolutionary "just-so" stories
	Explains certain phenomena by miracles	Yes: Divine	Yes: secular, e.g. abiogenesis
	Personal infinity	Yes: cannot be detected scientifically but God and people can communicate	No: cannot be detected scientifically or by any other means
	Incorporates infinity and/or purpose (outside the scope of empirical science)	Yes: eternal triune God of the Bible; infinite in knowledge and power; creates with purpose	Yes: singularity, multiverse and/or teleological processes
	Includes an explanation for the informational basis of life	Yes: God spoke the creation into being through the agency of the *Logos*, the Divine Word	Yes: abiogenesis to explain life and Neo-Darwinian evolution to explain all its diversity
	Starts with a "beginning"	Yes: God created the universe *ex nihilo* as declared in Genesis	Yes: the singularity was followed by the Big Bang at a point in the finite past
Foundations	Unprovable presuppositions	Drawn from interaction with the world around us and the Bible: God exists; God created the world, etc.	Drawn from human philosophy: God does not exist; the universe came into being by naturalistic processes, etc.
	Authority	Logical ideas which are consistent with science and the Bible.	Logical ideas which are consistent with science, reckoning all revealed religion as mythology and superstition.

We also draw readers' attention to the difference between the origin of the informational basis of life—the New Testament implies that the Divine *logos* was the ultimate source of all information in the physical world. In the case of polyinfinitism, abiogenesis and Neo-Darwinian evolution are the sources of biological information. Hence, the effect of the naturalistic worldview is to replace Christ.

Conclusion

In this chapter we saw that modern ideological atheism and secular humanism are implicit beliefs in a range of non-personal infinities, all of which are beyond the scope of science. This belief system, which we term infinitism (or polyinfinitism) also has an alternative in purposivism. These have several key similarities with religious belief, although secular humanists do not derive the usual benefit that religion provides—the corporate life associated with religious congregations. On the contrary, the rise in evolutionism has been accompanied by a significant loss of social capital, a problem that social researcher Robert Putnam has highlighted in America and politician Andrew Leigh has raised in the Australian context. This rise of secularism, and consequently atheism, has led to a less connected, more polarized society, attended by lower rates of marriage and partnering, and all the social benefits these bring (for example, happier childhoods overall and greater emotional stability in individuals).

Of course, most individuals in Western nations are not ideological atheists, but perhaps implicit atheists or agnostics. This partly flows from the working principle of modern science and humanities, which holds that everything should be examined and explained with the presupposition of *etsi Deus non daretur* ("as if God did not exist"). Thus, God becomes a concept to be avoided at all costs; He who should not be mentioned (if an academic wants a career). The mainstream paradigm is self-restriction, agnosticism or even deliberate ignorance, though not an aggressive atheism.

Unlike the atheists of centuries ago, contemporary ideological atheists are emboldened to the extent that they may engage in satanic discourse publicly. This does not mean that they believe in the existence of Satan any more than they do in the existence of God, but it shows that in the current era, their "fear of the people"—that is, the Christian populace—has evaporated.

Evolution is a key foundation and enabler of the civic and philosophical neopagan religious perspectives. This perspective, like the paganism of ancient Rome, is deeply incompatible with Christianity. Darwinism, and later Neo-Darwinism, have been major factors in the switch in the Western worldview from Christianity to neopaganism.

Finally, we saw that there is a substantial symmetry between the biblical and evolutionary origins accounts, so much so that many philosophical critiques of either account from the perspective of the other can be applied the other way. Those who regard the Christian origins account as a fairy tale have likely not understood that the exact same argument applies in reverse to the secular origins account.

In view of the damaging social effects of evolutionism, the outlook for Western civilization is for decline. We will outline an aspect of this in Part 4 concerning some recent developments in academia and society, but first, we turn our attention to an analysis of the biblical creation account.

References

Barker, D. (2016). *God: the most unpleasant character in all fiction*. New York: Sterling.

Beale, C. (2011). *Born out of Wenlock: William Penny Brookes and the British origins of the modern Olympics*. Derby: The Derby Books Publishing Company.

Bergman, J. (2012). *Slaughter of the Dissidents—The shocking truth about killing the careers of Darwin doubters* (2nd edn, Vol. 1). Southworth: Leafcutter Press.

Closer to Truth. (2020, September 4). Paul Davies—The Rise of Scientific Atheism. Retrieved February 21, 2025, from youtube.com/watch?v=gLp8Ame-FQ4

Darwin Correspondence Project, "Letter no. 1924". Accessed on January 9, 2025, from darwinproject.ac.uk/letter/?docId=letters/DCP-LETT-1924.xml

Dawkins, R. (2002). Is Science a Religion? Retrieved January 26, 2025, from skeptical-science.com/essays/science-religion-richard-dawkins/

Dawkins, R. (2003). *A Devil's Chaplain*. London: Weidenfeld & Nicolson.

Dawkins, R (2004). *A Devil's Chaplain*. London: Phoenix.

Dawkins, R. (2019). *Outgrowing God*. London: Transworld Publishers.

De Duve, C. (2013, November 30). Does the universe have a purpose? Retrieved from web.archive.org/web/20131130121113/http://www.templeton.org/purpose/essay_de_Duve.html

Dunphy, John J. (1983). A Religion For A New Age. Accessed August 17, 2025, from johnjdunphy.medium.com/a-religion-for-a-new-age-f812839c4cb8

Eternity News - Podcasts. (2019, September 30). Eternity News—Podcasts. Retrieved from eternitynews.com.au/podcasts/ep-5-social-capital/

Galton, F. (1883). *Inquiries into human faculty and its development*. New York: Macmillan and Co.

Gould, S. J. (2012, June 14). Stephen Jay Gould: "We derive benefits from appearing as a new priesthood". Retrieved from thedesignspectrum.wordpress.com/2012/06/14/stephen-jay-gould-we-derive-benefits-from-appearing-as-a-new-priesthood/

Grothe, J.D. (2007, August 10). "Peter H. Gilmore –Science and Satanism". Retrieved from pointofinquiry.org/2007/08/peter_h_gilmore_science_and_satanism/

Halloway, A. (2008, September 23). Reiss resigns as Royal Society stifles debate on evolution. Retrieved June 1, 2024, from creation.com/reiss-resigns-as-royal-society-stifles-debate-on-evolution

Herald, T. S. (2010, October 2). The Sydney Morning Herald. Retrieved from smh.com.au/national/trading-in-the-lectern-for-a-bully-pulpit-20101001-160xy.html

Hoyle, F., & Wickramasinghe, C. (1981). *Evolution from Space—A Theory of Cosmic Creationism*. New York: Simon and Schuster, Inc.

Mayr, E. (1982). *The Growth of Biological Thought: Diversity, Evolution, and Inheritance*. Cambridge, Massachusetts: Harvard University Press.

McGrath, A. (2011). *Why God won't go away*. London: Society for Promoting Christian Knowledge.

North, G. (2012). *Sovereignty and Dominion: An Economic Commentary on Genesis* (Vol. 2). Dallas: Five Point Press.

PewResearch.org. (2018, March 19). Share of married adults varies widely across U.S. religious groups. Retrieved November 14, 2024, from pewresearch.org/short-reads/2018/03/19/share-of-married-adults-varies-widely-across-u-s-religious-groups/

Potter, C.F. (1930). *Humanism: A New Religion*. New York: Simon and Schuster.

Reed, F. (2020, October 7). Stephen C. Meyer, Douglas Axe, Chuck Darwin and Me—Three Evolutionary Siege Howitzers and a Popgun. Retrieved April 27, 2024, from unz.com/freed/stephen-c-meyer-douglas-axe-chuck-darwin-and-me/

Reiss, M. (2008, September 12). Teachers should tackle creationism, says science education expert. Retrieved from theguardian.com/science/2008/sep/11/creationism.education

Ron Reagan: Not Afraid of Burning in Hell. (2015, December 11). Retrieved from youtube.com/watch?v=-h922B3Svzw

Ruse, M. (2017). *Darwinism as Religion—What Literature Tells Is About Evolution*. Oxford: Oxford University Press.

Smith, S. D. (2018). *Pagans & Christians in the City: Culture Wars from the Tiber to the Potomac*. Grand Rapids, MI: Wm. B. Eerdmans Publishing Co.

The Third Way: Evolution (n.d.). Rationale. Retrieved June 15, 2025, from thethirdwayofevolution.com

Turner, F.M. (1981). *The Greek heritage in Victorian Britain*. New Haven: Yale University Press.

Tushnet, M. (2016). Abandoning Defensive Crough Liberal Constitutionalism. Retrieved February 8, 2025, from balkin.blogspot.com/2016/05/abandoning-defensive-crouch-liberal.html

United Nations Convention on the Rights of the Child. (2023). Retrieved from unicef.org.au/united-nations-convention-on-the-rights-of-the-child

Part 3
A Young-Earth Creationist View of Genesis 1–11

15
Approaching Genesis

Earlier we quoted humanist philosopher Paul Kurtz on the profound influence that origins accounts have on shaping our understanding of who and what we are, and how we ought to live. Kurtz's assessment underscores the importance for Christians to assimilate the foundations of the Christian worldview in Genesis 1–11. The biblical narrative informs us of our identity and purpose, communicates details of the Divine plan of creation and redemption, and reveals important aspects of God's character. A proper understanding of Genesis 1–11 is very significant to Christians.

To emphasise this point, we draw readers' attention to the Apostle Paul's ministry in Athens (Acts 17:16–34), undertaken circa AD 50. The Greek philosophers that Paul addressed at the Areopagus (vv. 22–34) may have been aware of the proto-evolutionary thought posited by pagan thinkers from Thales to Lucretius (recalling Edward Clodd's work that we covered in Chapter 5) and may have held proto-evolutionary beliefs. It is possible that this episode was the first encounter between evolutionists and a Christian missionary—the beginning of "the battle for the beginning" (to borrow John MacArthur's phrase). After a brief preamble (vv. 22–23), Paul immediately challenges the thinking of the philosophers with presuppositions and concepts from the biblical creation account (vv. 24–26). Note that carefully: the Apostle Paul, God's "chosen instrument to take [His] name to Gentiles" (Acts 9:15) believed that the evangelisation of pagans began with Genesis—read in a straightforward manner. It behoves contemporary Christians, who live in an increasingly neopagan society, to follow this example.

This part of the book aims to show the coherence of the Genesis origins account from a young-earth creationist perspective, and to counter the typical criticism that the biblical origins account cannot be reconciled with contemporary science.

Authorship

Who was the author of Genesis? Genesis itself does not tell us, but the traditional view, dominant up until the latter part of the nineteenth century, held that Moses wrote Genesis—and indeed the first five books of the Bible.[1]

1 Jews call this part of the Bible Torah, and Christians call it the Pentateuch—literally "five scrolls or books"

While Exodus, Leviticus, Numbers and Deuteronomy[2] describe events contemporaneous with Moses, all the subject matter of Genesis precedes the time of Moses.

The traditional view that Moses was the main author of Genesis is backed by biblical evidence. Exodus, Numbers and Deuteronomy state that Moses wrote material relevant to these books,[3] and other references, including 1 Kings 2:3, imply that Moses had written "the law" or *Torah*:

> … and keep your obligation to the LORD your God to walk in his ways and to keep his statutes, commands, ordinances, and decrees. This is written in the law of Moses, …

Within the New Testament, 2 Corinthians 3:15 directly identifies Moses as author:

> Yet still today, whenever Moses is read, a veil lies over their hearts, ...

However, in writing the Torah, Moses almost certainly used older existing documents.[4] Genesis contains internal clues that older documents were used; Chapter 5, for example, begins with a statement suggesting that a genealogical source document or book, which may well have been authored by Adam himself, had been embedded by Moses:

> This is the document containing the family records of Adam. …

The composition of Genesis was not simply a case of oral history being passed down over millennia.

The family tree of the fifth chapter of Genesis is called a chronogenealogy which includes the ages of the individuals at their death and the age of the father at the birth of their offspring. Because these two details are provided for each member of the genealogy, an accurate chronology can be constructed.

Another interesting text is Genesis 10:19:

> The Canaanite border went from Sidon going toward Gerah as far as Gaza, and going toward Sodom, Gomorrah, Admah, and Zeboiim as far as Lasha.

Sarfati points out that this appears to derive from a historical document written while the doomed cities of Sodom and Gomorrah still existed, even though both cities had long been destroyed by the time Moses compiled the Torah.[5]

---

2 The last chapter of Deuteronomy contains a description of Moses' death, burial and subsequent mourning period. Jewish tradition suggests that these passages were written by Joshua.

3 Exodus 17:14, Numbers 33:2 and Deuteronomy 31:9 are representative examples—Sarfati lists more (Sarfati, 2018, p. 9).

4 Sarfati, 2018, p. 11.

5 Sarfati, 2018, p. 11.

We can also surmise that editors living in a later period than Moses made minor additions to the Torah. For example, the explanatory clause in Genesis 36:31 is likely to have been added at some point after the start of the reign of King Saul:

> These are the kings who reigned in the land of Edom *before any king reigned over the Israelites:* (emphasis added)

A reasonable view is that after Moses wrote Genesis, minor additions were made later by a priest such as Samuel.[6]

Summing up, the internal evidence of scripture implies that Moses compiled Genesis using existing documents, with some minor additions made at a later date, perhaps by Samuel, and that both Old and New Testament writers, and Christ Himself, considered Moses to have been the author.

Consistent with the *Chicago Statement on Biblical Inerrancy* (1978), we hold that the biblical text is inspired by God and therefore is completely trustworthy—although different interpretations are possible.

The Genesis Origin Account

The Genesis creation account is comparable to an airliner's cockpit voice recorder. It provides an account of past events in the form of a narrative, an account which aligns with measured data. Creationists argue that accurate scientific evidence, interpreted within a biblical framework, supports the Genesis account. Note the qualification carefully: the biblical framework is essential to our task. Indeed, creationists believe that interpreting Genesis within a non-theistic secular framework has been a historical problem for the church. Therefore, before we begin our survey of Genesis, we need to flag these secular presuppositions, which include philosophical uniformitarianism.[7]

The uniformitarian concept makes a lot of sense for atheists as a philosophy. Without a God to intervene in natural affairs, the atheist is likely to assume that the processes we see today are the only processes that have been operating in the past. The Christian, however, cannot accept this premise. Within the first four words of the Bible, God is mentioned, and His existence is assumed as a presupposition. Then, throughout the Old Testament, we see God occasionally intervening miraculously in various ways, just as the New Testament records Jesus intervening miraculously. Many of the Divine miracles of the Bible are by definition non-uniformitarian events.

---

6 See Rushdoony, 2002, pp. 235–236.

7 "… 'Methodological uniformitarianism' is merely a reformulation of the basic assumption of scientific methodology" (Alvarez, Hansen, Hut, Kauffman, & Shoemaker, 1988, p. 13). Christians in the past developed science based on this principle, and contemporary Christians continue to accept it. By contrast, philosophic uniformitarianism, sometimes called "… 'substantive uniformitarianism', or gradualism, is simply wrong" (Alvarez, Hansen, Hut, Kauffman, & Shoemaker, 1988, p. 13). That is, the extreme uniformitarianism of Hutton and Lyell is inapplicable to earth science.

Another difference between the biblical and secular origin accounts concerns the order in which processes are said to have occurred. The secular account is framed within a straightforward progression which we treated sequentially in Part 2: the emergence of matter and energy (Chapter 9), followed by the emergence of primitive life (abiogenesis, Chapter 10) and finally the evolution of higher life forms (Darwinism, Chapters 11–13). In comparison, the biblical account appears jumbled: by the time the sun and stars are created, for example, plant life has already appeared on the earth. In our presentation of the creation account, we will simply take the biblical text in the order it comes. We commence with an investigation of Genesis 1–2 (Chapter 14), then proceed to Genesis 3–6 (Chapter 16) and finally consider Genesis 7–11 (Chapter 17).

We will combine Genesis 1–11 with other relevant biblical texts from both the Old and New Testaments to present a unified interpretation incorporating descriptive scientific explanations where appropriate.

Origins research from a biblical perspective

Jesus Christ stated that "It is written, That man shall not live by bread alone, but by every word of God" (Luke 4:4, KJV). "Every Word of God" includes scripture in addition to the creative Word and the sustaining Word of God. The latter two are investigated by studying the Bible—and by studying science.

Not only did Jesus imply that God wants us to investigate our world, but the Old Testament contains similar hints. In the book of Deuteronomy, God tells the Israelites that He will bring them to a land "… whose rocks are iron and from whose hills you will mine copper" (8:9). The implication is clear: God is telling the Israelites to use their intelligence and knowledge of topography to identify promising sites for mining, dig out the copper and use it to make implements. As part of that, they needed to make fire burn sufficiently hot to melt copper; a bellows of some form might have been necessary. How to make that bellows? Leather tanning processes would come in handy. And so on. God expected scientific and technological development from His people, just as He expects it of us now. As Christians, our approach to origins research begins with the Bible, and the creationist movement operates in the same way.

Creationists propose models that seek to interpret scientific evidence within a biblical framework. An example is the model proposed by physicist Russell Humphreys to harmonize the immense age of the universe with the approximately 6,000 years of elapsed time since creation, a model Humphreys published in his 1994 book *Starlight and Time: Solving the Puzzle of Distant Starlight in a Young Universe*. Such models are then tested progressively, facing scientific criticism from within (or without) the creationist enterprise. A model which fails to withstand scrutiny is discarded. This is what happened with Humphreys' model; a flaw in the argument was found, and the creationist movement moved on to new models—a good example of the scientific method applied to origins. Readers will recall that the Copernican revolution began with Copernicus' suggestion of a helio-centric (sun-centered) model in which the planets, including earth, orbited the sun in circular paths. This model was somewhat effective but was improved upon by Kepler who proposed elliptical paths. In the same way, the starting point of creationism is the postulating of models that interpret evidence taking the Bible into account. Progress is made when such a model is either confirmed to be correct or falsified.

Because of the foundational significance of the Genesis origin account to Christianity, enemies of Christ and His church have historically sought to challenge belief in the authenticity and accuracy of Genesis. Not a few of the higher critics sought to undermine the traditional understanding of scripture and substituted alternate theories which were incompatible with a divinely inspired authoritative text.

In the following chapters we will present a case study that illustrates some of the processes of the creationist movement. We have chosen a relatively recent creationist book by earth scientist Dr. Ken Coulson, *Creation Unfolding* (2020), and will incorporate some of the concepts Coulson presented with our Genesis 1–2 narrative. Coulson is both a Christian who holds to a six-day creation belief, and an earth scientist, with several journal articles in both the secular and Christian scientific literature to his credit. As an earth scientist, Coulson focuses mostly on the geological aspects of the creation narrative and discusses cosmology only in passing (a single chapter).

Coulson's book has been thoughtfully reviewed by several creationist organizations: Creation Ministries International,[8] Answers in Genesis[9] and the Institute for Creation Research.[10] The reviews ranged from positive to sharply negative; our assessment is that none of the reviewers have fully drawn out the implications of Coulson's system, although Shaun Doyle's generally positive review goes furthest in this regard. Some of Coulson's key concepts reflect long-standing ideas within the creation movement, which he has further developed.

Our purpose is to demonstrate how new ideas can be assessed and developed with a view to permitting critique by specialists. The steps to reviewing a new creationist idea include:

1. Understanding what the author is proposing at presuppositional, theological (biblical), philosophical and scientific levels. Also, assessing whether the author has communicated clearly and avoided confusion and error.
2. Reviewing how the new ideas relate to existing work in the field.

8 Doyle, 2021.
9 Answers in Genesis, 2023.
10 Arledge & Thomas, 2020.

3. Considering the value of the new ideas—do they lead to new insights or developments or solve existing problems?
4. Ascertaining the implications of the new ideas to render them amenable to critique by specialists within the creation field, which could include biologists, earth scientists, cosmologists, philosophers or theologians.
5. Tentatively establishing, by the end of the critique phase, the advantages and disadvantages of the new ideas. If the disadvantages outweigh the advantages, the new idea should be assessed as unhelpful or even harmful. The overriding principle is that creationists hold to the Bible tightly, but less tightly to theories or explanations. We (as creationists) should feel no remorse in having to discard an erroneous or unhelpful theory. We should also expect a range of opinions and views to exist on particular theories and explanations—this is a healthy, and indeed, normal part of the scientific review process.
6. Accepting that the new ideas are of value if they lead to further new insights that are confirmed scientifically, theologically, or via some other relevant discipline.

In this chapter, we will cover the first three steps of review with respect to Coulson's new concept. Steps four through six are beyond our scope; they rely on the efforts of domain experts and generalists within the creationist movement, and should be reflected, over time, in technical journal articles and reviews within creationist literature.

Types of creation

The Genesis account contains two main types of creative act. The first is *ex nihilo* creation—*ex nihilo* is a Latin term meaning "out of nothing".[11] The second type of creation involves transformation of existing matter, such as causing the dry land to emerge and the waters of the earth to be gathered into one place. Creating Adam from the dust of the ground also involves transformation of existing matter. It is this second type of creation that Coulson's new concept applies to.

In brief, Coulson's concept is that the second type of creation occurs by what he terms "supernatural formative processes" (SFP). By this, Coulson means that in creating trees, for example, God did not just speak and—voila!—a stand of trees appeared. Instead, Coulson argues, God used natural processes, sped up supernaturally, to create. This argument, as applied to the plant

11 Ex nihilo creation brought the water-covered earth and the heavens into existence; only God can perform this type of creation, which is referenced in Hebrews 11:3: "By faith we understand that the universe was created by the word of God, so that what is seen was made from things which are not visible."

kingdom, had been proposed by creationists prior to Coulson. What is new in Coulson's approach is extending the concept to encompass various non-biological aspects of creation, such as earth science.

The crux of the argument is that items created by these processes did not merely have an appearance of maturity, they actually had a history of process—albeit accelerated—inherent within them.

The process aspect of SFP is scientifically interpretable. SFP is useful in re-evaluating a number of "old-earth" evidences within a "young-earth" biblical framework, while avoiding the problems of Gosse's *Omphalos*. It also suggests that God constrained the ways in which He acted during the creation week to benefit man's understanding of creation events.

Supernatural formative processes

Let's drill down further into Coulson's concept as it applies during the creation week. After the initial ex-nihilo creative act on the first day, the earth is "formless and empty" (Genesis 1:2). But by the sixth day, earth has extensive structure, with land, sea, river systems, mineral deposits such as gold (whose quality varies in different geographic locations), precious rock-like minerals such as beryl, onyx and bdellium (which may be a fragrant tree resin or gum) and complex ecologies supporting plant and animal life on land and in the sea. The existence of rivers implies that hills and valleys are in place, while the existence of plants implies that most of the vegetated land surface is covered in a layer of soil. Coulson argues that for man to have dominion over the earth (especially in relation to earth science, Coulson's specialty) God used supernatural formative processes to render the events of the creation week amenable to man's scientific enquiry. Coulson attempts to identify axioms (presuppositions) which make scientific analysis of events within the creation week possible, and to justify those axioms biblically.

Coulson begins by noting that several key authors from creationist ministries[12] have supported, in principle, the idea that God used naturalistic processes to fashion the earth and cosmos during the creation week, except that these processes were markedly sped-up. One of these is the creationist astronomer Danny Faulkner, whose cosmological *dasha* concept[13] shares some similarities to Coulson's ideas.[14] Thus, God's creation "appears to have a [full] history, and not just an appearance of age or maturity."[15] Coulson raises this sentiment to a hard axiom for events in the creation week after the ex nihilo act:

---

12 For example: Answers in Genesis, Creation Ministries International and the Institute for Creation Research.

13 Faulkner, 2013.

14 Coulson, 2020, p. 52.

15 Ibid., p. 43.

> God uses supernatural formative processes to create objects that themselves always, *without exception*, mirror naturalistic processes and thus retain *natural* histories.[16] (Emphases in original).

But this is not enough for science to be able to interpret changes occurring in the creation week. For this to work, Coulson deduces that God must have sped processes up uniformly—rather like viewing a video on fast forward. For example, compare the rate of any two natural events occurring today, such as the growth of grass versus the rate of land erosion due to river flow. For these two processes, a ratio of their rates of change can be determined. Coulson then proposes that for any such ratio calculated today, the same ratio would apply during the creation week. If we had a video replay of events during the creation week, they would appear to us as fast-forwarded, that is, they are undergoing *supernatural formative processes*. If this video were slowed down so that it approximated current conditions, there would be no surprises at the rates of any natural processes occurring on earth. Grass would be seen growing at expected rates; erosion by river flow would occur at expected rates, and so on.

There is one rate during the creation week which does not conform to Coulson's second axiom—the rotation of the earth, and thus the length of each day of creation. Days remain twenty-four-hour periods and they are not themselves sped-up or slowed. This affects available light and seasonal variation. Sped-up events completed during nighttime receive no light at all, while sped-up events completed during daytime receive uninterrupted light. As evidence favoring the reasonableness of this assertion in relation to events which normally require the action of the sun and seasons, Coulson posits the budding of Aaron's rod. In Numbers 17:8, Aaron's rod "sprouted, formed buds, blossomed, and produced almonds"; all this occurred overnight. Normally such changes occur over weeks and months, with growth requiring sunlight and seasonal variation. But this was not the case in this accelerated miracle, even though an otherwise naturalistic process had occurred. This can be considered another axiom: the rate of the earth's rotation was extremely slow in relative terms during most of the creation week, and events requiring sunlight or seasonal variation proceeded normally, whether sunlight was present or not. Ergo, the creation process was understandable, but miraculous at the same time.

Finally, Coulson considers how these sped-up rates might have changed during the creation week. He concludes that the initial accelerated rates of SFP were probably enormously high during the first four days but slowed to current levels at some time during the fifth day—when sentient animal life was first created—or during the sixth day, when mankind was created. This is Coulson's

---

16 Ibid., (Coulson quoting Deem approvingly).

final axiom—God controlled events in a manner which would achieve His purposes.

The net result of Coulson's axioms is that secular scientists, working under philosophical uniformitarian assumptions, can have significant success in science, even though the earth is not the billions of years old that it seems. Although Coulson does not provide much detail on how the rate of processes might have changed over time, he does suggest an exponential decline in this acceleration[17] during the fifth or sixth days, and logic suggests that the rate by which normal processes are accelerated must approach zero by the end of the sixth day. This is so that events are passing at current rates by the onset of the seventh day—at which point the current epoch begins. (We refer to the first six days of creation as the *creative epoch*, and the time interval from the beginning of the seventh day until the present as the *post-creation epoch*.[18])

Up until now we have considered creation events from an earth-centric perspective, as the Bible does. But imagine we were observing from a theoretical fixed point in outer space, looking down at the earth as it rotated on its axis each day. Now imagine (assuming Coulson's SFP concept is correct) that our observations occurred in a varying slow motion, in such a way that all of the process rates during the six days of creation would be seen to occur at a rate equivalent to what we see today. In that case, our observations of the earth's rotational speed would be such that it would be seen as increasing progressively from near zero on the first day. There would be a significant uptick in apparent rotational speed during the fifth or sixth days, but by the end of the sixth day, it would be observed rotating at the current rate.

Some may ask—if Coulson's approach is correct, does it imply that God has set up the world in a way which deceives scientists? Coulson doesn't believe so, and as evidence for this, he cites a corollary with Jesus' teaching in the New Testament. While Jesus' disciples listened to His teaching in order to learn, others approached it with an attitude of hostility. They wanted to catch Jesus out in His words, for the ultimate purpose of shutting down His ministry. Jesus therefore taught using parables, and when questioned as to why he used this method,

> He answered, "Because the secrets of the kingdom of heaven have been given for you to know, but it has not been given to them. … That is why I speak to them in parables, because looking they do not see, and hearing they do not listen or understand. …" (Matthew 13:11–13)

This fits with the supernatural formative processes posited by Coulson. All scientists approach the same evidence; for those that resist God's Word, their

17 Ibid., p. 54.

18 The post-creation epoch could be further broken down into a number of other periods: pre-fall; post-fall pre-Flood; the Flood; and the post-Flood.

stance can lead them into self-delusion. Others, approaching the Word of God with humility, draw very different conclusions from the same evidence. God responds differently to different people depending on their attitude. In 1 Peter 5:5 we read that "God resists the proud but gives grace to the humble".

At the same time, God provides equally for those that reject Him and those that accept Him. In Matthew 5:45, Jesus says "For he causes his sun to rise on the evil and the good, and sends rain on the righteous and the unrighteous." In the same manner, it makes sense to believe that God has established a system in which both types of people are able to conduct useful science and thereby earn their livelihoods.

Like all creationists, Coulson begins by holding to a Genesis-as-history interpretation of scripture, then analyzes scientific evidence in light of a six-day creation (and global flood) belief. The result is a synthesis which preserves the authority of scripture and successfully incorporates a number of aspects of contemporary earth science that are unrelated to the sedimentary and fossil record. Coulson's SFP concept, which relates to processes during the creation week, seems particularly useful for understanding the processes which caused the earth's crust to differentiate during its formation.

Furthermore, it may be possible to extend Coulson's SFP concept successfully to cosmology, perhaps in a more direct fashion than he attempted.

With these preliminaries in mind, we will proceed to our investigation of Genesis 1–11. Although we will make use of Coulson's SFP idea in relation to Genesis 1–2, our analysis will rely on concepts developed within influential creation ministries over the last few decades (with which Coulson's scheme is compatible). Coulson's approach is a development of several existing creationist ideas *in axiomatic form*. It is relatively new and untested. We are aware that, like any concept within the scientific and theological realms, this approach may be finetuned, modified, or even overturned in the future.

By presenting an interpretation of Genesis 1–2 based on Coulson's axioms, readers who follow this area of research over the next few years may see first-hand how the process of explanation, critique and analysis occurs within the multi-disciplinary creationist movement. Like Coulson's SFP concept, the ideas we present here may be discarded by the creationist movement, or they may become further developed and refined over time.

We present the following chapters in a spirit of humility: we believe they represent a biblically and scientifically credible interpretation of Genesis, but we remain open to other ideas. In fact, if our book promotes discussion that advances the creationist viewpoint beyond what we have presented here, we will have succeeded. However, any improved understanding must be:

1. Based on biblical presuppositions.
2. Consistent with biblical Christian doctrine.
3. Consistent with currently known scientific facts.

Regarding point (2), we note that a straightforward reading of Genesis implies creation occurring over a period of six literal, 24-hour days, and establishes a broad plan of history, where God plans to save the *kosmos* (John 3:16) to restore a good creation ruined by the Fall. Only in this scheme does the love and goodness of God, and the cross of Christ, make sense.

We commence our investigation with a day-by-day creation account.

References

Alvarez, W., Hansen, T., Hut, P., Kauffman, E.G., Shoemaker, E.M. (1989). "Uniformitarianism and the response of Earth scientists to the theory of impact crises", in *Catastrophes and Evolution: Astronomical Foundations (The 1988 BAAS Mason Meeting of the Royal Astronomical Society at Oxford, September 6th, 1988)*. Cambridge: Cambridge University Press.

Answers in Genesis. (2023, February 24). Is There Ambiguity in Creation? Retrieved March 11, 2024, from answersingenesis.org/young-earth-evolution/is-there-ambiguity-in-creation/

Arledge, S., & Thomas, B. (2020, July 31). Believe Anyway. Retrieved March 11, 2024, from icr.org/article/believe-anyway

Coulson, K. P. (2020). *Creation Unfolding: A New Perspective on Ex-Nihilo*. Phaneros Press.

Doyle, S. (2021, December). Creation Week, mature creation, and natural processes?—A review of Creation Unfolding: A new perspective on Ex Nihilo. *Journal of Creation*, 35(3), 41–46. Retrieved from creation.com/review-coulson.

Faulkner, D. R. (2013). A Proposal for a New Solution to the Light Travel Time Problem. *Answers Research Journal*, 6, 279–284.

Rushdoony, R. (2002). *Genesis*. Vallecito: Ross House Books.

Sarfati, J. D. (2018). *The Genesis Account—A theological, historical, and scientific commentary on Genesis 1–11* (Third edn). Powder Springs: Creation Book Publishers.

16

Genesis 1–2: the creation week

The creation comes into being in response to God's Word, of which the Son of God is the agent. "In the beginning God created the heavens and the earth." (Genesis 1:1) God literally spoke the world into existence from nothing. This is affirmed by many scriptures.[1] Our view is that this biblical description is consistent with modern science, including elements of Krauss's scheme, except it clearly identifies the personal infinity responsible for the existence of the cosmos.

The First Day

The earth was initially void, formless and covered deeply with water. It was unstructured and had no features of maturity (for example, probably lacking the differentiation of core, mantle and crust). The Spirit of God hovered over the waters, and there was darkness (Genesis 1:2–3).

While the meaning of "earth" is straightforward, the word "heaven" needs some elaboration. We know from other parts of the Bible that there are ultimately three heavens. The first heaven is the sky, or atmosphere. The second heaven refers to outer space, and the third heaven is the abode of God and the heavenly host (the large number of angelic creatures that are with God in heaven). The opening verse of Genesis certainly includes the creation of the third heaven, which was created complete and good along with the angels and other heavenly creatures. This third heaven is mentioned by the apostle Paul in the New Testament (2 Corinthians 12:2–4). Scripture records some instances in which the heavenly realm became visible to humans, such as the chariots of fire seen by Elisha and his servant in the Old Testament (2 Kings 6:17), and the martyr Stephen, standing before the Sanhedrin, looking up and seeing Jesus standing at the right hand of God in the third heaven (Acts 7:55–56).

Then God said, "Let there be light."(Genesis 1:3) Since this light resulted in day and night, it is widely accepted that the light came from a particular location adjacent to the rotating earth. One understanding is that the light emanated from the Spirit of God; Psalm 104 describes God as being surrounded by light as a cloak (Psalm 104:2), and Revelation 21:23 declares that the heavenly

---

1 Psalm 33:6, Isaiah 44:24, John 1:3, Romans 11:36, 1 Corinthians 8:6, Ephesians 3:9, Colossians 1:16, Hebrews 2:10, Hebrews 11:3, 2 Peter 3:5, Revelation 4:11.

city "... does not need the sun or the moon to shine on it, because the glory of God illuminates it ...". This light, which is denoted by the extra-biblical term "Shekinah glory",[2] is a physical manifestation of God, who is Spirit. In the Bible, the Shekinah glory is typically blinding light, fire or cloud.[3] Darkness was separated from the light. God saw that it was good, and this marked the end of the first evening and morning (Genesis 1:5).

How do we account for the earth's internal structure and surface features (a sphere comprising a core, mantle and crust)? On the first day, the earth was initially "without form" (Genesis 1:2), but it probably contained radioactive elements which decayed, releasing thermal energy. Using Coulson's SFP concept, these processes were sped up enormously—processes which under current conditions take billions of years were occurring within hours. We may hypothesize that radioactive decay could have created sufficient heat to melt part of the earth into magma. In this molten state, lighter materials, like quartz, floated upwards towards the surface of the earth, while heavier elements, including iron and nickel, sank towards the earth's center. This movement of thousands of cubic kilometers of matter released gravitational potential energy, adding to the heating via friction. A process of progressive melting and fractionation under gravity led to iron and nickel concentrating toward the center of the earth, forming the core, which solidified under the intense pressure. Lighter rocky material rose to underlie the waters above, forming patches of granitic continental crust (that is, parts of the crust that would eventually surface as land), separated by basaltic suboceanic crust.

The Second Day

On the second day God separated the waters below from waters above to create an intervening space referred to as the *firmament*, or more accurately, *expanse*. The Hebrew word contains the important concept of "thin layer". There remains some uncertainty as to what the separation of the waters refers to. An old interpretation, accepted by theologians such as Augustine and John Calvin and some contemporary experts, is that the separated waters above the expanse simply refer to atmospheric clouds.[4] In this interpretation, the expanse as atmosphere was created on the second day.

A more recent idea, called the "canopy theory", posited that a canopy of water vapor existed at the outer limit of the atmosphere pre-Flood, but this idea has generally been abandoned by the creationist movement.

2 The word "shekinah" is derived from Hebrew; the term "shekinah glory" is used in both Judaism and Christianity.

3 IVP Reference Collection, 1996, pp. 1090–1091.

4 Halley, 2023.

Others have proposed that these waters are a layer of icy particles surrounding the entire universe.[5] In this scheme, a thin layer of outer space (the expanse) was created on the second day, with a layer of water at the periphery of outer space. In our view, there is no reason for such a layer of water to remain at this location during the subsequent expansion of space. Instead, the water was probably progressively dispersed throughout the entire volume of outer space during its expansion, like a wave thinning as it runs up the shore.

We saw on the first day that the third heaven, the abode of God, was created in addition to one other heaven. This other heaven was either the atmosphere or outer space. Whatever interpretation is applied to "the expanse" on the second day establishes, by exclusion, the identity of the other heaven that was created on the first day. Until evidence emerges to tip the balance one way or the other, Christians are likely to choose that scheme which resonates at an individual level.

On the second day, God also began stretching out the expanse of outer space. Originally a thin layer, space underwent an enormous expansion. The observant reader of Genesis 1 will note that the second day is the only day on which God does not assess His work as "good". This could indicate that the stretching of the expanse was not completed that day, but continued until the end of the sixth day, when God described His work as very good (rather than just "good" as seen at the end of the first, third, fourth and fifth days). The expansion of the second heaven is described in several passages in the Old Testament, such as Isaiah 40, Psalm 104 and Jeremiah 10. The necessity for this rapid expansion of space becomes evident on the fourth day, when God created the stars, which are placed in the expanse.

Genesis is explicit that on the second day, a different type of creation occurred from that of the first day. Unlike ex nihilo creation, this second type of creation involves rearrangement of existing matter—namely the waters.

What transpired beneath the earth on the second day? The processes that began on the first day—fractionation and differentiation of the core, mantle and crust—likely continued into the second day, generating large amounts of heat. Was it this heat, combined with surface geological activity, that caused the separation of the waters? The separation of waters implies there was much less water covering the earth by the end of the second day.[6]

---

5 For example, Danny Faulkner suggested this water, or ice, may have been a source of the cosmic microwave background radiation (Faulkner, 2016).

6 See "Pre-Flood Earth Structure" in "Catastrophic Plate Tectonics: A Global Flood Model of Earth History", Austin, et al., 1994, p. 611.

The Third Day

On the third day the water on the earth was gathered into one place so that dry land, called Earth, appeared (Genesis 1:9–10). The text implies that there was a single ocean at this stage. We can again invoke Coulson's SFP to consider how this refashioning and rearrangement of the earth's surface occurred.

Processes that began on the first day had, by the third day, probably led to the formation of zones of thicker, lighter continental crust which remained submerged beneath the waters. At this stage, tectonic plate movements pushed together multiple areas of continental crust to form a Pangea-like supercontinent. At the same time, the denser, thinner oceanic crust pushed deeper into the mantle below, lowering the sea level. In this process, land emerged from the waters, and tectonic plate movements thrust up hills and mountains. Soils and sediments were generated, perhaps due to the erosive action of water during the uplift of dry land. Some of these sediments, having been deposited, hardened into rock.

The Bible states that a mist went up across the face of the earth to water it (Genesis 2:6).

> Then God said, "Let the earth produce vegetation: seed-bearing plants and fruit trees on the earth bearing fruit with seed in it according to their kinds." And it was so. … And God saw that it was good (Genesis 1:11–12).

While some types of vegetation were growing by the third day, the grass and herbs did not spring up until the sixth day (Genesis 2:5) suggesting that created seeds lay dormant on the third day until after Adam was created. The fruit trees did spring up and were certainly bearing fruit by the sixth day, and possibly earlier. These verses imply that plants sprang up from seed.

By God speaking the vegetation into existence, the biblical account suggests that information was introduced into organisms. We have already considered (in Part 2) the vast amount of information which is found in the DNA within cells. For life to occur and be sustained, the information in DNA is insufficient in and of itself. A whole machinery of metabolism and reproduction is also required, and what makes these processes possible are information-rich three-dimensional molecules, such as ribosomal RNA, and a variety of three-dimensional protein structures which serve as enzymes. Recall that enzymes are—or can be considered—precisely targeted nanotechnology machines which act on molecules. The amount of information which underpins these structures and processes is prodigious, and the biblical record implies that this information was provided by the Word of God during creation. John 1:3, referring to the Logos—the Divine Word—states that "All things were created through Him, …" The imprimatur of the infinite is explained by John 1:3, but not by secular origins science—the universe detectable by scientific means is finite.

As well as living cells, which are organized into the three-dimensional structures of plants and trees, a supportive ecosystem is required to provide the nutrients and habitat suitable for growing and reproducing. To this end, God put in place symbiotic relationships between plants and the animals that would later appear. Plants were an essential part of the environment to render ecosystems habitable for birds from the fifth day, and animals and humans from the sixth day. Plants store energy from light by the process of photosynthesis. This process generates carbon-based biomass for shelter and food, and releases oxygen from carbon dioxide into the atmosphere. On the other side of this equation are organisms such as birds and animals, which require oxygen for efficient metabolism. Animals, birds and plants are co-dependent; at a biochemical level, they metabolize energy-rich molecules such as sugars and fats into carbon dioxide and water. Animals also replenish the carbon dioxide in the air and may assist in plant fertilization and the dispersion of seeds across the environment.

Coulson's analysis suggests that the growth of this vegetation was extremely rapid. Would these trees have appeared to have passed through a normal, extended period of biological growth and maturation? Coulson argues that they did; these trees and shrubs would have matured according to normal criteria. If someone were to assess the age of these trees using philosophical uniformitarian principles, they would mistakenly conclude that some of the trees were very old.

In the Bible, "death" refers to the death of sentient animals or humans, which also have breath and lifeblood. Microscopic organisms such as bacteria are understandably not mentioned. The death of plants, or at least plant material, must have occurred before the fall of man, since these were given to man and beast as food. Thus, it can be speculated that multiple generations of vegetation, including algae, may have lived and "died" during the rapid processes that took place on the third day. Following on from this, simple organisms such as bacteria would also be required to break down dying vegetable matter, and these bacteria, or simple non-sentient organisms such as worms or insects, may also have undergone death and decomposition. Plants which could cope with poor, rocky soils likely sprang up early on the third day. Multiple generations of these decayed and composted, rapidly improving the soil quality. At that point, those plants which required high quality soil sprang up.

The light source available to plants was the first-day light; as we speculated above, this could be the Shekinah glory of the Spirit of God.[7] Even this was only available to a proportion of the plant life because of the slow rotation of the earth relative to the accelerated pace of plant growth.[8] God still had not created the sun and stars; the expansion of space, still in progress, was inadequate for these to attain sufficient distance from the earth. At close range, the radiant heat from these celestial bodies would have burned up the earth. (This recalls the Transfiguration (Matthew 17:1–8), in which Jesus' face shone like the sun without causing any deleterious effects on the disciples, and the angel of Revelation 18:1, who illuminated the earth "by his splendor.") We again note Coulson's suggestion that some of the plant life was growing miraculously on the third day in the absence of any light in a manner analogous to Aaron's budding staff.[9]

The Fourth Day

On the fourth day God spoke the lights of outer space into existence. The sun and the moon were created to rule the day and the night. God also made the stars for signs, seasons, days and years. "And God saw that it was good" (Genesis 1:14–19). The Bible tells us that the heavens declare the glory of God. They are beautiful, expansive and majestic.

By the fourth day, outer space was sufficiently expanded to enable the sun and other stars to be placed at a significant distance from the earth, where their radiant heat did not damage the earth or its nascent life. Since vegetation was created on the third day, it is evident that sunlight was required to support growth, which the newly created sun supplied from the fourth day onwards. The mention of seasons signifies that the axis of the earth's rotation was tilted relative to the sun by the fourth day.[10]

---

7 An alternative view is that God may have created a temporary light source, fixed in relation to the rotating earth, which was no longer required after the sun was created on the fourth day.

8 Under SFP, multiple generations of plant life may have lived and died in regions that were experiencing night conditions.

9 The growth of flora that was located on the "night" side of the earth is, in our opinion, a potential challenge for Coulson's SFP concept. Coulson has proposed a miraculous answer to this problem, supported by a theological argument (the budding of Aaron's staff). Alternative explanations should be considered, however, because the default position of the creationist enterprise is a preference for naturalistic explanations where possible. An alternate explanation might note that the shape, size, and position of the pre-fourth day light source determined the duration of the night. The night could have been as short as a few hours depending on those parameters. Following from this, most locations on the surface of the earth on the third day cycle from day → night → day with a smaller subset cycling from night → day → night. Any area in daylight generates biomass via photosynthesis. As it moves into night, the biomass could serve as a substrate for growth of fungi and species of plants which can parasitize fungi. This permits an ecology of flora to survive the night period. Other organisms can live in conditions devoid of sunlight, for example, various types of bacteria in deep ocean hydrothermal vents can support a complex vent food chain.

10 Sarfati, 2018, p. 189.

Also noteworthy is the enduring value of the stars as "signs" for celestial navigation. For millennia mariners have used the stars to determine their bearings; navigators on Second World War-era aircraft used sextants on night flights. The advent of GPS in the twentieth century might have rendered celestial navigation obsolete, except that the United States military, realizing that GPS is dependent on functional electronic systems that may not be available in wartime, resumed training its personnel in celestial navigation in 2015.

Coulson argues that the movement of the tectonic plates (mentioned above) largely ceased at the end of the fourth day of the creative epoch. In the current epoch, movement of the tectonic plates is associated with natural disasters such as earthquakes, volcanoes and tsunamis, which did not affect human and animal life until after the fall of man. But the pre-fall earth likely entered a period of tectonic quiescence during the fourth day. The deceleration of process rates occurring at some point after the fourth day implies that tectonic activity also must have decreased (everything was occurring more slowly by the fifth or sixth days compared with prior days).

It is also likely that the earth's magnetic field was established by this stage to protect the atmosphere from the solar wind. Earth's magnetic field is believed to result from the geodynamo mechanism of circulating elements of the molten outer core of the earth.

The Fifth Day

On the fifth day, God created swarms of sea creatures, including giants of the deep. The waters teemed with this new life, an abundance that was to be later reflected in the fossil record. The mighty aquatic creature, Leviathan, mentioned in Psalm 104 and Job 40, was likely a now-extinct giant crocodile such as *Sarcosuchus imperator*.[11] The sea creatures which creep, such as crabs and lobsters (in addition to others which are now extinct but revealed in fossil layers), were also created on the fifth day. That the waters "swarmed", or in some translations "teemed"[12] with every living creature which moves perhaps suggests that the accelerated process rates—including reproduction of fish and other sea animals—continued at vastly higher rates than in the post-creation epoch.

Birds were created to fly across the face of the expanse—that is, in front of the expanse of outer space, and to multiply on the earth. This included the original ancestral kind of the now-extinct species *Archaeopteryx*. God blessed the sea creatures and the winged animals to multiply.

---

11 Booker, 2005.

12 For example, New International Version.

What becomes more apparent from the fifth day are the elements of common design which God used across various kinds of animal life. For example, most kinds of fish, except for rare examples, rely on gills to efficiently extract the oxygen dissolved in water and to excrete carbon dioxide. In order to function, gills require blood and a pump to circulate it—a heart. Gills provide an advantage to creatures which cannot access the surface of the water, which occurs if the surface of water freezes over in winter. Some types of crab can keep their gills moist even on land, enabling them to respire when not submersed. Gills are highly efficient, yet they appear to place a limit on the size of organism they can support; the largest gilled creature is the whale shark, with a body mass of around 20 tonnes.

God designed other sea creatures such as whales, dolphins and seals with lungs. Like gills, lungs require circulating blood pumped by the heart to function. Lungs can support a very large aquatic biomass of up to nearly 200 tonnes, in the case of the blue whale. In addition, lungs support more complex brains. The lunged sea creatures tend to be more sociable and trainable than the gilled creatures.

Birds' lungs are supplemented by additional structures which allow a highly efficient and unique flow-through system (rather than the 'in and out' or 'bellows' lung system that humans possess). This design supports the very high metabolic rate required for flight and can enable flight at high altitudes where oxygen levels are rarefied.

In designing organisms, God has clearly reused useful anatomical, physiological and molecular structures across different kinds. However, some are unique variations on a general theme, such as the flow-through design of birds' lungs. Certain similarities are necessary. For instance, since the originally created animal and bird kinds broadly had the same food sources, we would expect to see similarities in their digestive processes, and consequently the design of their digestive tracts. Following from this we would expect significant sections of the DNA blueprints of kinds with matching features to be very similar. This is indeed what we find.

And God saw that it was good (Genesis 1:20–23).

The Sixth Day

On the sixth day, God created cattle, creeping things and beasts after their kinds. This includes the herbivorous dinosaurs, such as the behemoth, described in Job 40, which had a tail like a cedar tree. Every green herb was given for food for these animals and the winged creatures.

There is enormous diversity among the kinds of land animals. Cold-blooded kinds have a metabolic rate somewhat dependent on their surroundings. In cold weather, their metabolism slows, making them extremely energy efficient. However, on the downside, such creatures are sluggish under these conditions. Warm-blooded kinds such as humans maintain their temperature accurately over a wide range of environmental conditions, but this comes at the cost of higher energy use. For this reason, warm-blooded animals require a higher caloric intake for a given body mass, irrespective of the energy density of the food they consume.

Insects use thin tubes called spiracles to direct oxygen to body cells. The system works well on a small scale but does not translate effectively to larger organisms. As we saw with whales and dolphins, larger land animals have lungs to oxygenate their larger body mass. Clearly, God designed the structure of the animals to follow the functional requirements.

Based on structural similarities and differences between the many species of plants and animals, Carl Linnaeus, the eighteenth century Swedish taxonomist, categorized living organisms into a nested hierarchy (groups within groups) known today as the Linnaean taxonomy. This taxonomy partitions organisms into categorical ranks of kingdoms, phyla, classes, orders, families, genera and species. A species is a population which can naturally interbreed to produce fertile offspring. It is understood that the kinds mentioned in Genesis likely correspond roughly with families or genera in the Linnaean taxonomy (although Wieland's axiomatic definition of kinds, as introduced in Chapter 8, is to be preferred). Because many land animals share anatomical, functional and biochemical elements, we should see similar DNA sequences among different kinds. This is borne out by the evidence.

God created the kinds with an ability to adapt to their environment. This ability enables the animals and birds to live virtually all over the earth in different climatic and ecological conditions. Variation and speciation were an inherent ability of God's originally created kinds, with one seeming exception: mankind. Humans exhibit variation—for example, recognizably different ethnolinguistic groups and nations have emerged over time—but not speciation.

God noted that there was no man to till the ground. Then God formed Adam out of the dust of the ground and breathed into his nostrils the breath of life, at which point he became a living being (Genesis 1:24–31). Adam was not created ex nihilo, but by the refashioning of existing matter. The narrative also indicates Adam has a "composite nature: material and non-material".[13] Coulson's SFP synthesis is again useful here. Usually dust comes from windblown soil, which is formed over long time periods by the breaking down of rock (and subsequent enrichment with composted plant matter). "The dust of the ground" here is readily understood as what we might call "soil", particulate matter of which the earth under our feet is composed. On the sixth

13 Sarfati, 2018, p. 291.

day, soil was present for Adam to be formed, and God created him by refashioning this matter into the body of a sexually mature man.[14]

God planted a garden in Eden from which flowed a river which itself split into four rivers. The river provided water for the garden. God caused beautiful trees with edible fruit to sprout from the ground, including the tree of the knowledge of good and evil. God placed Adam here to live and tend the garden, commanding him to eat fruit from any of the trees except for the tree of the knowledge of good and evil. If he were to eat of the latter, on that day he would die. God presented a subset of all animals to Adam, who named them. The subset consisted of all livestock, animals of the field and all birds (Genesis 2:7–20).

Because Adam needed a helper and none of the animals were suitable, God caused Adam to fall into a deep sleep. God removed a rib and fashioned it into a woman whom he presented to Adam (Genesis 2:21–23).

Bible commentator Matthew Henry, writing in the eighteenth century, opined that woman was created out of Adam's side to be equal to him, under his arm to be protected by him and close to his heart to be loved by him. He stated that if man is the head, woman is the crown. Adam is refined from dust, Eve is twice refined, one remove further from the earth.[15]

Adam called her Woman because she came from man. Both Adam and the woman were naked and not ashamed (Genesis 2:23–25). The man and woman together were made in the image of God (Genesis 1:27). God blessed them to be fruitful and to multiply and fill the earth. (This was a command that God expected to be obeyed; when mankind disobeyed at Babel, God deliberately dispersed people across the face of the earth.) As a representative first couple, Adam and Eve were given the task of subduing the earth and ruling over all the fish in the sea, and over the creeping land animals and the birds. And God saw that the creation was very good (Genesis 1:28).

Before moving on we need to consider the phrase "in the image of God". What does this mean? It certainly does not merely refer to the power of speech, nor intelligence above that of the animals. Angels possess both of these attributes yet are not made in the image of God. It may refer to man's dominion over the animals, his creativity, or the fact that he has a spirit which was breathed into him at creation. Genesis also identifies man as a tiller of the ground, a gardener. Once again this relates to the image of God, a God who created the garden of Eden. Animals and angels do not till the ground or create gardens.

---

14 There is a wordplay in the Hebrew on 'adam and 'adamah, where the latter means "clay". The impression this gives is that God has formed the structure of a man from "clayey-soil" and then animated it (Sarna, 1989, p. 17). An English equivalent to adam/adamah is homo/humus (Sarna, 1989, p. 17).

15 Henry, 2008, p. 8.

But man is unique among all creatures in being both a spiritual being and one who begets, as shown in Figure 16.1. This is contrasted with the animals, which beget, but are not spiritual; and the angels, which are spiritual, but do not beget (angels do not marry). The distinguishing characteristic of man, the attribute that sets him apart from all other creatures, is that he is (at least partly) a spiritual being who begets and names his offspring. Animals do not name their offspring because they do not have the power of speech; angels do not have offspring to name. Man is therefore like God in this respect, and is the only part of creation like this.

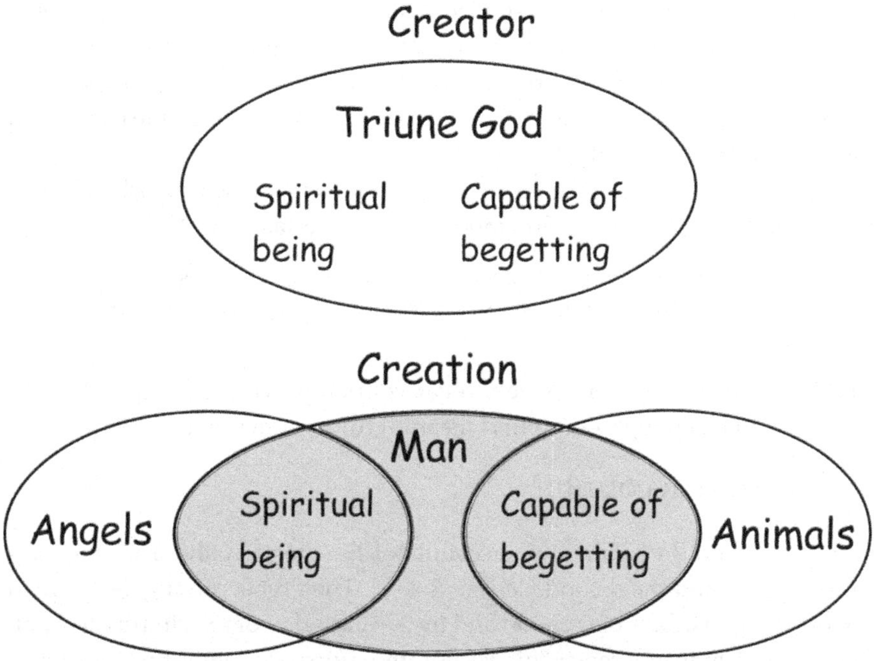

Figure 16.1: The uniqueness of mankind in creation

There are two more key differences to note regarding man and the animals. The first concerns hereditary differences. Whereas all humans, including Eve, are effectively descended from a single person (Adam), animals were created in larger numbers from the outset, both male and female. Hence the unity of *Homo sapiens* is more absolute than that of the animals. Man has an unmistakable corporate essence represented at the head by Adam (and now, for those who believe, by Christ).

The second key difference is that God planted a garden and placed the man in it. By design, man belongs in a planned environment. Many animals are creatures of the wild.

The end of the sixth day marks the end of the creative epoch in terms of the Anno Mundi[16] calendar. By the end of the sixth day, God had brought the entire creation to the point of maturity which He determined was appropriate for the commencement of human history. As Genesis 2:1 says of the sixth day, "So the heavens and the earth and everything in them were completed."

The Seventh Day

On the seventh day, God ceased from his work and rested. He did this to set a pattern for mankind to follow (Exodus 20:8–11). He blessed the seventh day and set it apart as a day of rest (Genesis 2:1–3). The seventh day is notable because it marks the start of the post-creation epoch—God's work of creation is finished. The same laws of nature that are operating today were also operating on the seventh day, except that today there are also the effects of the curse such as disease and thorns and thistles.

We note in passing that Genesis presents a positive view of work, which is echoed later in the Bible. King Solomon said in Ecclesiastes 3:12–13:

> I know that there is nothing better for them than to rejoice and enjoy the good life. It is also the gift of God whenever anyone eats, drinks, and enjoys all his efforts.

In line with this, modern society regards unemployment as a problem for those unfortunate enough not to find meaningful remunerative work.

Two creation accounts?

Genesis 1 and 2 contains two accounts of the creation, the first spanning Genesis 1:1–2:3, and the second Genesis 2:4–25. The first account gives a broad overview of the six days of creation and the seventh day of rest. In this account, the creation of man references his corporate nature. Recalling the concept of "the One and the Many", God is one, and Man, though (eventually) comprising many individuals, also has a corporate (unitary) nature. All of mankind are ultimately derived from Adam, the first man. Genesis 1:26 emphasises this aspect of creation, as God says:

> Let us make man in our image, according to our likeness. They will rule the fish of the sea, the birds of the sky, the livestock, the whole earth, and the creatures that crawl on the earth.

The first sentence reveals that Man is in a sense 'one'—analogous to God. The next sentence makes clear that the oneness of man comprises a multiplicity of human persons. But this account focuses on the "one" and is less personal.

---

16 The Anno Mundi calendar counts forward from the first day of creation.

The second account in Genesis 2 focuses solely on the sixth day, but provides significantly more detail than the first account, and, notably, expands on Adam as a personal being ("the many"). God talks person to person with a specific human—Adam. Because of the different characters of the two accounts, two different Hebrew terms are used for God: *Elohim* is used exclusively in the first (a more general word for God, which is a plural noun in Hebrew). The second introduces the term *YHWH* (the personal name of God revealed to Moses by God on the occasion of the burning bush in Exodus 3). The first three times the word 'God' appears in Genesis 2 still uses Elohim (vv. 2, 3), and thereafter in the chapter the two are combined, as *YHWH Elohim*, translated 'LORD God' (v. 4). These two narratives represent different aspects of the same account and anticipate the trinitarian doctrine which is fully revealed in Jesus Christ.

The historicity of Adam and Eve

In an online article, the high-profile evangelical leader Tim Keller (1950–2023) pointed out that the issue of Adam and Eve's historicity is one of four difficulties that Protestant Christians have raised (even though the early church fathers were unanimous about the historicity of Adam and Eve):

> One way to reconcile what current science says about evolution is to propose that the account of Adam and Eve is symbolic, not literal, but what does this do to the New Testament teaching of Romans 5 and 1 Corinthians 15 that our sinfulness comes from our relationship with Adam?[17]

In our estimation, Paul's teaching in Romans 5 is crystal clear. Verse 12 states that sin entered the world through one man (Adam), and thus "death spread to all people …". This statement clearly identifies Adam as a man. Paul later declares that "death reigned from Adam to Moses" (Romans 5:14). This phrase includes Moses, whose historicity is not in doubt, as the second of two men who marked out the time during which death reigned, further indicating that Adam occupied a definite place in history. The mention of Adam with another historical man also brings to mind the genealogy of Christ given in Luke 3, which includes Abraham and David and, going backwards, concludes (Luke 3:38) with "son of Enos, son of Seth, son of Adam, son of God." In this genealogy, Adam is listed with over 70 other historical men.

The language in 1 Corinthians 15:22 follows this pattern. Paul writes that "For just as in Adam all die, so also in Christ all will be made alive." Adam is again mentioned alongside a historical man, this time Jesus Christ. The New Testament teaches repeatedly that Adam is a historical man—as historical as

---

17 Keller, 2012.

Abraham, Moses, David and Jesus Christ, to cite just a few of the men that Adam is named alongside.

The New Testament also provides strong evidence as to the historicity of Eve: "And Adam was not deceived, but the woman was deceived and transgressed." (1 Timothy 2:14) The reference to the woman is clearly Eve; and the text leaves us in no doubt that she was a real woman.

Beyond this, the New Testament also teaches that Adam was "a type" of Christ (Romans 5:14), a reference which is one of the supporting scriptures for the doctrine of typology. (This doctrine interprets certain Old Testament people or events as "types" which are fulfilled in the New Testament—for example, the sacrificial system of the Old Testament prefigured, and was superseded by, the sacrifice of Christ on the cross). The creation of Eve from Adam typologically prefigured the relationship between Christ and the church. Christ is repeatedly described in the New Testament as a bridegroom, or husband (Mark 2:19, 2 Corinthians 11:12) and the church as bride (Ephesians 5:22–32, Revelation 21:9). The mode by which God created Eve from Adam's side, in Genesis 2:21–22, is fulfilled in the death of Christ on the cross, as shown in Figure 16.2. This is not a new-fangled interpretation; it was taught by John Chrysostom (347–407) who served as Archbishop of Constantinople in the patristic period.

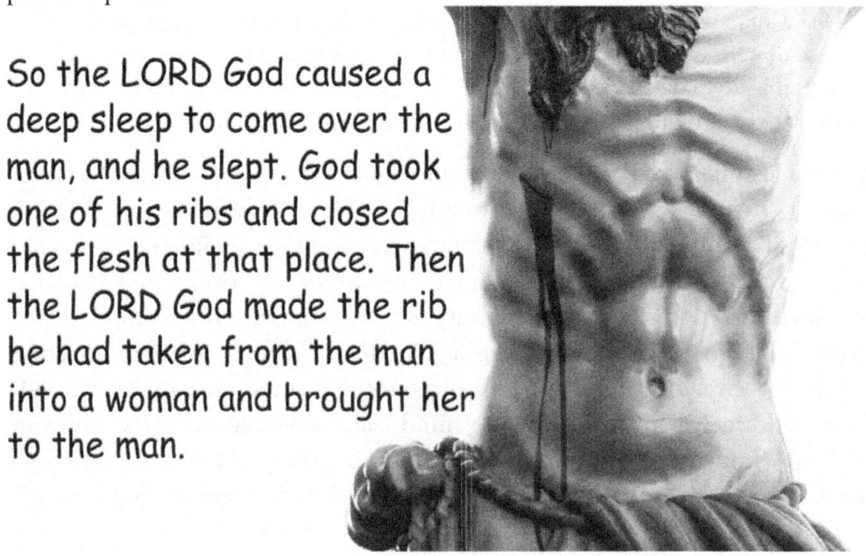

Figure 16.2: Christ's side was pierced, bringing forth "blood and water"

The blood and water that emanated from Christ's pierced side (John 19:34) are emblems of communion and baptism, the two sacraments recognized by the Protestant churches. We may say then that Eve derived from Adam in the same way that the church derives from Christ. In fact, this typology shows in big-picture terms what the church age is about. In the same way that God formed Eve from Adam then brought her to him, God is also in the process of forming Christ's bride, the church. When the bride is complete, God will bring her to the Man, who is God Incarnate (Revelation 21:2). An implication of this typology is the centrality of the creation account to Christian life and practice. Every time a Christian is baptized, and every time a Christian partakes in communion, they do so in a way which recalls the creation of Eve from Adam.

This typology also implies that the church is regarded biblically as a corporate whole. Humanity, like God, can consist corporately as a multi-personal unity. Scripture is replete with references in both the Old and New Testaments to corporate responses (for example, Judges 20:1, Ezra 3:1, Acts 2:1, and compare Acts 19:29). The Bible makes this meaning clear when it states, "Let us make man in our image, according to our likeness ..." (Genesis 1:26). It uses a plural article for God and man. The corporate and individual natures of man are expressed in the two accounts of the creation. Thus, humans as corporate Man reflect both the Trinity (multiplicity) and oneness (unity) of God. As we noted, the two representatives of corporate Man in history are Adam and Jesus Christ.

Marriage and blessing

In Genesis 2:24 we read:

> This is why a man leaves his father and his mother and bonds with his wife, and they become one flesh.

In Matthew 19:4–5, Jesus discusses these very verses from Genesis:

> "Haven't you read," he replied, "that he who created them in the beginning made them male and female," and he also said, "For this reason a man will leave his father and mother and be joined to his wife, and the two will become one flesh?" So that they are no longer two, but one flesh. Therefore, what God has joined together, let no one separate."

In so doing, Christ treats Genesis as literally true and provides biblical definitions for sex and marriage. The biblical religious perspective holds that there are two human sexes—male and female. Biblical marriage is monogamous, involving a single male and a single female, and should only be dissolved by death or serious covenantal infractions. Christ clearly says (v. 4) that this was God's plan from the beginning.

Genesis 1:28a can be understood as both a command and a blessing for corporate man:

> Be fruitful, multiply, fill the earth, and subdue it.

This is the original (and unfinished) mandate God gave to man—to fill the earth and subdue it.

The God of Genesis

So, what is the nature of the infinity underlying the Christian worldview? Both old-earth creation proponents and young-earth creationists have a measure of agreement on this. For example, John Lennox highlights several features of God deduced from the first chapters of Genesis in his book *Seven days that divide the world,* which we summarize here:[18]

1. God exists—He is the only non-derivative, uncreated, self-existent being.
2. God is the eternal creator—ultimately everything which exists (excluding Himself) was created by God so He owns, understands, knows and controls everything.
3. God is distinct from his creation—not only is He transcendent and separate from the creation, He also transcends time.
4. God is personal—therefore we are accountable to Him.
5. God is a fellowship—this refers to the Trinity; God does not need other beings for fellowship. Also, by implication, both the individual and the corporate, the "one and the many", are equally ultimate.
6. God has a goal in creation—therefore history has meaning.
7. God creates by His Word—the world is therefore full of information, and this information is coherent. It is understandable, at least analogically, by those made in His image.
8. God is the source of light—without God we live in darkness.
9. God's creation is good. (It is sin and the curse on sin which have brought disaster, pain and the death of humans and sentient animals.)
10. God has set a pattern by which man performs six days of work followed by a seventh day of rest—both work and rest are important. Unemployment is viewed as a misfortune, while working seven days per week is unsustainable, if not unhealthy. This weekly pattern also draws attention to the fact that God has created time, and that He gives each of us a certain amount of time in our lives.

This concludes the creation week narrative. As an origins account, it must align with scientific findings to be plausible. It is well known, however, that secular science regards the universe as billions of years old. How are we to square this with the Bible, which, when read in a straightforward manner, gives an age of the earth as around 6,000 years? We turn to this crucial issue in the next chapter.

18 Lennox, 2021, pp. 127–150.

References

Austin, S., Baumgardner, J., Humphreys, D., Snelling, A., Vardiman, L., Wise, K. (1994). Catastrophic Plate Tectonics: A Global Flood Model of Earth History. *Proceedings of the Third International Conference on Creationism.* (pp. 609–621). Pittsburgh: Creation Science Fellowship, Inc.

Booker, P. (2005). A new candidate for Leviathan? *Journal of Creation* (formerly *TJ*), 19(2). Retrieved from creation.com/images/pdfs/tj/j19_2/j19_2_14-16.pdf

Faulkner, D. (2016). Thoughts on the rāqîa' and a Possible Explanation for the Cosmic Microwave Background. *Answers Research Journal*, 9, 57–65. Retrieved June 21, 2025 from answersresearchjournal.org/raqia-cosmic-microwave-background/

Halley, K. (2023, April 20). The 'waters above' in Genesis 1—a brief survey of competing interpretations. Retrieved from creation.com/waters-above-survey

Henry, M. (2008). *Matthew Henry's Commentary On The Whole Bible—Complete and Unabridged.* Peabody: Hendrickson Publishers.

IVP Reference Collection. (1996). *New Bible Dictionary* (Third edn). (I. H. Marshall, R. R. Millard, J. I. Packer, & D. J. Wiseman, Eds.) Nottingham: Inter-Varsity Press.

Keller, T. (2012, 2 23). Creation, Evolution, and Christian Laypeople. Retrieved June 3, 2024, from biologos.org/articles/creation-evolution-and-christian-laypeople

Lennox, J. C. (2021). *Seven Days that Divide the World: The beginning according to Genesis and Science* (10th Anniversary edn). Grand Rapids: Zondervan Reflective.

Sarfati, J. D. (2018). *The Genesis Account—A theological, historical, and scientific commentary on Genesis 1–11* (Third edn). Powder Springs: Creation Book Publishers.

Sarna, N. (1989). *The JPS Torah Commentary: Genesis.* Philadelphia: The Jewish Publication Society.

17
The conundrum of the ages

The inferred ages of the earth and cosmos are issues in which a straightforward reading of Genesis and secular science are at odds with each other. In Chapter 6 we saw the importance of this topic. Reverend Driver commented in 1897 that "read without prejudice or bias, the narrative of Genesis i. creates an impression at variance with the facts revealed by science." We also saw that, in the 1960s, Ronald Numbers "slid down the proverbial slippery slope toward unbelief" after seeing evidence that *he thought* implied the earth was at least 30,000 years old.

Before we consider how these seemingly intractable differences might be resolved, we explain how one famous Christian reading Genesis in a straightforward fashion deduced the approximate age of the earth. This Christian was none other than the Irish archbishop of Armagh, James Ussher, who published his estimate, 4004 BC, in 1650. This deduced age became widely known in the English-speaking world following the Church of England's decision to include it in the marginal notes of King James Bibles from 1701.

A biblical estimate of the age of the earth

Ussher's chronology, written in Latin, was a substantial work of scholarship which relied on Ussher's detailed analysis of evidence from the Bible and other historical sources. We will briefly discuss the method by which Ussher established his chronology and review some criticisms of his work.

Ussher's aim was to carefully and logically improve the chronology of ancient biblical history using biblical and non-biblical sources.[1] This was no easy task because it required synchronizing the Old Testament calendar with multiple pagan calendars from Egypt, Ancient Babylonia and other places. The Old Testament source Ussher relied on was the Masoretic text, the standard Hebrew version[2] regarded by Jews and Protestants as very accurate. (The earliest surviving copy of the complete Hebrew Masoretic text dates to around AD 1000). Ussher applied an Anno Mundi (AM) calendar, counting forward from the first day of creation. Anno Mundi is Latin for "year of the world". So, AM 930 would be 930 years after creation.

---

1 Barr, 1984–1985, p. 575.

2 Ibid., p. 584.

The Gregorian calendar, in use since 1582, has a fixed point at AD 1 which is an approximation of the year of Christ's birth. ("AD" stands for "Anno Domini", Latin for "in the year of our Lord", namely Christ. The year prior is 1 BC or 1 "before Christ". There is no zero year in this system.)

Genesis 5 and 11 present important data for the dating exercise. Chapter 5 begins with the sentence "This is the document containing the family records of Adam" (Genesis 5:1a), indicating that a written genealogy, dating to the pre-Flood era, was incorporated by Moses at this point. The genealogy of Chapter 5 is a chronogenealogy; that is, it lists the age of the father at the birth of the next offspring in the list, regardless of whether that offspring is a son, grandson, or great-grandson. This enables a chronology to be established in years. This will be in Anno Mundi terms, since the "birth" year of Adam was the year of creation week. A chronogenealogy from Adam to Noah's children Shem, Ham and Japheth is found in Genesis 5. Genesis 11 contains a follow-on chronogenealogy of Shem to Abraham (then known as Abram) and Lot.

To synchronize the Bible-based Anno Mundi calendar with the Anno Domini calendar, a historical event with a known date in both calendars had to be found. Ussher used the year that King Nebuchadnezzar II of Babylon died (and his son Evil-Merodach began to reign), 563 BC, as his major synchronization date. Ussher was able to identify this date in the Anno Domini calendar because of his extensive knowledge of both Greek and Roman historical writings. Today it is considered that Nebuchadnezzar II died in 562 BC, which is an indication of how accurate Ussher's analysis was, at least on this point. Ussher synchronized this with an Old Testament detail in 2 Kings 25:27 which identifies that Evil-Merodach released Jehoiachin, the King of Judah, from prison in his first year as ruler.

From here Ussher traced his chronology back through the Old Testament books of 2 Kings and 1 Kings to the reign of King Solomon, and from there back to creation. But as mentioned, this was not a simple task, and in places a degree of interpretation was required. For example, consider the three verses Genesis 5:32, 7:11 and 11:10 placed consecutively (King James Version):

> And Noah was five hundred years old: and Noah begat Shem, Ham and Japheth. In the six hundredth year of Noah's life [the Flood began]. This is the history of the generations of Shem: Shem was an hundred years old, and begat Arphaxad two years after the flood.

To the modern reader and to Ussher this was confusing; the first two sentences imply Shem was 100 years old at the Flood whereas the second implies he was 100 two years after the Flood. Ussher concluded that Japheth was born first and Shem two years later.

However, some modern commentators believe that in the context of Genesis 1–11 it probably means:[3]

> Noah was five hundred years old when he became the father of Shem. Later he had Ham and then Japheth. In the six hundredth year of Noah's life [the Flood began]. This is the history of the generations of Shem: Shem was 100 years old [at the Flood], two years after which he became the father of Arphaxad.

There is another example which involves much more than a matter of two years. Here we consider Genesis 11:26, 11:32 and 12:4 together, again using the King James Version:

> And Terah lived seventy years, and begat Abraham, Nahor, and Haran. And the days of Terah were two hundred and five years: and Terah died in Haran. … and Abram was seventy and five years old when he departed out of Haran.

The modern commentator, Donovan Courville (died 1996), reads this as:

> Abraham was born when Terah was seventy years old, and afterwards, Nahor and Haran were born. Abram left Haran when he was seventy-five years old. Terah remained in Haran and died when he was two hundred and five years old.[4]

Ussher's interpretation differed from modern scholarship here too, and can be summarized as: Nahor and Haran were born when Terah was seventy, but Abraham was born sixty years after that. Terah died after which Abraham left Haran.

Ussher moved Abram's birthdate 60 years forward. His understanding is believed to be heavily influenced by Acts 7:4, where Stephen, the first martyr of the Christian faith, said about Abram:

> Then he left the land of the Chaldeans and settled in Haran. From there, after his father died, God had him move to this land in which you are now living.

Stephen implies that Abraham left Haran after Terah his father was dead, and Ussher took this to be true. But some recent commentators do not accept this reasoning, implying that Stephen's words may have been misinterpreted or mistaken. (After all, Stephen was not an apostle, nor does the text state that he was speaking in the Spirit at this point[5]—and some commentators might place emphasis on these considerations.) Either way, this example shows the kinds of decisions that bear on the analysis and the difficulty in making these decisions.

---

3 Rushdoony, 2002, p. 117.

4 Courville, 1971, pp. 139–140.

5 Although it does later in the speech (v. 55)

A further issue concerns the starting point for Israel's sojourn in Egypt. The duration of the sojourn is not in doubt; it is specified in Exodus 12:40 as 430 years. However, two major interpretations of the starting point of this period have emerged, sometimes referred to as the short and long chronologies (or sojourns).[6] Ussher and authorities such as Courville preferred the short chronology, which views the start of the time in Egypt as being Abram's entry into Canaan (and later Egypt) after he left Haran. This is the time in which God gave a promise to Abram and is consistent with Galatians 3:17. Egypt was the main power near Canaan and thus entering Canaan is considered synonymous with entering Egypt. The promise to Abram specifically states that his offspring would be afflicted for four hundred years, after which the exodus would occur. The affliction is understood to have begun with Ishmael's mocking Isaac on the day he was weaned (Genesis 28:9). This is approximately 30 years after the original promise given to Abram. Under this scheme, Jacob and his sons enter Egypt 215 years after the promise to Abram. They, and the nation of Israel after them, spend 215 years in Egypt. For more than 70 years they are treated very well but later, under the aegis of a new pharaoh, the children of Israel are enslaved and treated harshly.

By contrast, the long chronology identifies the entry of Jacob and his sons into Egypt as the start of the 430-year sojourn, which adds an extra 215 years into the chronology. Courville considered this interpretation to be inferior.

The timespan from the Exodus to the third year of the reign of Solomon is 480 years, as mentioned in 1 Kings 6:1. But from here onward dating becomes more difficult; in the period of the kings some of the reigns may overlap. The upper limit is about 430 years to the destruction of the temple if the reigns are placed end-to-end, but Ussher gives about 423 years for this period. From there, Ussher has 26 years passing until the death of Nebuchadnezzar.

Courville notes that Ussher made a mistake of logic toward the end of his analysis. Ussher's working showed that Christ was born in about 4000 AM; he subtracted that from 4 BC (which is the year Herod died and close to when Christ was born) to arrive at his date of the creation of the world—4004 B.C., a calculation that Courville felt was too precise given the imprecision of the data that Ussher was working with. However, Courville considered that overall, Ussher's scriptural analysis and chronology were quite acceptable and within the same ballpark of other authorities using the Masoretic text.

There are other ancient versions of the Old Testament whch differ from the Masoretic text. In the Septuagint,[7] the ages of fathers at the time of their children's births in Genesis are mostly larger by 100 years. (It is possible that

6 Ibid., pp. 8–9.

7 A couple of centuries before the time of Christ, many Jews spoke only Greek instead of Hebrew. For this reason, 70 Jewish translators produced a Greek version of the Hebrew Old Testament. It was called the "Septuagint", meaning "70" and is often abbreviated as "LXX", the Latin representation of the same number.

the Jewish translators may have altered these ages to increase the length of the history to align Genesis with the timelines of the Egyptians and Phoenicians.[8]) For the Orthodox churches that usually rely on this text, the date of creation is calculated at about 5508 B.C., significantly earlier than those based on the Masoretic text.

Other documents containing parts of the Old Testament, such as the Samaritan Bible, are also taken into account by experts.[9] (Written in Hebrew, this text was discovered by Western scholarship in Damascus in 1616.)

But what of the Jewish interpretation of their own scriptures? In Daniel Gordis' highly regarded *Israel: a concise history of a nation reborn*, Gordis concludes his acknowledgements with the place and date:

> Jerusalem, May 2016, 5 Iyar 5776, the sixty-eighth anniversary of Israel's independence[10]

indicating the date of creation as 3760 BC, some 244 years later than Ussher's estimate.

However, most Protestant Christians would agree that a straightforward reading of the Bible places us in the seventh millennium AM, based on the more accurate chronogenealogies of the Masoretic text.[11]

Summing up, internal evidence from the biblical texts, as understood by commentators such as Ussher and contemporary young-earth creationists, does not allow the 30,000-year timeline that proved a stumbling block for Ronald Numbers, let alone the billions of years that secular science claims for the age of the earth.

Scientific estimates of the age of the universe

We now turn to consider the age of the universe from a scientific point of view. In a philosophically uniformitarian scheme, the distance between the stars and the earth is relevant to the minimum age of the cosmos. Determining those distances is crucial. Clearly, analysis of the distance of stars and estimating the age of stars falls under the forensic branch of science—an applied science. This means that the evidence is circumstantial in nature and that multiple lines of evidence are required to establish knowledge more firmly.

---

8 Barr, 1984–1985, pp. 582–583.

9 Barr, 1984–1985, p. 583.

10 Gordis, 2017, p. 432.

11 The chrono genealogy in early Septuagint (LXX) manuscripts imply that Methusaleh outlived the flood by 14 years. This does not align with other biblical references which only include Noah, his wife, and his three sons and their wives—a total of eight (1 Peter 3:20)—among those who boarded the ark. This issue may have cascaded from an earlier textual error in a proto-LXX document, where the age of Jared when Enoch was born was wrongly copied as 62 years instead of 162. This was corrected in later LXX manuscripts (Cosner & Carter, 2015, p. 102). The Masoretic text does not have this issue, and is mostly more reliable (Cosner & Carter, 2015, pp. 103–105).

In short, some or all of this information may change in the future, although it appears to be relatively consistent and reliable currently.

Scientists use two main methods to estimate the distance to stars within our own galaxy. One is called *trigonometric parallax;* this assesses the apparent change in the position of a star when viewed at opposite extremes of the earth's orbit and converts that change into a distance measurement. The other method is known as *proper motion* and relates, under specific circumstances, the apparent velocity of a star to its distance from earth.

There are two commonly used astronomical measures of distance, the *astronomical unit* which is the average distance of the earth to the sun, and the *parsec,* which is a little over 3 light-years, where a light-year is the distance light travels in a year.[12]

To measure the distance to extremely remote stars, a key consideration is the degree to which starlight is redshifted. Here, secular science infers a distance from earth to the star using the ΛCDM Big Bang model. But as we saw in Chapter 9, Big Bang models, including ΛCDM, are arguably built on top of several secular, Genesis 1-denying presuppositions. Therefore, these inferences are unacceptable from a creationist perspective. For example, the Hubble space telescope detected a star in 2018 that was believed at the time to be the most distant star ever found. Dubbed Icarus, the enormous blue star was estimated to be 9 billion light years away.[13] However, this distance was inferred based on the presuppositions of ΛCDM, and as a result, creationists who disregard ΛCDM, would in turn disregard this inferred distance.

Creationists have not yet agreed on an alternative cosmological model for the expansion of the universe which can map redshifts to distance measures. In the absence of such a model, creationists must fall back to geometric calculations of star distances which do not rely on a cosmological model with secular presuppositions. For example, a trigonometric technique, which does not rely on ΛCDM, was used to calculate the 430 million light year distance to a galactic feature in 2020.[14] (We use this example because, at the time of writing, this was the most distant astronomical measurement we could find that does not rely on ΛCDM.) Under a philosophical uniformitarian interpretation, light from that galaxy has taken 430 million years to reach earth.

Light travel problem

This type of finding leads to what is known as the light travel problem for biblical creation. If the earth was only created some 6,000 years ago, how could

12 Weinberg, 2008, pp. 14, 15, 17.

13 NASA Hubble Mission Team, 2018.

14 The feature was a water megamaser found in the barred spiral galaxy indexed as 6264 in the New General Catalogue (NGC 6264) (Pesce DW et al., 2020).

light have traveled for 430 million years? This issue invalidates simplistic and naive attempts to defend young-earth creation, such as the claim that the vast distances are not real; the science underpinning these measurements, and the measurement themselves, are credible. (If all the stars were bunched together so that the furthest was only 6,000 or 7,000 light-years away, the radiant heat would likely burn up the earth.)

One solution put forward by creationists is the concept that God created beams of light emanating from distant stars 'on their way' to reach earth. While reasonable at first glance, this idea has a major problem. An observer on earth sees things 'happening'—rotating, colliding, exploding, flaring up, even transmitting information about the chemical composition of stars. The "created light beam" concept implies that any information gleaned from objects more than six to seven thousand light-years away—the vast majority of stars—never left its source but was fictitious. It would evidence events that never happened at the source. Earth would be surrounded by a gigantic light show of unreality.

Danny Faulkner, the creationist astronomer whom we mentioned earlier, pointed out that there is a more immediate light travel issue to be considered in Genesis. The stars were made on the fourth day yet at least some had to be visible by the seventh day at the latest (stars were created for signs and seasons). How, then, did Alpha Centauri, the closest star outside our solar system, become visible in the night sky within three days, given it is 4.3 light-years from earth? If the closest star was not visible on the seventh day, then neither were any other stars. This short-distance light travel problem is at least as significant as the large-distance light travel problem.[15]

Some creationists have considered relativistic effects as part of the solution to the light travel problem. Hence, we will need to understand Einstein's theory of relativity in broad terms.

Relativistic effects: satellites and muons

The Newtonian axiom that time was an independent variable,[16] and the same in all reference frames, was upended by Einstein's theories of relativity. According to Einstein, time depended on the relative velocity of an object with respect to an observer, and also any difference in potential between the observer and an object within a gravitational field. Hence time was not universal; the passage of time could differ between objects. This counter-intuitive result has

15 Faulkner, 2013, p. 279.

16 Empirical science requires measures of distance and time, and in the Newtonian era, which was based on Euclidian geometry, both were deemed to be the same everywhere, and under all conditions. In the Einsteinian era, time and Euclidean distance can no longer serve as absolute measures. Instead, the speed of light in a vacuum serves as a theoretically stable constant. In secular cosmology, the constancy of the speed of light in a vacuum throughout all time and space is typically taken as axiomatic (that is, assumed to be true as a presupposition), an application of the philosophy of uniformitarianism (North, 2012, p. 336).

been very well established empirically. It's worth illustrating these effects with a couple of real-world examples.

Very accurate time measurements on satellites orbiting earth show that both special and general relativistic effects come into play (although they partially cancel each other out, as they work in opposite directions). Satellites move relatively quickly, and the special relativistic effect due to that velocity very slightly slows the passage of time on the satellite with respect to earth. However, the general relativistic effect, due to the greater gravitational potential at the altitude of the satellite's orbit compared with an observer on earth, tends to increase the speed of time. Consider a satellite such as a GPS transmitter, which spends its time in a geostationary orbit (an orbit which takes 24 hours to complete, thereby keeping the satellite stationary above a position on the earth; this corresponds to an altitude of around 20,000 kilometers). The special relativistic effect due to the satellite's velocity causes a time loss of around 7 microseconds per day but working counter to that is the general relativistic time gain of +45 microseconds per day due to the higher gravitational potential. Taken together, these effects net out to a time gain of 38 microseconds per day. To maintain synchronization with an earthbound clock, the satellite's clock must constantly compensate for these effects. Clearly, these time differences are small—it would take over 70 years for the satellite's clock to gain a single second—but nonetheless, the effects are real. If GPS positioning systems did not deliberately compensate for them, the position information would accumulate error by about 400 meters per hour, meaning the system would very quickly become useless.

Another interesting effect, but much larger, concerns muons (also known as mu-mesons), which are particles similar to electrons. Muons are created in the upper atmosphere at an altitude of about 10km when cosmic rays—extremely high-speed atomic particles emitted from the sun—strike the earth's atmosphere. These highly unstable muons only exist for about 2 microseconds. Travelling at near the speed of light, or 300,000 km per second, we would expect these to cover about 600 meters in the 2 microseconds they have. However, some of them reach the earth. How can this be? The answer is special relativistic effects. At close to the speed of light, time slows down for the muon, and, from our perspective as earthbound observers, the muon lasts considerably longer before decaying, long enough to travel 10km in this lengthened time of survival. From the point of view of the muon, it lasts only the expected 2 microseconds, but the distance it travels is foreshortened due to special relativistic effects. The muon travels more than 16 times further than Newtonian physics would have indicated. Einstein's theories check out.

The key concept we emphasize here is that the passage of time occurs at different rates in different locations in the universe, depending on velocity, acceleration and differences in gravitational potential. When an object is moving rapidly in relation to an observer at rest, special relativity tells us that the observer sees time for that object passing more slowly than for themselves. When an object and observer are both in a gravitational field, time passes more quickly for objects at a higher gravitational potential compared with the observer, from the observer's point of view.

Biblical cosmologies

Faulkner pointed out that creationist models relying heavily on relativistic effects have not been successful in resolving the light travel problems to gain unanimous support, despite their mathematical sophistication. Faulkner reviewed a number of models that have been put forward over the decades and categorized them into seven different types, with the proliferation of models being an indication of an ongoing problem. He notes:

> I see that most of these solutions to the light travel time problem have advantages and disadvantages. If there were one solution that worked, there would not be such sharp disagreement.[17]

At this point Faulkner put forward his "Dasha solution",[18] in which he noted, like Coulson later, that both process, and the creative purpose of God are important. (We will not describe the dasha solution in more detail here, except to note that we view Coulson's SFP idea as a reworking and development of Faulkner's key concepts.)

Coulson himself does suggest a cosmological explanation, but it is less in line with the SFP concept which he applied to earth science with some success. For that reason, our interest is in categorizing other putative cosmologies that seem to be consistent with the SFP concept.

Ages according to secular science

The secular position on the age of the cosmos, given as 13.8 billion years, relies entirely on the notion of a current epoch stretching back in time to the beginning of the big bang.

Secular scientist John Gribbin provides an account of how the age of the universe was ascertained in his book *13.8*.[19] Commenting on NASA's WMAP probe (short for Wilkinson Microwave Anisotropy Probe), Gribbin relates that data from the first year of observations "pinned down the age of the Universe as 13.4 plus or minus 0.3 billion years". After nine years of data, "the WMAP data on their own implied an age of 13.74 plus or minus 0.11 billion years ... Adding in data from other kinds of observations changed these numbers to...

17 Faulkner, 2013, p. 281.

18 The name "dasha" relates to a Hebrew word taken from Genesis 1:11.

19 Gribbin, 2015.

13.772 plus or minus 0.0059 billion years". Then, following the retirement of WMAP, the European Space Agency's Planck satellite took up the mantle. "More detailed analysis of the Planck data... released in February 2015, implied that ... the age of the Universe is 13.799 billion years."[20]

The presupposition of philosophical uniformitarianism leads to the interpretation of the data as supporting 13.8 billion or so years. This is not an observed fact, but rather an interpretation conditioned by philosophical uniformitarianism. Secular scientists are not measuring time; they are inferring "time" from the amount of process that has taken place, interpreted in the context of the secular ΛCDM Big Bang model.

Turning to the age of the earth, secular science currently gives an age of 4.5 billion years. Various measures are used in calculating this value, including modeling based on philosophical uniformitarianism and radiometric dating. (This latter method remains scientifically problematic in establishing absolute time values in the opinion of creationists.) The 4.5-billion-year value appears to be a minimum age for the earth given available evidence and secular scientific modeling, rather than a specific age.

We reject both the secular "age" of the cosmos (13.8 billion years) and the earth (4.5 billion years) as both are underpinned by philosophical uniformitarianism. In the discussion that follows, we will add the suffix "PU" to secular timescales, signifying that they are based on "philosophical uniformitarian" principles and as such, remain outside a young-earth creationist worldview. Using this notation, for example, 13.8 billion years (PU) refers to 13.8 billion years inferred using uniformitarian presuppositions.

A biblical analogy

Coulson's SFP tells us, in effect, that with respect to science, "time" is not the critical aspect during the creative week, but rather, the amount of process that has occurred. This is analogous to an example from the Bible that we considered earlier in the book. Imagine if two individuals arrived on the scene shortly after Jesus healed the long-term paralytic (Matthew 9:1–8). One of these individuals subscribes to naturalistic, miracle-denying thinking, and holds that, since they have never seen a miracle, miracles are not possible. The other believes that miracles might be possible; this person will weigh all evidence carefully to determine whether a miracle has plausibly occurred. Since neither was on hand to observe the event, their presuppositions will be critical as to how they understand what happened, even though they both agree on the amount of process that has taken place (long-term bedridden to walking). The bystanders explain to them that the man had been a long-term paralytic. The uniformitarian looks at the man and infers that several months of rehabilitation

20 Ibid., pp. 214–217.

have occurred. This calculation assumes that the event occurred in a naturalistic fashion; therefore, average rates of healing of other patients and the assumption that this case must have occurred in similar fashion, leads to the multi-month conclusion.[21] The miracle-believer, who believes the eyewitness testimony that Jesus healed the man, understands that the mechanism of the miracle is the acceleration of natural processes. If the individual wishes to know when the miracle took place, they must ask witnesses; it is not possible to infer the timing. (Asking when the miracle occurred is analogous to Ussher analysing scriptural and other sources to calculate a date of creation. The miraculous nature of the creation means that its timing cannot be inferred but must be revealed by God.) If the bystanders reveal, for example, that the healing occurred several minutes ago, the miracle-believer might accept this account if it is well corroborated. By contrast, the uniformitarian regards this reported timing as implausible and dismisses the eyewitness account out of hand. Importantly, whether miracles are possible or not is a religious or philosophic question, not scientific.

Similarly, in origins, what matters is the amount of process. Some creationists (such as ourselves) broadly agree[22] with secular scientists that the amount of process that has occurred is potentially amenable to scientific enquiry; the main disagreement lies in the way these processes unfolded—a Coulsonian interpretation involves massively sped-up processes during the creative epoch, followed by "normal" rates after that.[23] Under this scenario, the biblical account of creation around 6,500 years ago is entirely consistent with the scientific evidence. The ages of the earth and cosmos given by secular science, which confounded the likes of Philip Henry Gosse in the late nineteenth century, were and are calculated under uniformitarian assumptions, which do not apply in Christian origins. Modern Christians need to be clear on this point.

The uniformitarian approach considers the evidence and concludes that processes have occurred over billions of years. We pose the question: if we apply Coulson's SFP to secular "ages" deduced from astronomy, how does this fit with the creation week in relation to cosmology? Clearly, there are several ways to apply SFP to this question, of which we will consider three. (We reiterate we are using the secular ages here as rough measures of the amount of process which has occurred.)

The material that follows is challenging and some readers may wish to skip to the Conclusion of the chapter.

---

21 In a similar way, secular scientists infer the age of the universe, they were not there to observe its formation.

22 By "broadly agree" we mean that the amount of process which has occurred may be able to be estimated by scientific methods; we reject all origins science results based on philosophical uniformitarianism.

23 There is also a question about rates during the Great Flood; the apostle Peter identifies an analogy between the waters of the great flood and the creative epoch in 2 Peter 3:5–6, perhaps implying in some sense the second period is also a type of creative epoch. In this chapter, we do not incorporate this notion into our analysis.

Assuming synchronized rates

The first type of analysis assumes that the SFP process rates on the earth and in space are synchronized.

When applying SFP to the creation week, the acceleration of processes is enormous at the start of the week. But before the end of the sixth day, processes had decelerated to the same rate they are today. Coulson suggests that processes could have been initially billions of times faster than current rates through the third day. Rates may then have slowed to hundreds of times faster than current rates during the fourth day,[24] indicating that the acceleration of process rates decayed exponentially. More specific details are needed to model this with a degree of precision, such as: the magnitude of the initial acceleration of process rates; when did the exponential decline in rates begin; or did they follow some other deceleration pathway? For the purposes of our analysis, we will assume that rates had slowed to current rates by the start of the fifth day (this is somewhat arbitrary), and for simplicity, we will further assume the accelerated rates were constant, then suddenly stepped down to rates we see today. This gives us four days of constantly high rates, followed by three days of "normal" rates.

We will use the 430 million light year measurement, mentioned earlier, as a lower bound for the amount of process occurring in the cosmos during the creation week. That is, *at least* 430 million years (PU) worth of process has taken place. We also recall that the expansion of outer space occurred between the second through sixth days inclusive.

- Under our SFP analysis, process rates were accelerated from the first through fourth days inclusive.
- Because of our SFP window, the expansion on the fifth and sixth days was unaccelerated and therefore we can exclude it from our analysis.
- Therefore, 430 million years (PU) worth of process has occurred in three days, which is 143 million years (PU) worth of process on each of the second, third and fourth days.
- The first day was also accelerated. We multiply 143 million years (PU) worth of process by 4 (the number of accelerated days), which results in 572 million years (PU) worth of process on earth.
- Note that the amount of process on the earth (572 million years PU) exceeds that of the universe (430 million years PU). This is not a problem in a biblical scheme, because a straightforward reading of Genesis implies that the earth is older than the celestial bodies.

---

24 Coulson, 2020, p. 54.

- This particular lower bound estimate of the amount of process occurring in the first four days of creation equates to an average acceleration over current rates by a factor of at least 52 billion times.

Non-synchronized rates

Our other two types of analysis differ from the first in that they allow non-synchronized rates for earthly and cosmic processes. These imply different expansion models for outer space while remaining consistent with Genesis 1. They permit more process to occur in outer space than occurs simultaneously on earth.

First, for illustrative purposes we assume that 30 billion years (PU) worth of process occurred in outer space from the second through fourth days inclusive, and 10 billion years (PU) worth of process occurred on earth from the first through fourth days inclusive. Note that both these figures are higher than the ages given by secular science (13.8 billion years and 4.5 billion years respectively), demonstrating that any amount of process can be incorporated into this type of modelling. Rates are not synchronized between earth and outer space; the SFP we apply to the earth occurs at a different rate to the SFP we apply to the rest of the cosmos. Again, we assume that the accelerated process rate is constant from the first through fourth days.

- Therefore, in the time the earth rotates four times on its axis (first through fourth days), 30 billion years (PU) worth of process has occurred in outer space, working out to 10 billion years (PU) worth of process on each of the second, third and fourth days.
- Simultaneously, 10 billion years (PU) worth of process has occurred on earth, working out to 2.5 billion years (PU) worth of process on each of the first, second, third and fourth days.

In this analysis, the amount of process in outer space exceeds that occurring on the earth, even though the earth was created before the celestial bodies. Further, the rate of process in outer space is 3.65 trillion times faster than current rates, while the rate of process occurring on earth is roughly a trillion times faster than current rates.

Our second type of non-synchronised analysis allows process rates in outer space to vary in proportion to the distance from the earth. In this analysis, the greater the distance from earth, the higher the process rates which occur. In particular, the speed of light in a vacuum is much higher at vast distances from earth than it is close to earth. Using the same figures from the previous example:

- In the time the earth rotates four times on its axis (first through fourth days), 30 billion years (PU) worth of process has occurred in outer space at the furthest extent of the visible universe, working out to 10 billion years (PU) worth of process on each of the second, third and fourth days in this region.

- Simultaneously, 10 billion years (PU) worth of process has occurred on earth, working out to 2.5 billion years (PU) worth of process on each of the first, second, third and fourth days. The area of the cosmos adjacent to the earth also undergoes this amount of process.

All these types of cosmic SFP illustrate that it is possible to incorporate billions of years (PU) of process into creationist models. Further, the latter two types of cosmic SFP demonstrate that it is possible to analyse systems which preserve the order of creation of the earth and cosmos found in Genesis 1 while simultaneously allowing more process to have occurred in outer space than on earth (these differential process amounts are interpreted by secular science as indicating that the earth is younger than the cosmos).

In all these analyses, we are suggesting that, during the first through fourth days of creation, the speed of light in a vacuum was multiple times that of what we currently observe. These analyses also define a day as one rotation of the earth, in line with a straightforward reading of Genesis.

Given that God is rational, the laws of physics during the creation week may have comprised a superset of which the current laws of physics are a subset. It may be possible for creationists to develop models for some of the miraculous events of the creation week that shed more light on how the processes unfolded.

Fitting the model

In discussing the models above, we decided on values somewhat arbitrarily, simply to illustrate the concepts. There are many parameters that may be adjusted, which would yield a suite of models. The adjustable parameters include:

1. The parameter for setting the initial earth conditions—the state of the composition of the earth substance after creation ex nihilo on the first day.
2. The process rate parameter—to what degree were processes accelerated from the first day onward? Was this rate constant, or did the parameter vary during the creation week?
3. The time at which process rates decreased to current rates—adjustable from the fourth through sixth days.
4. The initial cosmological conditions—what was the state of expanding outer space at the commencement of the fourth day?
5. When allowing non-synchronized earthly/cosmic rates, the process rate for outer space could vary with respect to location and/or time.

Scientists adjust the parameters of a model to improve its degree of fit with existing data. This is known as "fitting the model". It is analogous to someone adjusting the settings on a graphical equalizer to improve the quality of sound reproduction. In the case of any biblical model, any adjustments to the model must be compatible with both scripture and science. But even if a model can be tuned, the real test of a model's value is whether it leads to new predictions that are later confirmed and validated.

If one of these schemes were to pass initial inspection by creationist generalists, then it would need further collaborative work by theologians, earth scientists and cosmologists to refine and analyze the details and investigate the strong or weak points. Clearly, the creationist enterprise is a team effort requiring both generalists and specialists in science, philosophy and aspects of Old Testament and New Testament theology. And just like any other field of science, there can be false starts and dead ends.

Our hope is that, with further work on a biblical cosmology, Coulson's axiomatic SFP scheme, or one like it, will be found to harmonize with science and Genesis. While relativity may be part of that explanation, the concept of SFP could be more important. For instance, our view is that SFP, applied to cosmology, means that during most of the sped-up period of the six days of creation (the first through fourth or fifth days), the speed of light in a vacuum was potentially orders of magnitude higher than the contemporary speed of light. In theory, this is helpful in dealing with at least part of the light travel problem at both smaller and larger distances.

Before concluding, two other points need to be emphasized. First, all types of cosmologies we have briefly described have two characteristics which are entirely consistent with scripture—that the earth was created before other celestial bodies, and that the earth initially had a unique and preeminent place in the cosmos. Secondly, all analyses utilized non-uniformitarian process rates. These points contradict the presuppositions underlying secular cosmologies, but this type of analysis may align with scripture and the evidence of modern science simultaneously.

In science, ideas such as those we presented here need to be analyzed carefully from different angles before too much weight is placed on them. Proposing them is helpful, because it may allow them to be falsified. Falsification of ideas (to rule them out) is an important part of scientific progress. Only when ideas survive repeated attempts at falsification should any level of credence be attached to them.

Conclusion

Secular scientists must fall back to philosophical uniformitarianism in their interpretation of current data, which results in inferred ages of 13.8 billion years for the universe and 4.5 billion years for the earth. However, a philosophical uniformitarian interpretation is incompatible with scripture. Coulson's axiomatic scheme, on the other hand, and other creationist interpretations, interpret precisely the same data in a manner which is compatible with the Genesis account. Coulson's scheme underlines the fact that earthly and cosmic ages are deduced from biblical revelation, and the corresponding amount of process is investigated by scientific inquiry. For the Christian, SFP moves the *age* of the earth and cosmos from the realm of science into the realm of revelation. Under this kind of analysis, Ussher's chronology, or similar, becomes crucially important in origins science.

The Bible tells us that faith will always be required to accept the account of creation found in the first two chapters of Genesis. As the writer to the Hebrews puts it, "By faith we understand that the universe was created by the word of God, …" (Hebrews 11:3a). This is not the same as the 'blind faith' with which Christianity is often caricatured, but a reasonable faith, one that makes sense of reality.

References

Barr, J. (1984–1985). Why the world was created in 4004 B.C.: Archbishop Ussher and Biblical Chronology. *Bulletin of the John Rylands University Library of Manchester, 67*(2), pp. 575–608. Retrieved March 6, 2024, from escholar.manchester.ac.uk/api/datastream?publicationPid=uk-ac-man-scw:1m1647&datastreamId=POST-PEER-REVIEW-PUBLISHERS-DOCUMENT.PDF

Cosner, L., Carter, R. (2015). Textual traditions and biblical chronology. *Journal of Creation*, 29(2), 99–105.

Coulson, K. P. (2020). *Creation Unfolding: A New Perspective on Ex-Nihilo*. Phaneros Press.

Courville, D. A. (1971). *The Exodus problem and its ramifications* (Vol. 1). Loma Linda: Challenge Books.

Faulkner, D. R. (2013). A Proposal for a New Solution to the Light Travel Time Problem. *Answers Research Journal, 6*, 279–284.

Gordis, D. (2017). *Israel: A concise history of a nation reborn*. New York: HarperCollins Publishers Inc.

Gribbin, J. (2015). *13.8—The quest to find the true age of the universe and the theory of everything*. London: Icon Books Ltd.

NASA Hubble Mission Team. (2018, April 2). Hubble Uncovers the Farthest Star Ever Seen. Retrieved March 6, 2024, from science.nasa.gov/missions/hubble/hubble-uncovers-the-farthest-star-ever-seen

North, G. (2012). *Sovereignty and Dominion: An economic commentary on Genesis*. Volume 2. Dallas: Point Five Press.

Pesce DW, Braatz JA, Reid MJ, Riess AG, Scolnic D, Condon JJ, Gao F, Henkel C, Impellizzeri CMV, Kuo CY, Lo KY. (2020). The Megamaser Cosmology Project. XIII. Combined Hubble constant constraints. Retrieved October 13, 2025, from arxiv.org/pdf/2001.09213

Rushdoony, R. J. (2002). *Genesis*. Vallecito: Ross House Books.

Weinberg, S. (2008). *Cosmology*. New York: Oxford University Press.

18
Genesis 3–6: the fall and judgment

In addition to differences between the biblical and secular origins accounts identifiable in the first two chapters of Genesis, the third chapter of Genesis communicates another fundamental divergence. The Bible reveals that Adam and Eve, the progenitors of mankind, were tempted and fell into sin. Their original goodness was lost; they became corrupted, alienated from God and subject to death, a condition inherited by all their descendants.

Secular humanists do not acknowledge the innate moral corruption of fallen man. They see no problem with humans retaining personal autonomy or basing morality on the shifting sands of human reason. But this is a fatal mistake. The unrestrained autonomy of fallen beings must ultimately end in disaster. Christians recognize this as a basic truth. Fallen humanity must be removed from control and replaced with the unfallen Jesus Christ as Lord. Within this worldview, Christians acknowledge the Bible, rather than human reason alone, as the authoritative source of truth and morality.

A second divergence concerns the role of death. Genesis teaches that death of sentient animals did not occur before the fall of man (and mankind). This is incompatible with the purported death of the unfit occurring prior to the fall, indicating that evolution cannot have been involved in the creation of the sentient animals or man. In essence, mankind has not evolved from lower animals.

In this chapter we investigate the fall of man and the events that flowed from that to the end of the pre-Flood period, a span of over 1,600 years. We will also examine some aspects of God's response to the fall of mankind. We will investigate these chapters from a biological point of view—admittedly not an approach that most Bible commentators take, but one which provides a useful counterpoint to the secular origins account.

The fall of Man

Adam and Eve were created without any genetic errors (as implied by Genesis 1:31). When Eve was created from tissue removed from Adam's side, the male Y chromosomes may well have been removed from the cells, and the X chromosomes duplicated, making them biological twins. Whatever way it was achieved, Eve was Adam's biological descendant, and as perfect as he was. Their perfection should not be taken to imply that they were genetic "super-people", but they had no harmful dominant or recessive genes. In addition to their error-free genetic heritage, Adam and Eve were both delegated rulership over all the earth. They were children of God, made in His image.

For reasons which the Bible does not explicitly state, God allowed the devil,[1] a powerful but fallen angelic creature, to enter the physical creation. The devil had aspired to be like God, but now found himself confronted with man, a class of creature made in the image of God and who had been given judicial authority on earth. This was a mortal threat to Satan's agenda. The devil, perhaps taking possession of a serpent, tempted Eve to sin—an encounter which had profound ramifications. The spiritual and physical welfare of Adam and Eve—and by extension, the entire human race—hung in the balance. This well-known narrative is presented in Genesis 3.

When the serpent tested Eve on her knowledge of evil, she made a slight error. She added to what God had said by stating she was forbidden to touch the tree of the knowledge of good and evil, on pain of death. In fact, God had prohibited Adam only from eating the fruit.

The serpent responded (Genesis 3:4–5):

> No! You will not die, ... In fact, God knows that when you eat it your eyes will be opened and you will be like God, knowing good and evil.

Since Eve already had theoretical knowledge of good and evil based on the blessings and the prohibition of God, what was the serpent offering her? The answer is found in the clause "you will be like God". A god does not just know what good and evil are—a god determines and establishes them.[2] This is more than earthly wisdom; this is enlightenment, a promise of becoming a god and determining good and evil.

Satan was effectively offering initiation into a new worldview, one which was subtle, dangerous, and man centered. That the initiation rite was to commit a capital offense[3] had several implications. This was a one-way event from which there could be no turning back, for God had warned Adam, "on the day you eat from it, you will certainly die".[4] The new worldview presupposed that God was not trustworthy, that His Word was incorrect, and that the pursuit of knowledge was independent of God's Word. Indeed, God's Word was to be empirically tested by breaking it. The moral position of the new worldview was that ends justify the means. Committing a capital offense could be justified if the end was perceived as "good". Henceforth, man would develop his own ways of living independent of God. The initiation rite was disobedience against an express commandment of God. (From a Christian worldview perspective,

---

1 The New Testament equates the serpent with the devil, or Satan, in several places, e.g. Revelation 12:9

2 Rushdoony, 2002, p. 33.

3 As we discuss in the text below, Adam was not deceived; from his point of view, he was committing a capital offense.

4 This is a literal translation of the original Hebrew.

secular humanism includes many of these implications identifiable within the temptation narrative.)

When incited to commit a capital offense, Adam did not resist. In eating the forbidden fruit, he sinned both as an individual and as the representative of corporate man. The Bible tells us "For just as in Adam all die, ..." (1 Corinthians 15:22), meaning that Adam decided this question on our behalf.

However, any human would have failed at this point. Jesus told the Apostle Peter that "Satan has asked to sift you like wheat" (Luke 22:31), demonstrating that the spiritual force the devil can bring to bear in temptation is beyond what man—in his own strength—can resist. However, Jesus reprised the "first man" role on our behalf (Matthew 4:1–11), relying on scripture to maintain His integrity in the face of Satanic temptation.

The trial

After Adam and Eve had eaten the forbidden fruit, they realized they were naked. Their eyes were opened—not only to their new condition, but to whom God is and His holiness. At this realization, they felt shame at their nakedness. They sewed fig leaves together as crude coverings.

God walked through the garden later that day (Genesis 3:8). (This appearance, known as a *theophany*[5] is a visual manifestation of the presence of God.) On hearing God approaching, Adam and Eve hid among the trees.

God called to Adam, "Where are you?" Adam replied that he had hidden himself because he was afraid due to his nakedness. God then asked him if he had eaten of the forbidden fruit. Adam replied:

> The woman you gave to be with me—she gave me some fruit from the tree, and I ate.
>
> God asked the woman, "What is this you have done?"
>
> She replied, "The serpent deceived me, and I ate."[6]

We note that although Adam's response was true, implicit in his reply was a suggestion that the woman and God were partly responsible for his sin. Yet we know from 1 Timothy 2:14 that Adam was not deceived. He was not prepared to take full responsibility for his act.

Eve, on the other hand, spoke truly, because she was deceived, as the New Testament confirms.

5 The word theophany is a combination of two Greek words, meaning *God* and *appearance*.

6 Partially quoted from Genesis 3:9–13

The curse

God then cursed the animal kingdom, especially the serpent, which henceforth would move on its belly, a posture of humiliation.[7] The serpent's offense was to tempt Eve to eat the forbidden fruit; the punishment was to eat dust. The serpent would also suffer a state of permanent hostility with humans Genesis 3:14–15. God's final words to the serpent are found in Genesis 3:15:

> I will put hostility between you and the woman, and between your offspring and her offspring. He will strike your head, and you will strike his heel.

In the King James Version, "her offspring" is translated "her seed". The reference to seed of the woman is a highly unusual phrase in the Old Testament, which typically references the seed of a man (for example Genesis 21:12; 1 Chronicles 16:13; Psalm 89:3–4). Why then is "the seed of the woman" used here? The majority Christian view is that this is an allusion to Christ, who would be "born of a virgin" and thus seed of the woman, rather than seed of a man. In the New Testament, when Christ performs his first miracle at the wedding in Cana (John 2:11), He refers to his mother as "woman", identifying Mary as "the woman" of Genesis 3:15 and himself as the seed, born of a virgin.[8] Thus, Genesis 3:15 is referred to as the *protoevangelium*, meaning "first gospel".

The woman was cursed with increased pain in childbearing and condemned to marital discord (Genesis 3:16). Clearly this would have adverse effects on her role of dominion and exerting influence within the family.

Because of Adam's sin, the ground was cursed to bring forth thorns and weeds, which would render gardening and agriculture more difficult and labor intensive (Genesis 3:17–19). Agriculture would require Adam's sweat. There would be more fatigue and slower progress; exercising dominion would be more difficult. At this point, herbs and seed-bearing plants began to grow upon the earth—accompanied by weeds.

Finally, the man (and by implication mankind henceforth) was cursed to die, and to return to dust. The mortal remains of the body after death would decay.

The New Testament reveals that the effect of the curse was that the creation began to groan. Not only was the earth cursed to bring forth weeds, but its tectonic movements would cause volcanoes, tsunamis and earthquakes. All of creation was cursed. The diets of some animal kinds were altered to omnivory and carnivory.[9]

---

7 Sarna, 1989, p. 27.

8 Taylor, 1961, p. 306.

9 The fossil record supports the view that carnivory (and by extension omnivory) existed pre-Flood—some animals are fossilized in the process of eating other creatures.

God had said that on the day Adam ate of the forbidden fruit, he would "surely die" (Genesis 2:17). Here, death has a threefold state—judicial, progressive and final. Judgment was passed by God leading to condemnation. Therefore, Adam and Eve were placed on death row on that day. From that time forward, progressive death was occurring; the aging process had begun. Adam would eventually die aged 930 years (Genesis 5:5); Eve also died. An additional understanding is that in the Bible death means separation. Adam and Eve were separated from God, and God's garden, on that day.

After passing judgment, God rejected man's fig-leaf coverings, and provided clothing made from animal skins instead. This is a picture of God ordaining the shedding of blood on man's behalf as a covering for sin. God also prevented access to the tree of life (Genesis 3:21–24). This was a logical necessity since Adam and Eve had been condemned and placed on death row—"he must not reach out, take from the tree of life, eat, and live forever" (Genesis 3:22).

Note also that Adam called his wife Eve, "because she was the mother of all living" (Genesis 3:20). This tells us that all humans have descended from Adam and Eve, who were the first humans and, as such, our representatives. Both the blessings and the curses delivered to Adam and Eve apply to all their descendants (Romans 5:12).

The family of Adam and Eve

Eve became pregnant by Adam and gave birth to her firstborn, Cain. Sarfati notes that the famous German reformer, Martin Luther, translated Eve's cry on the birth of Cain as "I have gotten a man: YHWH",[10] suggesting that Eve not only understood the protoevangelium, but believed it to have been immediately fulfilled. Even if her theology can be considered close to the mark, Eve's timing was out by about 4000 years. Other experts have supported Luther's translation.[11]

Eve soon learned that her son was not at all like YHWH, and thus when she named her second-born, she called him Abel (vanity). Cain became a gardener like his father, whereas Abel kept sheep. Prior to the Flood, humans remained vegetarians, so Abel's sheep would have provided milk, skins and wool.[12]

Adam and Eve had other sons and daughters too. The Bible does not record how many children the first couple had, but Jewish tradition holds that Adam and Eve had 33 sons and 23 daughters.

Abel brought an offering to God of the best of his sheep and was accepted by God, possibly by a display of fire consuming the sacrifice[13] (Genesis 4:4).

---

10 Sarfati, 2018, pp. 406–407.

11 Ibid., pp. 407–408.

12 Sarna, 1989, p. 32.

13 Ibid., p. 33.

Cain brought an offering of "fruit of the ground". Not only was this not a blood sacrifice, but the text does not even indicate the offering was first-fruits or best of the crop.[14] Cain evidently believed that all ways lead to God.[15] When God rejected this offering, Cain was distressed. Not being able to strike out at God directly,[16] he struck at the image of God instead by murdering his brother Abel, causing the first human death in the world[17] (Genesis 4:8).

Again, there was a trial, in which God questioned Cain about Abel and then passed judgment (Genesis 4:9–12). By asking about Abel, God identified himself as the God of Abel. This implies the resurrection of the dead because, using the logic of Christ (Mark 12:27): "He is not the God of the dead, but of the living".[18] Cain was cursed to be a nomad whose agricultural efforts would not be rewarded. After telling God he could not bear his punishment, God put a mark on Cain to protect him from attack (Genesis 4:13–15).

Cain moved away and started a family (Genesis 4:16–17). It is likely that he married a female descendant of Adam and Eve, most probably a daughter rather than a granddaughter. (Such consanguineous marriages, we note in passing, were necessary for the first generation, but were not harmful because human DNA did not initially contain harmful dominant or recessive genes. These are mostly caused by heritable mistakes in copying DNA (mutations) and only became possible after the fall. Mutations tend to accumulate as fresh errors arise in subsequent generations to add to those already present in a population.

Assuming for argument's sake that the number of children born to the early, long-lived humans was in line with the 33 sons and 23 daughters ascribed to Adam and Eve by Jewish tradition, the population could have exceeded 500 million as Adam's life came to a close (assuming the early people started having children at age 150, and each couple had 20 surviving children that also went on to have 20 surviving children). Comparing this figure to the modern United States, which has a population of around 350 million, built up over a period of some 400 years, it is possible that Adam would have lived to see a very substantial antediluvian civilization. The biblical record of Cain going to an existing city, Nod, is therefore easily supported.

Unfortunately, this early civilization became marred by extreme violence. One of Cain's descendants, Lamech, was also a violent murderer (Genesis 4:18–19, 23–24).

Some have questioned the long lifespans of that period.[19] An analogy from today is relevant. The genetic disease *progeria* causes symptoms of premature

---

14 Sarfati, 2018, p. 410.
15 Ibid., p. 410.
16 Rushdoony, 2002, p. 44.
17 Sarna, 1989, p. 33.
18 Sarfati, 2018, p. 418.
19 Other ancient sources corroborate long life spans, but to an immoderate extent, such as the Sumerian King list (241,200 years) and the ten antediluvian regents of Berossus (432,000 years) (Sarna, 1989, p. 41).

aging, with the average age of death being 13 years.[20] On average, a person with progeria has a lifespan about six times shorter than an unaffected person. This is analogous to the preflood individuals comparing their lifespan to ours—by comparison it is as if modern humans have progeria. The difference is that progeria is caused by one specific genetic mutation, whereas it is likely that modern human lifespans are conditioned by multiple genetic mutations.

The Bible also mentions the development of technology and culture by Cain's descendants, demonstrating that at this time knowledge and expertise were passed down within families and tribes (Genesis 4:20–22). One of Cain's descendants, Tubal-Cain, is noteworthy as "an instructor of every craftsman in bronze and iron". This calls into question the notion of a "stone age"; the Bible states that men were crafting bronze (an alloy of copper and tin) and iron in the pre-flood period. It is likely that metalworking has been practiced, in differing places and at differing times, for almost as long as people have been on the earth.

Following the narrative of Cain, the Bible records that Adam and Eve had a son called Seth. This son was significant because he would go on to establish a godly family line. People began to call on the name of God at about the time Seth had his son Enosh (Genesis 4:25–26).

Violence and corruption

At this stage of history, anarchy ruled the earth. There was no capital punishment for murderers; civil government had yet to be ordained. In the absence of any check on violent humans—be they murderers, rapists, thieves or kidnappers—violence and aggression were survival benefits. The advantage lay in the fact that violent individuals could kill off the non-violent with impunity. This state of affairs presaged one of the darkest periods in human history as a negative feedback loop of increased violence took hold.

We see an example of this in the story of Cain and Abel. The Bible is silent on whether Abel had children of his own before Cain killed him; it is possible that he didn't. Cain, on the other hand, went on to have a family, the genealogy of which is listed in Genesis 4. Cain's great-great-great grandson, Lamech, killed a man for wounding him. This hints that the spreading violence was passed along family lines.

Further insight into the pre-diluvian period was given by Jesus Christ, who, in reference to His second coming, stated:

> As the days of Noah were, so the coming of the Son of Man will be. For in those days before the flood they were eating and drinking, marrying and giving in marriage, until the day that Noah boarded the ark. They didn't

20 Health Research Fund, 2024.

> know until the flood came and swept them all away. This is the way the coming of the Son of Man will be.[21]

These antediluvian humans sought the same enjoyments and celebrated the same sort of life events—and were marred by the same spiritual blindness—that the people at the time of Christ's second coming will have.

The Bible tells us that "the sons of God" were marrying the "daughters of mankind" (Genesis 6:2). One popular interpretation holds that the godly line of Seth, "the sons of God", were intermarrying with ungodly tribes. However, another prominent interpretation of this verse is that the sons of God were angelic beings who left their assigned place to pursue strange relations with human women, who then bore children[22 23]. In the New Testament, both 2 Peter 2 and Jude shed light on this aspect of Genesis 6, providing support for the interpretation that "the sons of God" were indeed fallen angels. This section of Genesis also mentions the Nephilim, who were "men of renown" (sometimes wrongly translated as "giants").[24] These men of renown, or "heroes", are otherwise not identified in the Bible, but may correlate with the ancient heroes of other cultures, such as Gilgamesh, who was simultaneously a figure found in ancient Mesopotamian mythology, and a king included in the Sumerian King List. Certainly, the Nephilim were familiar to the contemporaries of Moses. The writer of Genesis points out that these Nephilim were present both *before* and *after* the fallen angels united with human women, indicating that the Nephilim were not produced by angelic/human unions, but were fully human. In opposition to some ancient mythologies, this clearly tells us that these heroes were not in any sense divine.[25] Nevertheless, the mingled line of angelic and human beings produced offspring which were corrupted. Any corruption affecting humans, and more particularly the unadulterated line of Seth, was a dangerous development because the future redeemer was to come from this line. It was at this time that God fixed the date of the flood at one hundred and twenty years into the future (Genesis 6:3).[26]

God became increasingly grieved with the violence and corruption on earth. Due to this violence and lawlessness, it is highly probable that God caused climate change, localized floods, geological catastrophes, degradation of the environment and pestilence in the same way that He did post-flood. During this period, carcasses of dead animals and humans were unlikely to have become fossilized, but simply would have rotted away, leaving virtually no trace of the antediluvian world. Indeed, the apostle Peter stated that "... the

---

21 Matthew 24:37–39.
22 Sarfati, 2018, pp. 473–480.
23 Sarna, 1989, p. 45.
24 Sarfati, 2018, pp. 482–484.
25 Gentry, 2019.
26 Sarfati, 2018, pp. 480–482.

world of that time perished when it was flooded" (2 Peter 3:6), indicating that all living land-based creatures outside of the ark died (evidence of these deaths persists in the fossil record).

God's choice of Noah and the timing of the flood

God delayed the coming near-extinction event as long as possible to give people time to repent. The state of the violence on the earth reached the point that God regretted that he had made man at all. However, we read in 2 Chronicles 16:9:

> For the eyes of the LORD roam throughout the earth, to show himself strong for those who are wholeheartedly devoted to him.

God's eyes found Noah. Noah was the head of the last remaining godly family of the line of Seth. Noah and his wife had three sons, each of whom also had a wife (Genesis 6:5–13).

The Bible clearly states that Noah was "perfect in his generations" (Genesis 6:9, KJV). We can take this to mean that Noah was descended from the godly line of Seth, with no intermarrying with lawless, ungodly tribes. He was probably more or less as genetically favored as Adam because of the blessing of God. We can reasonably speculate that he had few if any harmful recessive or dominant genes, and that Noah's wife was also genetically pure Sethite.

The chronogenealogy also provides a hint as to how God chose the exact timing of the flood. Methuselah, the longest-lived human recorded in Genesis, died in the same year as the flood, presumably of natural causes. It appears that none of the godly line of Seth died in the flood—all died before, except for the final Sethites, Noah and his family, including Shem, who were preserved in the ark. The chronogenealogy from Genesis 5, presented in Figure 18.1 (next page), indicates this (additional information regarding Seth's lifespan is sourced from Genesis 11).

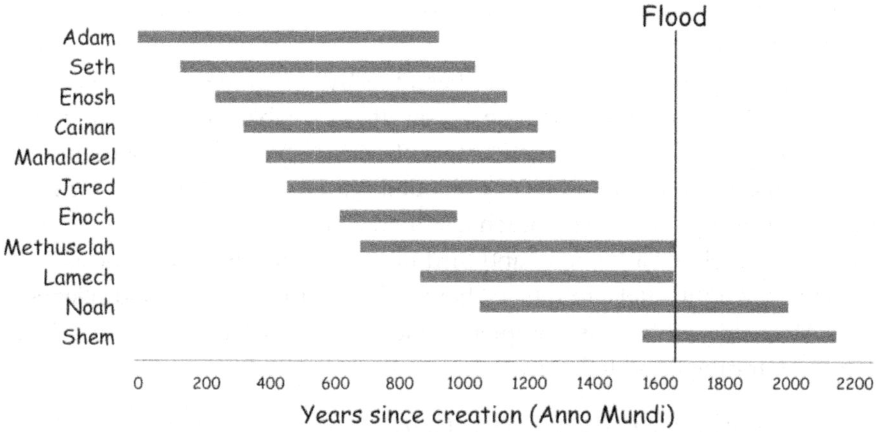

Figure 18.1: Chrono genealogy from Adam to Seth (Genesis 5)

The ark

God determined to save Noah from the coming near-extinction event for humans, land dwelling animals, and birds. He told Noah to construct an ark, or ship (Genesis 6:18) of gopher wood according to a design given by God. It was to have three decks divided into rooms and be covered with a roof. The hull was to be covered with pitch both inside and out, with a door located in its side. The craft was more than 120 meters long, about 22 meters wide, and about 13 meters high, enclosing an enormous volume. This massive logistical project and engineering feat took Noah and his three sons, Shem, Ham and Japheth, less than 100 years to complete, and possibly closer to 70 years.[27] It is possible Noah was wealthy and used hired help for building the ark and accumulating dehydrated food supplies.

Notably, the ark had neither a steering mechanism nor sails; while this simplified construction, it also meant that the ark would drift at the mercy of the floodwater currents and wind.[28] As the ark neared completion, Noah's family must also have gathered and preserved the wide variety of foods necessary for different animals and humans, work which may have been largely carried out by the four wives. These stores were then moved onto the ark.

The biblical text indicates that seven mating pairs of each clean animal and bird kind were taken aboard the ark prior to the onset of the flood. Of the unclean animals, including dinosaurs of various kinds, only a single breeding pair was taken aboard. Presumably, juvenile specimens of the larger kinds, such

27 Sarfati, 2011.

28 Sarna, 1989, p. 55.

as some of the dinosaurs, were taken onto the ark to preserve space and minimize food consumption.

Conditions in the ark would have been mostly cold and dark as very little sunlight would have shone into the vessel. It is possible that some animals, perhaps many, would have entered a state of dormancy, such as hibernation in mammals and brumation in the cold-blooded reptiles. In that state, their food requirements would have been much less than usual.

It took a week for all animal and bird kinds to enter the ark (Genesis 7:4). The animals sought safety in the ark having been prompted by God (Genesis 6:20b). After this work was completed, God closed the door of the ark in preparation for the flood to begin.[29]

Kinds, speciation and extinction

Prior to the flood, some speciation of the original animal kinds, as well as plants, would likely have taken place. This could have occurred at least partly from selective breeding; Abel distinguished better juveniles from his flock for sacrifice (Genesis 4:4); he might also have selectively bred his flocks, resulting in new varieties of livestock.[30]

Clearly, the pairs (or seven pairs for clean animals and birds) that boarded the ark were descended from an original kind. However, we cannot say for certain that all these pairs exactly represented the original created kinds; speciation may have occurred prior to the flood.

Since God directed the choice of the pair and supernaturally commanded them to make their way to the ark, individuals may have been selected for maximal variability (heterozygosity), allowing them to adapt and speciate according to conditions after the Flood.

It is possible that some entire land-dwelling kinds, meaning all the descendants of a particular created kind, became extinct prior to the Flood. If so, they would almost certainly be unknown to paleontology, since it is very unlikely they would have fossilized. (Fossilization generally requires rapid burial in order to prevent scavenging and decay.)

Pre-flood speciation of aquatic or sea animals would have been possible, and even likely. Nonetheless, the fossil record indicates that many types of these did not survive the catastrophic environmental upheavals, including the

---

29 Author John Woodmorappe investigated the design and practicalities of the ark from a scientific perspective in his 1996 book *Noah's Ark: A feasibility study*. Woodmorappe covered a wide range of topics including the identity of the animal kinds on the ark, space requirements, quantities of food and water required, waste management, heating, ventilation, illumination, ancient animal husbandry, the design and engineering of the ark, human resource requirements, implications of animal crowding, and how various dietary requirements may have been accommodated. His conclusion was that the biblical account of the ark was scientifically defensible, and "could have happened exactly as the Bible teaches". We encourage those that wish to explore this topic in more detail to read Woodmorappe's book.

30 Lamb, 2007.

massive undersea volcanism of the global deluge. Where such extinctions involve all of the species descended from an original aquatic kind (for example, the trilobites), the kind itself became extinct. The only evidence that it ever existed is found in the fossil record.

Conclusion

The consequence of mankind's sin was devastating. Human relationships often descended into violence, and the natural environment and animals were not spared adverse effects. The violence among humans had a counterpart among the animals in carnivory; natural disasters further afflicted living organisms.

The flood that was about to commence was not just a near-extinction event. The Bible implies that the geological structure of the earth was also to be destroyed and refashioned (Genesis 6:13b); the flood and its aftermath would bring an end to the single-ocean geography of the earth's surface, and usher in the period of dispersed continents and peoples that we know today. The Garden of Eden, mentioned in Genesis 2, was destroyed by the flood and no longer exists.[31]

References

Gentry, P. (2019, November 18). Were the sons of God in Genesis 6 fallen angles? Who were the Nephilim? Retrieved June 23, 2025, from youtube.com/watch?v= qKtHwc3mMY8

Health Research Fund. (2024). Progeria Life Expectancy. Retrieved March 11, 2024, from healthresearchfunding.org/progeria-life-expectancy/

Lamb, A. (2007, July 28). Sheep and goats? Since when? Retrieved from creation.com/if-sheep-and-goats-are-the-same-kind-then-were-they-both-on-the-ark

Rushdoony, R. J. (2002). *Genesis.* Vallecito: Ross House Books.

Sanders, L., & Carter, R. (2018, June 15). Where was Eden? Part 1—examining pre-Flood geographical details in the biblical record. Retrieved November 17, 2024, from creation.com/eden-1

Sarfati, J. (2011, March 12). Timing of events in Noah's life. *creation.com*. Retrieved March 10, 2024, from creation.com/timing-of-events-noahs-life

Sarfati, J. D. (2018). *The Genesis Account—A theological, historical, and scientific commentary on Genesis 1–11* (Third edn). Powder Springs: Creation Book Publishers.

Sarna, N. M. (1989). *The JPS Torah Commentary—Genesis.* Philadelphia: The Jewish Publication Society.

Taylor, E. K. (1961, May). The Woman of Cana. *The Furrow, 12*(5).

Woodmorappe, J. (1996). *Noah's Ark: A Feasibility Study.* Dallas: Institute for Creation Research.

31 The text in Genesis 2:10–14, compiled post-Flood, indicates that Moses knew where the Garden of Eden was located in relation to contemporary geography. However, commentators have suggested that Moses relied on earlier written records in composing that part of Genesis. The names of the rivers near Eden were probably generic and were sometimes applied to other rivers post-Flood. (Sanders & Carter, 2018).

19

Genesis 7–11: Flood and confusion

They deliberately overlook this: By the word of God the heavens came into being long ago and the earth was brought about from water and through water. Through these things the world of that time perished when it was flooded (2 Peter 3:5–6).

So wrote the apostle Peter in the New Testament period. With these words Peter affirmed the universal destruction of the Great Flood. The sin-tainted world of Genesis 3–6, an era that spanned around 1,650 years,[1] was lost to history. The only remains of that period are the sparse record of these early chapters in Genesis and an array of folklore left to us from various ancient cultures. The Flood was a profound dislocation—ecologically, geologically and historically; it was and remains the major non-uniformitarian[2] event since the creation. The world that emerged from the receding floodwaters differed in some important respects from the antediluvian world of the ancients.

Yet in other ways the world remained the same: the tendency toward sin still marred mankind, and God was again confronted with a rising tide of human rebellion. God responded again—but in a different way that would have far-reaching consequences. God confused the languages at Babel, which scattered mankind across the earth and set the stage for the multiplicity of nations, with their different languages and cultures, to emerge. By the end of Genesis 11, there is a clear line of sight from the world as it then existed to the world we live in today. The human origins account is essentially complete.

The Flood narrative is an essential component of the biblical origins account, and as such has been targeted heavily by secular critique. In the late eighteenth century, the main thesis of Hutton's *Theory of the earth* was that geological features were better explained by uniformitarianism than by catastrophes such the worldwide Flood of Genesis. Readers will also recall Darwin's approving reference to Charles Lyell shaking "the faith in the Deluge &c". From the Christian viewpoint, the first decisive step in the creationist movement was Morris and Whitcomb's defense of the global deluge in *The Genesis Flood*, published about a century after *Origin*.

---

1 Sarna, 1989, p. 41.

2 Non-uniformitarian in the historical philosophical sense, as pertains to earth science, not in the methodological sense, which pertains to the scientific method.

Without the global Flood, the Genesis account cannot be reconciled with the fossil record—its vast extent, and the disease and carnivory that it records. In this chapter we investigate this major event and discuss implications for the origins debate.

Biblical context

The apostle Peter mentions that eight souls survived the Flood (1 Peter 3:20), correlating exactly with the Genesis account. In the opening quotation of this chapter, Peter is referring to both the creation account in Genesis 1, and the worldwide Flood event of Genesis 7–8, "Through these things the world of that time perished when it was flooded". Peter has harsh words for those who reject this history: they are "scoffers", "following their own evil desires" (2 Peter 3:2–7). The clear message of the New Testament is that the worldwide Flood of Noah's day was a historical event, and that Christians should believe the biblical account.

Peter explicitly makes a connection between the Flood and baptism in the New Testament. In 1 Peter 3:20–21, he writes about how God waited patiently in the days of Noah while the ark was being built, and that only a few people, eight in all, were saved through water. He goes on to confirm that

> Baptism, which corresponds to this, now saves you (not as the removal of dirt from the body, but the pledge of a good conscience toward God) through the resurrection of Jesus Christ …

Clearly, every Christian baptism is done in a manner which recalls the Genesis Flood.

The Flood (Genesis 7–8)

The Genesis account of the Flood clearly states that the flood was global in extent. The tallest mountains were covered; the world was in a sense returned to the watery state of Genesis 1:2.

The floodwaters had two main sources. The first is familiar to us today—precipitation—in this case forty days and nights of torrential rain. The second is not so familiar—the "fountains of the deep" (Genesis 7:11)—subterranean water under pressure.

The text tells us that "the floodgates of the sky were opened". This expression indicates an astronomical amount of rain and has been alternatively translated as "the windows of heaven were opened",[3] perhaps intimating that localized flooding began to occur in areas of land before complete inundation occurred. The phrase "windows of heaven" is only used a handful of times in

---

3 Sarna, 1989, p. 55.

the Bible. In Genesis, these "windows" open as the flood begins, and close as the flood concludes. After this, they are only mentioned in the conditional sense (for example, "... even if the LORD were to make windows in heaven ...", as in 2 Kings 7:2, 19 and Malachi 3:10) The Bible is telling us that the rain of the floods had an intensity that has not been repeated. They were part of a one-off, never to be repeated event.

After the fortieth day the rain was restrained. The rising waters had become immensely deep, such that even the highest mountaintops were submerged to a depth of about 6 meters (Genesis 7:20). God then caused a wind to develop, and the waters began to subside; rain ceased, and the fountains of the deep were closed.

Psalm 104:6–9 gives us extra information about the recession of the flood waters. It clearly describes enormous geological change occurring as the flood waters receded:

> ... mountains rose and valleys sank—to the place you established for them.

This can reasonably be taken to imply that the sinking down of valleys, especially undersea valleys, assisted in causing the flood waters to recede. The combined effects of the sinking valleys and geological uplift is said to result in the waters not being able to pass certain limits in the future:

> You set a boundary they cannot cross; they will never cover the earth again.[4]

One hundred and fifty days after the Flood began, the ark rested on the mountains of Ararat. A further 220 days passed before God told Noah it was safe to disembark with his family and the animals. Not only did the waters need to subside, but the ground needed to dry and harden, vegetation germinate and recover, and the earth's surface stabilize geologically. By this stage all humans, land animals, and birds formerly outside of the ark had long since perished.

As judgement for sin and violence, the Flood had a miraculous aspect, but creationists also believe it has a naturalistic explanation and is amenable to scientific enquiry (it belongs in the type 3 category of miracle we described in Chapter 8).

Explaining the naturalistic aspects of the Flood

The most prominent model creationists use to explain the world-wide Flood is Catastrophic Plate Tectonics (CPT), described in a series of papers published from 1986 to 2019. We present a very brief outline of CPT here, and provide a more detailed summary in Appendix E.

---

4 Psalm 104:9.

The model built on the work of French scientist and geographer Antonio Snider-Pelligrini (1802–1885), a creationist, who published *The Creation and its Mysteries Unveiled* in 1859. Snider-Pelligrini hypothesized that all of the earth's continents were originally connected together before spreading apart during the time of Noah's Flood.[5] [6]

It is a well-known scientific fact that the oldest layers of the sea floors are geologically younger than the oldest layers of continents. From this important fact, and the science of plate tectonics, geophysicist John Baumgardner deduced that rapid ocean floor replacement (that is, the old ocean floor descended into the mantle of the earth and new ocean floor was produced as this occurred) could provide a physical mechanism to explain the global Flood. In this model the flood onset, severity and offset relate to geological events.

To establish the scientific plausibility of the concept, Baumgardner utilized computer modeling incorporating contemporary physics. He simulated various aspects of the Flood, including the movement of tectonic plates, the generation of the flood waters, sediment deposition, and how the flood onset may have been triggered.

CPT provides a scientifically plausible explanation for the Genesis Flood and correlates well with a range of geological facts. Within a creationist framework, CPT represents an interesting and important contribution to origins science. Nevertheless, modeling work on CPT is not exhaustive, and some major problem areas of the model have not yet been adequately resolved. A range of other flood models have been proposed by creationists.

Evidence for the Flood in the geological record

Creationists have different perspectives on the fossil record and the geological column inferred by secular science. Some creationists generally accept the principles while others are critical of them. First, we will explain what stratigraphy and the geological column are.

Modern stratigraphy began when Nicolas Steno enunciated several presuppositions aimed to permit analysis of layers found in geological rock formations. A key principle was that of superposition, the idea that rock layers lower down in a formation were formed earlier than those higher up. Steno also posited that sediment layers are originally horizontal and extend laterally for a given distance. Under this scenario, two rocks found a distance apart that exhibit similar layers may have originally been part of the same rock. The separation is explained by later erosion or damage.

---

5 Austin, et al., 1994, p. 610.

6 Scientific priority goes to Abraham Ortelius of the Netherlands who proposed continental drift in 1596 (in Thesaurus Geographicus).

As the field of stratigraphy developed, a universal geological column was developed which described a specific vertical ordering of deeper rock layers. The ordering depended upon the type of rock, the types of fossils it might have contained, and the types of the neighboring rock layers. For example, if a stratigrapher identified rock layers A, B, and C in a particular location (but no others), and later identified layers B, C, and D in a different location, the geological column would comprise layers A, B, C, and D, even though not all these layers were present in any one location. The geological column is therefore a theoretical construct, informed by geological evidence sourced from many locations and agreed by secular scientific consensus. It is used by secular scientists to reconstruct geological history, for example, assigning relative ages to fossils. It is also used in relation to oil, gas, and mineral exploration, among other applications.

The geological column can be several kilometres deep in places and similar across continents. Roughly half of the column consists mostly of mud derived shales, with the balance split between carbonate rocks (for example, chalk and limestone), and sandstones. All of these are potentially related to water-borne sediments, consistent with a worldwide flood.

Some creationists are critical of the role of fossils in modern stratigraphy because they are interpreted in the philosophical uniformitarian, atheistic, Flood-denying paradigm of modern science. Other creationists, who generally support the principles of stratigraphy, observe that widespread deeper layers exist which not only are found across individual continents but across all continents, suggestive of global formative processes—exactly the sort of processes that would have been in operation during the six days of creation and the year-long Flood period. The tremendous geographical extent of some of these formations fits in very well with a global catastrophic event. Indeed, there are a number of specific features of these deep layers and general findings of science which fit particularly well with the Flood narrative.[7]

First, multiple discontinuous fossil sequences have been found. This occurs where a fossilized species is found in some deeper layers of the geological column but is absent from shallower layers. Secular scientists then infer that the species has been extinct for millions of years—but it is later found to be living. An example of a discontinuous fossil sequence is the coelacanth, an "ancient" species of fish which, judging by the geological column, disappeared from the fossil record for "65 million years". Then, in 1938, a fisherman caught one off the coast of Africa. This was followed in 1998 by another coelacanth catch, this time off Indonesia.[8] The coelacanth example is easily explained if the bulk of the geological column was laid down in a short space of time, such as

7 Batten, 2019.

8 Erdmann, Caldwell, & Moosa, 1998.

creationists believe happened during the global Flood. But in an evolutionary scheme, in which the fossil record is layered over deep time, the disappearance of a "living fossil" from the record for "65 million years" is deeply problematic. Why is it not found in the higher layers, given it is still alive?

The coelacanth is not the only example of a discontinuous fossil sequence. Another example is the Wollemi pine, which secular scientists believed was absent from the fossil record for "150 million years". However, a stand of Wollemi pine was found in the Blue Mountains, near Sydney.[9] A third example is the rodent *Laonastes aenigmamus*, which was thought to be extinct for 10 million years or more, but was observed in a Laos market about to be placed onto a kebab skewer.[10]

A second type of evidence for the recency of fossils in the geological column concerns what is known as *amino-acid racemization*. In living organisms, all amino acids exist as the levo isomer form (as explained in Chapter 10). After organisms die, it is known that levo isomers in the remains begin converting to the dextro form. It is estimated that after 20 million years, about half of the levo form will have converted to dextro, resulting in a 50:50 mixture (which is referred to as *racemic*). However, the Fig Tree Chert of South Africa,[11] estimated to be 3 billion years old, only contains amino acids in the levo form. This is consistent with fossils laid down in a recent flood, but not with the deep time scenario of secular science.[12]

Another line of evidence against the deep time interpretation of fossils in rock strata are the findings of soft tissue, including elastic blood vessels in dinosaur bones. If these really were millions of years old, soft tissues should have decayed and disintegrated. Under favorable circumstances, soft tissues can survive for thousands of years,[13] but modelling suggests it is extremely unlikely for soft tissues to survive millions of years. Material resembling intact osteocytes (bone cells) which contained appropriate protein signatures, and intact DNA and Histone H4 have also been found in dinosaur fossils. Similar to soft tissue substances, these should have disintegrated if they were millions of years old.[14] Instead, they are compatible with an interpretation of burial in a world-wide flood approximately 4,000 years ago.

In Appendix F, we list other evidence consistent with a world-wide flood 4000–5000 years ago.

---

9 Oard, 2011.

10 Catchpoole, 2008.

11 Chert is a siliceous rock, of which flint is an example.

12 Wieland, 1993.

13 Sarfati, 2020.

14 Ibid.

How the Genesis Flood explains the fossil record

Some marine fossils can be explained by catastrophic geological events affecting the sea floor. These events may have caused underwater landslides, trapping marine animals under layers of mud and carbonates early in the flood process.

On land, many animals and humans would have sought higher ground as the floodwaters rose. Landslides may have buried specimens which were unsuccessful in relocating to more elevated locations. Tsunamis, related to catastrophic tectonic plate movements, may have carried other organisms out to sea or trapped them in debris on land. The fossil record contains evidence that some sediments were laid very quickly, capturing and fossilizing animals in dynamic poses—reminiscent of what was seen in Pompeii[15]—such as the ingestion of other animals. Animal footprints, and even dung, were also fossilized, suggesting relatively rapid burial.

Within slurries of submerged sediments, various degrees of hydrological sorting may have occurred.[16] Hydrological sorting refers to the layering that occurs as sediments and organic materials settle in water. This phenomenon offers a different explanation for layering from the secular account. Whereas secular scientists regard geological layers as being laid down gradually across eons—meaning that deeper layers were laid down far earlier—hydrological sorting is a quick acting process. Factors such as current flows and the viscosity of sediment slurries, as well as body shape and density, affect which layers the water-borne objects come to rest in. Humans may have been the most capable of avoiding the floodwaters by seeking higher ground, which is a possible explanation for the relative lack of human fossils in the geological record.

As the floodwaters receded, we can imagine some of the likely effects. The rapid water flows would have carved out huge canyons. A substantial amount of marine life likely became trapped on emerging land surfaces, with a proportion of these becoming buried due to mud slides and other geological activity. Large numbers of land-dwelling animal carcasses and skeletons, as well as marine animals, must have become buried under mud flows and settling sediments. Some of the strata laid down in the Flood were likely disrupted and damaged by later geological upheaval in the post-Flood period. These effects will have rendered some of the contemporary fossil record challenging to interpret.

---

15 Casts of bodies buried by the volcanic ash in Pompeii (79 AD) demonstrate a range of lifelike poses.

16 Note that hydrological sorting was not ubiquitous. In some places, fossilized footprints have been found. These would not survive the process of hydrological sorting.

Life after the Flood

After the Flood ended, Noah offered burnt offerings of clean animals. God blessed Noah and his sons and their descendants. In the same way that God had instructed Adam to be fruitful, he now repeated this to Noah, saying "be fruitful and multiply, filling the earth" (Genesis 9:1).

Unlike the original creation of Adam and Eve as the parents of the entire human family, the immediate post-Flood period featured several couples who could procreate. Again, marriages between relatives must have taken place,[17] but there was less need for sibling marriage, and the offspring from Noah's sons probably married their cousins. Noah was also "perfect in his generations"; perhaps his genome had relatively few errors. (As history progressed, and the number of mutations grew in the human gene pool, a point was reached where consanguineous marriages became noticeably harmful. By the time of the ancient Israelites, the Law of Moses explicitly forbade consanguineous marriages.)

God's blessing also stipulated that from this point forward, "every moving thing that lives" was for food, with the caveat that man was to avoid eating the blood. Man was no longer to be a vegetarian. Henceforth animals would fear humans, who were now lawfully meat-eaters.

Finally, God made a covenant with humans, land animals and birds that never again would a worldwide flood cause complete destruction of the earth. Never again would there be a global near-extinction event. God chose the rainbow as the sign of that covenant, reminding us that the regularities of nature will continue until the end of time. But there is another implication—never again would mankind be allowed to degenerate into utter lawlessness and violence on a global scale—the sort of condition that prompted God to ordain the Flood event. Global anarchy was gone. God instituted the death penalty for killing a human, on the grounds that man is made in God's image (Genesis 9:6), and thereby established civil government. This was a corrective against the runaway multiplication of violence that marked the pre-Flood era. The balance was tipped away from violent and aggressive people towards peaceful, law-abiding citizenry.

Human lifespans in the immediate post-Flood period remained lengthy, which would have helped these early families replenish the earth's population. Within a century of the Flood, lifespans began to decrease in an exponential

---

17 Nowadays, marriage between relatives that are cousins or closer (known as consanguineous marriage) is illegal in many countries. The contemporary danger is that any children born from such unions are at increased risk of inherited diseases caused by an accumulation of harmful recessive genes in the population. Most individuals today carry such genes, but if they marry someone who is unrelated, their recessive genes are unlikely to match their partner's, leaving a good chance that the children will not manifest such diseases. However, if the parents are very closely related, there is a risk that their offspring will inherit two copies of the same harmful gene and hence manifest a genetic disease.

fashion, eventually levelling off at about 70–80 years (refer Figure 19.1). This is biologically plausible; a steady accumulation of slightly harmful mutations in the human gene pool could have progressively decreased life expectancy. Adverse environmental or dietary changes may also have played a part. Similar effects may have applied to animal lifespans in the post-Flood period, although the Bible does not specify this.

Noah had taken seven pairs from each of the clean animal kinds (characterized later in the book of Leviticus), but only one pair from the unclean kinds. As noted above, one animal of the clean kinds was sacrificed after Noah's family returned to dry ground, leaving six pairs of each clean animal kind intact. Thus, the clean kinds had a reproductive pool several times as large as the unclean kinds, perhaps making them slightly less prone to extinction. Humans had a reproducing population four times as large as unclean kinds. The archeological evidence demonstrates that many species and kinds have become extinct since that time. Notably, the dinosaurs, representing unclean kinds, are all extinct. This may have been due to a combination of factors such as the actions of humans, or to ecological changes post-Flood. Creatures have been known to go extinct for as long as people have been taking notice of such things, and the reasons will differ for every species. Certainly, the restricted gene pool of the unclean kinds may have made them more susceptible to extinction.

Some species of plants and animals have developed post-Flood via processes which probably involve mutations, natural selection and hybridization. These effects can occur quickly, especially when humans direct the processes, such as in dog breeding.

Rapid speciation after the Flood

In the previous chapter, we noted that most pre-Flood species and varieties of land animals and birds would have been lost. Following the Flood, we would expect that rapid variation and speciation occurred as the animals adapted to new habitats.

But as animals became more adapted and specialized, typically by losing genetic information, the rate of change of this variation probably slowed. In the first place, there was less need for change as the animals became adapted to their environment, but crucially, they also had less genetic capacity for change as they became more specialized.

Human lifespans, which are given in the Genesis chronogenealogies, give us a hint as to how this might have played out. Figure 19.1 shows the exponential reduction in human lifespans in the post-Flood period.

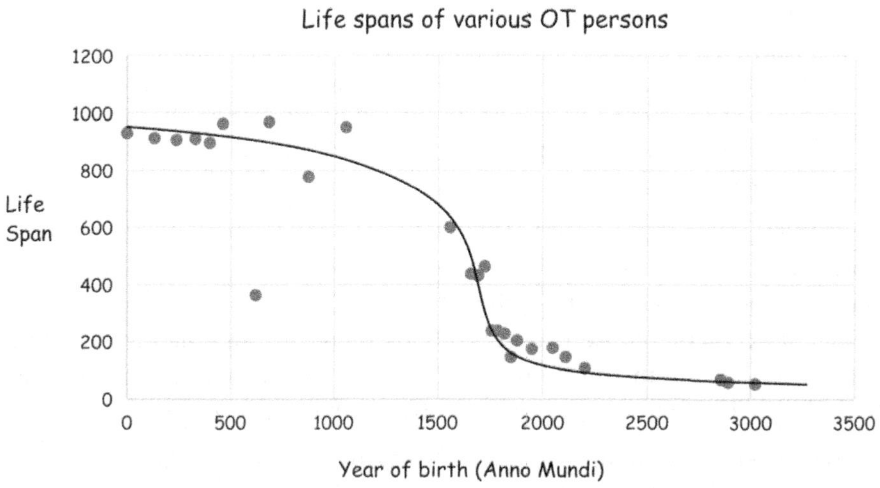

Figure 19.1: Exponential reduction in human lifespans after the Flood

We hypothesized earlier that an accumulation of genetic mutations caused the reduction in lifespans during the post-Flood period. These mutations damaged working genes, reducing the information in the human gene pool. What was happening in humans was likely replicated in animals—mutations were rendering some genes nonfunctional, which may have led to improved adaptation,[18] but they simultaneously reduced the capacity of these species for further adaptation.

By the time of Darwin, very little speciation was occurring. This explains why many of his scientific contemporaries believed that species were fixed.

Comparing the Genesis account with the Epic of Gilgamesh

A specific ancient flood account is described in the *Epic of Gilgamesh,* a Babylonian flood story. One of the best-preserved copies of this account was found in the ruins of the library of the Assyrian king Ashurbanipal (died 631 BC), where a destructive fire had literally "fired" the clay tablets, preserving them from disintegration.[19]

The Epic of Gilgamesh was first translated by an Englishman, George Smith, who became a pioneering Assyriologist by studying in his spare time after work. In his book *The Chaldean Account of Genesis* (1876), he stated:

18 Reduced genetic information leading to increased adaptation is termed devolution. For examples of devolution, refer to Chapter 6.

19 The Gilgamesh Epic is written on twelve tablets, with the flood account on the eleventh tablet.

> The great value of the inscriptions describing the Flood consists in the fact that they form an independent testimony in favor of the Biblical narrative at a much earlier date than any other evidence[20] ... In the account of sending forth the birds there is a difference in detail between the Bible and the Inscriptions which cannot be explained away; this and other similar differences will serve to show that neither of the two documents is copied directly from the other.[21]

Smith regarded the Epic of Gilgamesh as supporting the biblical account. He provides evidence—key differences between the two accounts—indicating that one has not been copied from the other.

Secular scholars, whose presuppositions deny the supernatural and the historical accuracy of the Bible, interpret the Epic of Gilgamesh in an entirely different way. The following is representative of the secular perspective:[22]

> The accepted cause of similarities between the two versions in biblical scholarship is that the *Gilgamesh Epic* account of the Babylonian Flood story is the source of the Genesis Flood Story. The view, which assigns a Mesopotamian background for the biblical story, stems from the perception of the earlier date of the Babylonian Flood accounts, and the idea that the Flood story "fits the geo-hydrological conditions of Mesopotamia more than those of Palestine".

So, is the biblical account based on the Epic of Gilgamesh? Japanese postgraduate researcher, Nozomi Osanai, investigated the relationship between the Epic of Gilgamesh and Genesis. To do this, Osanai first learned the Akkadian language so she could read the Epic in its original language. She compiled her findings into a thesis which earned her an M.A. from the Wesley Biblical Seminary in the U.S. in 2004. This thesis is available on the Creation Ministries International website,[23] and the following discussion draws heavily on Osanai's work.

It is true that original compilations of the Epic may predate extant copies of Genesis, and that there are numerous similarities between the two accounts. But an analysis of the internal evidence reveals major difficulties with the idea that the Genesis account is based on the Epic, for three reasons: first, the Genesis account is internally coherent, whereas the Epic leaves readers unsure of why the story progresses the way it does; second, the Genesis account includes realistic details that the Epic lacks; and third, the Genesis account includes the "Table of Nations", which lists the names of ancient people groups that separated after the Flood. We will examine each of these in turn.

---

20 Smith, 1876, p. 286.

21 Ibid., p. 289.

22 Pryke, 2019.

23 creation.com/comparative-study-of-gilgamesh-and-genesis-thesis-by-nozomi-osanai-introduction

Coherence of the accounts

The Genesis account presents a clear motivation for the flood. A monotheistic God "regretted that he had made man on the earth," owing to the great wickedness of mankind, and therefore purposed to "wipe mankind, whom I created, off the face of the earth, together with the animals, creatures that crawl, and birds of the sky ... ." Among the people on earth, "Noah, however, found favor with the LORD." (Genesis 6:6–8). God told Noah to build an ark which would save him and representative animals from the flood. Then, in Genesis 7 and 8, Noah builds the ark and is delivered from the flood, with his family and the animals. The rest of the earth perished. God tells Noah and his sons to "be fruitful and multiply and fill the earth" (Genesis 9:1b). Thus, the original command to Adam to fill the earth is repeated to Noah and his sons. The narrative is internally coherent.

On the other hand, the Epic only implies that the flood was due to man's sin; a fair interpretation is that the flood is a capricious decision of a council of gods. The hero of the Epic, Utnapishtim (the person analogous to Noah), intentionally deceives the people, as instructed by the god Ea, and has them build the ship for him. After the flood, the gods bestow immortality on Utnapishtim, but it is not clear why. The epic lacks clear reasons for why the gods decided to send the flood, how Utnapishtim was chosen as the hero of the story, and why he was accorded immortality afterward.

Details

Although the Epic implies a global flood, it only identifies rain as the source of the floodwaters. This rain fell for about a week (translations have varied between six days and six nights to seven days and seven nights), which seems insufficient to cause a global deluge. Osanai points out that if the flood were not global, it would have made more sense for Utnapishtim to migrate to high ground rather than commence a major ark-building project.

The Genesis account cites multiple sources for the floodwater. It emanated from "the fountains of the great deep" and "the windows of heaven". Rain was on the earth for forty days and forty nights. The flood is clearly described as global, with the waters topping the mountains to a depth of 15 cubits (more than 20 feet, or 6 meters).

The Gilgamesh Epic describes the dimensions of the ark that was built for Utnapishtim in vague terms. A reasonable interpretation is that Utnapishtim's ark was a cube of side 60 meters (almost 200 feet)—about the height of a 20-storey building—with seven decks. This gives an unlikely image for an ancient shipyard. Crucially, a floating cube of this size would be highly unstable; one wonders if the writer of the Epic had any nautical experience.

By contrast, Noah's ark was built to a design specified by God (Genesis 6:14–16). It was of gopherwood, covered inside and out with pitch, with rooms occupying its three decks. The ark was 137 meters (450 feet) long, 23 meters (75 feet) wide, and 14 meters (45 feet) high.[24] These dimensions result in stable flotation.

In summary, the details of the Epic do not appear to identify an adequate water source for a global flood, and the design of Utnapishtim's ark is removed from nautical reality—a large, cube-shaped structure would be unstable and very difficult to build.

Table of nations

The Epic does not include a reference to the repopulation of the earth after the flood. By contrast, Genesis 10 presents what is commonly known as the "Table of Nations", or a listing of the nations descended from Noah. Many of the names mentioned in Genesis 10 have come down to us today. For example, Genesis 10:14 mentions the Philistines. This people group occupied the land of Canaan from the days of Abraham through to the kingdom years of ancient Israel. They gave their name to the region where they lived; the Romans knew the area as *Palaestina,* which has given rise to the modern term, Palestine.

Reviewing all these names is beyond our scope, however it is worthwhile to look at several other examples. The late creationist writer, Bill Cooper, researched where the people groups from the Table of Nations lived:[25]

> Noah's son Japheth (Genesis 10:2) was the progenitor of the Indo-European peoples. He was worshipped by the Greeks as *Iapetos,* and the Romans as *Iupater,* which is Anglicized as *Jupiter.* One of Japheth's grandsons, Tubal (10:2) was the progenitor of a people group which in Roman times lived in the area of present-day Georgia, before migrating northeast to Siberia. The name of Tubal lives on in the the city of Tobolsk, which is situated alongside the Tobol river. Another of Japheth's grandsons, Ashkenaz (Genesis 10:3) was the father of a people group that settled in Armenia, although some traditions also associate him with Germanic races. The name Ashkenazi is applied today to Jews of Central European origin.

A final example from Japheth's grandsons is Dodanim (Genesis 10:4). Dodanim's descendants settled in Asia Minor, in the vicinity of Troy. These coastal regions are known to this day as the Dardanelles, a region in present-day Turkiye.

The evidence today

Osanai presents an analysis of flood accounts from 200 people groups. Many of these accounts share features in common with the Genesis account. We

---

24 The ark was comparable in length to the nine-masted Chinese treasure ships built in the 15th century (Nunn, 2016).

25 Cooper, 1990.

present the first five features from Osanai's list, with a percentage showing how many of the worldwide flood accounts incorporate each feature:

1.	Is there a favored family?	88%
2.	Were they forewarned?	66%
3.	Is the flood due to the wickedness of man?	66%
4.	Is the catastrophe only a flood?	95%
5.	Was the flood global?	95%

The global flood of Genesis is a part of humanity's collective memory. This is exactly what we would expect given that all people on the earth today have descended from Noah or his sons.

The Tower of Babel (Genesis 10:21–11:9)

Genesis 11 records that the entire human population settled in a valley[26] in Shinar and started to construct a city and a tower.[27] Their purpose was to avoid spreading over the face of the earth, in defiance of God's earlier command to "fill the earth". The leader of this project was Nimrod, a mighty hunter who had descended from Ham via Cush. Genesis 10:8b,10 provides some details of Nimrod's reign:

> He began to be a mighty one in the earth ... The beginning of his kingdom was Babel, Erech, Accad and Calneh in the land of Shinar.

Nimrod is often interpreted as "rebel"; he is regarded as a tyrant and a hunter of men, implying that he enslaved many.[28]

God observed this rebellious behavior and confused the languages so that the people could no longer understand each other. For this reason, the city, which remained unfinished, was known as Babel, and the people were scattered over the face of the earth to ultimately become different nations and ethnicities. Nimrod moved into Assyria and built the cities of Nineveh, Rehoboth Ir, Calah and Resen, almost certainly with slave labor.

The Genesis narrative states that the division of the earth occurred in the lifetime of Peleg (Genesis 10:25). This text is generally taken to refer to the dispersion of various language groups occurring at Babel. This probably also inaugurated the spread of populations across the now separated continents.

Tectonic plate movements, principally during the flood, may explain the division of the earth into these continents, although continental drift continues very slowly even today. This is implied by Genesis 11:9, which describes God scattering the people "over the face of the *whole* earth" (emphasis added). Various animal kinds also spread to different continents. The animals spread to

---

26 Sometimes translated as a plain (Sarna, 1989, p. 81).

27 Sarna, 1989, p. 81.

28 Rushdoony, 2002, p. 93.

various places, some by human intervention, and their fauna became sequestered on them. It became impossible for some of these animals, such as koalas on the Australian continent, to migrate elsewhere under ordinary conditions.

The near permanent geographic separation of these early human populations was initially due to different language groups. But variations between separated human populations occurred, partly driven by differing ecological environments and climates, leading to the development of distinct human races and nations with their own physical characteristics and cultures.

The remains of various human groups identified as Neandertals, Denisovans and *Homo erectus* may date from this early post-flood period. They were fully human—an analysis of Neandertal and human skulls showed that Neandertals "had the capacity to perceive and produce human speech." Neandertals deliberately buried their dead and used the same stone tool technology as *Homo sapiens*.[29] Neandertals wore jewelry: pendants of a variety of materials, with holes drilled in them so they could perhaps be worn as necklaces, have been attributed to this extinct people group.[30] Moreover, they developed plied string—scientists found a piece of ancient 3-ply string in a cave in France believed to have been occupied by Neandertals.[31] So why did they die out? Neandertals, and other people groups which we know today as Denisovans, did not "die out", as much as they were "bred out" by more successful people groups that they encountered. Over a number of generations, they ceased to exist as distinct people groups, and the only way to find evidence of Neandertal ancestry is to look in the DNA of individuals. The DNA of people today whose recent ancestry is non-African will likely comprise about 2% Neandertal genes.

The range of animals in particular geographical locations also changed due to local extinctions and the effects of the human population. The Old Testament mentions lions and bears in the territory of ancient Israel. These were regarded as a menace by shepherds (for example, 1 Samuel 17:34–36) and at some point after the reign of Solomon (Proverbs 22:13), they were likely hunted to local extinction. On the other hand, humans transported various species from one location to another. An example is the importation of apes and monkeys to ancient Israel by King Solomon's ships (1 Kings 10:22). Evidently these animals were not native fauna in Israel at that time.

29 Oard M. J., 2022.
30 Sanders and Carter, 2020.
31 Cox, 2024.

The effect of nations

By forming different races and nations, God had put in place the conditions by which His original commandment to man to "fill the earth and subdue it" would be fulfilled. At the same time, God had fashioned for himself a range of general mechanisms by which he could tilt the table in favor of law-abiding communities and against lawless communities—mechanisms which, acting at a societal level, remain operative to this day.

In the post-Babel era, God used nations as a means of punishing certain communities and nations. For example, Israel would be used by God to bring the recalcitrant Canaanites to near extinction once the measure of their sins was filled. Israel itself also was to suffer for its apostasy under the Babylonian exile, as well from 70 AD when the Romans destroyed Jerusalem. It is almost as if God were fashioning population groups to serve as instruments of judgment by which to punish evil nations and cities, as affirmed in 2 Chronicles 15:6:

> Nation was crushed by nation and city by city, for God troubled them with every possible distress.

The notion of fragmentation and scattering of disobedient communities is explicit in the Babel narrative. It is also found in other parts of the Old and New Testaments. If we consider humans in two populations, obedient and disobedient communities, fragmentation and scattering is a mechanism by which God weakens those latter communities in history. Conversely, Christ explicitly prayed for unity in the church (John 17:20–23), and Paul appealed for it (1 Corinthians 1:10) to strengthen the new corporate man in Christ (see Romans 5:12–21). The concepts of fragmentation and unity are motifs that run through the Bible.

Competition among ethnolinguistic groups and nations meant that each was forced, over time, to improve their law codes. Law codes that favored violence and slavery slowly lost out to those populations with better law codes. The Israelites had laws which were proportionate, impartial and favored private property and economic freedom. Not only did the law of Moses punish murder, rape, theft and kidnapping, it punished delinquency in male youths, thereby strengthening the family. All of these measures helped improve the moral character of the nation.

Genealogy of Shem

Following the Table of Nations, a final genealogy, that of Noah's son Shem, is presented in Genesis 11. (The non-biblical word "Semite" is derived from Shem's name.) One of Shem's descendants was Eber, from which the biblical word "Hebrew" is derived. Abram and his wife Sarai were descendants of Eber. Abram, later renamed Abraham, is noted in the New Testament as the father of our faith as Christians. God promised Abraham that his descendants would be a blessing to all the families (Genesis 12:3) and all the nations of the earth (Genesis 22:18). It was Abraham's descendants, via Isaac and Jacob, who not

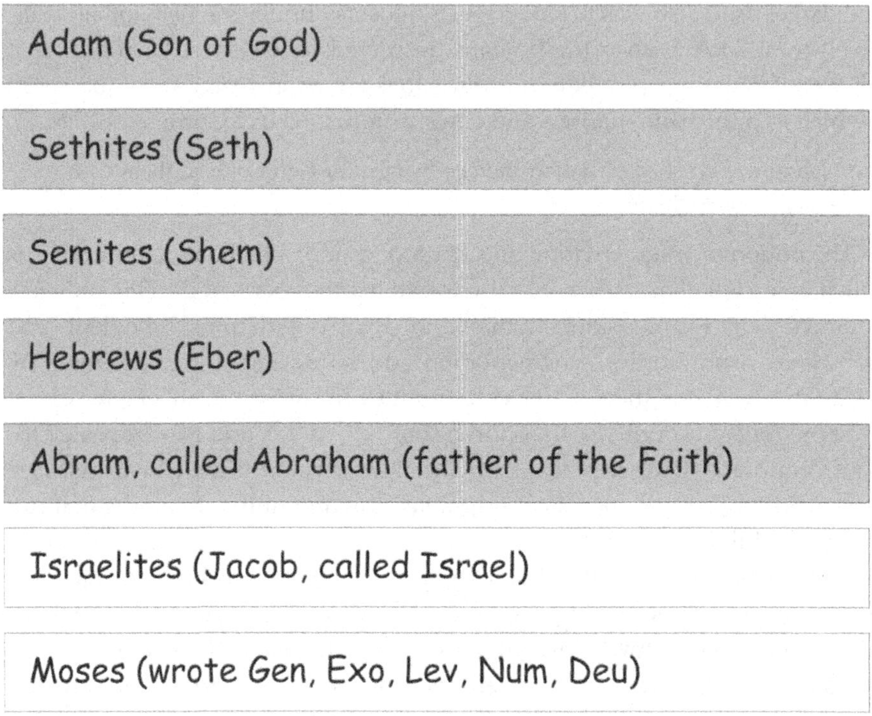

Figure 19.2: Abbreviated genealogy of Abraham and Moses

only brought forth the Word of God as scripture, but also the incarnate Word of God, Jesus Christ. These events were to be the most important in human history, blessing Christian believers from all nations.

Figure 19.2 shows an abbreviated genealogy of key people in God's redemptive plan. Genesis 1–11 covers the shaded area of the diagram (although most of the narrative concerning Abram/Abraham occurs later in Genesis). This period, of some 2,000 years, is roughly half of the time period covered in the entire Old Testament. It is a short account of a long and crucially important period of human history. The significance of Moses, the final person in this geneaology, is that he was the author of Genesis.

Conclusion

Genesis 1–11 provides coherent explanations for many features of the world around us. These include the fact that man is a conscious being, his fallen nature and the problem of sin, the origin of the secular humanist worldview, the pre-eminence of man above the animals, the cause of most of the fossil record, the rise of nations and languages, and a hint that Abram (later Abraham) would be a person of immense historical importance.

The secular humanist account ignores some of these, is unable to account for others, and, in the case of consciousness, offers mere conjecture. Regarding the flood, secular scholars interpret flood accounts left to us by virtually all people groups as myths or folklore, and insist that they describe only localized events. The worldwide abundance of fossils and the wide geographic extent of sedimentary rock layers coincides well with the Genesis flood account. Therefore, pointedly denying the global extent of the flood has long been a key task of the secular worldview. In the modern sense it originated with Hutton and Lyell, but the Apostle Peter mentions that the same spirit of denial was prominent in his day.

The Bible leaves the Christian, regardless of educational level or the specific scientific era their life falls under, in a position of far greater discernment. This allows Christians, and Christian communities, to lead wiser lives—lives that are in tune with God and His creation.

References

Austin, S. A., Baumgardner, J. R., Humphreys, D. R., Snelling, A. A., Vardiman, L., & Wise, K. P. (1994). Catastrophic Plate Tectonics: A Global Flood Model of Earth History. *Proceedings of the Third International Conference on Creationism. 3*, pp. 609–621. Pittsburgh: Creation Science Fellowship, Inc.

Batten, D. (2019, September 18). Age of the earth—101 evidences for a young age of the earth and universe. Retrieved March 18, 2024, from creation.com/age-of-the-earth

Catchpoole, D. (2008, May 14). The 'Lazarus effect': rodent 'resurrection'! Retrieved March 19, 2024, from creation.com/the-lazarus-effect-rodent-resurrection

Cooper, B. (1990). The early history of Man: Part 1 - the table of nations. Retrieved August 14, 2024, from creation.com/images/pdfs/tj/j04_1/j04_1_67-92.pdf

Cox, G. (2024, November 4). Neanderthals: A very human race. New discoveries are changing perceptions. Retrieved November 17, 2024, from creation.com/neanderthals-are-human

Erdmann, M. V., Caldwell, R. L., & Moosa, M. K. (1998, September 24). Indonesian 'king of the sea' discovered. Retrieved March 18, 2024, from nature.com/articles/26376

Nunn, W. (2014). Amazing ancient Chinese treasure ships: Giant sea-going wooden vessels challenge Ark skeptics. *Creation* 37(1): 12–13.

Oard, M. (2011, May 2). Are fossils ever found in the wrong place? Retrieved November 17, 2024, from creation.com/fossils-wrong-place

Oard, M. J. (2022, June 17). Neanderthals becoming more modern with time. Retrieved November 17, 2024, from creation.com/neanderthals-modern

Pryke, L. M. (2019, November). The Influence of Gilgamesh on the Bible. Retrieved March 25, 2024, from bibleinterp.arizona.edu/articles/influence-gilgamesh-bible

Rushdoony, R. J. (2002). *Genesis*. Vallecito: Ross House Books.

Sanders, L., Carter, R. (2020, January). The sophisticated Neandertal: Is the Flintsones a more accurate picture of Neandertals than evolutionary documentaries? *Creation* 42(1):12–13.

Sarfati, J. D. (2020, April 23). DNA and bone cells found in dinosaur bone. Retrieved March 18, 2024, from creation.com/dino-dna-bone-cells

Sarna, N. M. (1989). *The JPS Torah Commentary—Genesis*. Philadelphia: The Jewish Publication Society.

Smith, G. (1876). *The Chaldean Account of Genesis*. New York: Scribner, Armstrong & Co.

Wieland, C. (1993, September 25). Shaking hands on a recent creation. Retrieved March 18, 2024, from creation.com/shaking-hands-on-a-recent-creation

Part 4
Outlook

20
The failure of the evolutionary worldview

The X Club of nineteenth century Britain promoted a naturalistic worldview which its members hoped would displace belief in the God of the Bible. In this battle of worldviews, Darwinian evolution was one of the main weapons in their armory. "The student of science," wrote Thomas Huxley, "who is satisfied that the evidence upon which the doctrine of evolution rests, [which] is incomparably stronger and better than that upon which the supposed authority of the Book of Genesis rests, will not trouble himself further with these theologies ...".[1]

The pyrrhic victory of the X Club

These words have proven to be prescient, even if the hoped-for change in worldview only arrived decades after Huxley's death in 1895. It was the twentieth century Cold War between the West and the Communist bloc that provided the setting for the incorporation of Darwinism into state-funded compulsory education systems. In 1957, Russia had shocked the West when their Sputnik satellite beat the Americans into orbit. This prompted a rethink in the United States about the effectiveness of the education system. The National Science Foundation reformed its curriculum, including revisions to the biological sciences. It was in this context that the Epperson v. Arkansas (1968) decision was made. Huxley's "student of science"—or in reality, hundreds of millions of students of science over the ensuing decades—was now to be taught not to "trouble himself further with [Genesis-based] theologies".

Epperson was more significant than a routine court decision. Firstly, the fact that the United States Supreme Court decided the case meant that the ability to teach the evolution of man from lower animals was protected by law in all fifty states. Further, such education carried the prestige and imprimatur of the state and was seen as a national priority. Students were being told, essentially, that without world-leading science education, which included human evolution in the biological sciences, the United States was doomed to fall behind the Communist bloc. The rest of the Free World, which looked to the Unites States for leadership, took note.

---

1 Huxley, 1893, pp.147–148.

The Epperson decision can be summarised: *We may teach our children that human evolution from lower animals is settled science, with the implication that the Genesis 1 account is myth.*

This was a development which the X Club would surely have approved. A centuries-long anti-Christian offensive had reached its denouement. Christianity would no longer be reviled merely by a clique of the Western elite; Darwinian evolution would finally drive a presuppositional shift in the masses, and Christianity would switch rapidly from the acknowledged basis of the civilisation to an increasingly abandoned faith. The X Club had apparently won the war of the worldviews and achieved one of its primary objectives.

However, if the primary purpose of the X Club was to secularize society, their thinking was fundamentally flawed. The evolution idea, as we saw in Chapter 14, is of ancient pagan origin. While incompatible with Christianity, the origins evolution idea is not necessarily secular; it serves as a foundation for a wide range of pagan religious perspectives. To counter the Bible by Darwinian evolution, which was the X Club's modus operandi, does not secularize a society, but rather replaces Christianity with neopaganism. In that sense, Western de-Christianisation since Epperson is directly associated with the rise of neopaganism; in fact, we regard these two phenomena as different sides of the same coin. Having overlooked the neopagan implications of a broad-based belief in evolutionary origins science, the X Club may have altogether neglected the dissonance between paganism and the notion of objective truth.

Paganism frequently downplays the concept of objective truth, with many adherents being more interested in the subjective truth of personal experiences. Thus, for example, the ancient Greeks enthusiastically incorporated the gods of the surrounding nations into their pantheon. This attitude stood in direct contrast to the Christian worldview, which is intensely interested in, and protective of, objective truth (consider John 14:6) and refuses to syncretise other religious perspectives. Crucially, the Christian ethos, which favours the pursuit of objective truth, provided fertile ground for a sustained scientific enterprise.

Before we explore these considerations more fully, we pause to evaluate the extent to which the West has been de-Christianised since Epperson.

Tell-tale signs of rising paganism

Taking Australia as a bellwether for the wider West, religious data reveals the time at which de-Christianization began in earnest. Christian affiliation, as captured in national censuses from 1901 to 2021, first showed a step decrease in the Depression era (1930s), before recovering slightly over the next several decades (refer Figure 20.1[2]). Overall, the first half of the twentieth century was marked by a relatively minor decrease in Christian affiliation, certainly not the kind of decrease which amounted to a wholesale change. However, beginning in the 1970s, a long-term, structural decline in Christian affiliation commenced, timing which matches the Epperson decision of the late 1960s. From this point forward, a significant proportion of the population gradually shifted from Christian affiliation to "no religion". We argue that the post-1968 "no religion" category can be understood as neopaganism in its various forms. Also noteworthy is the slow but steady growth in the "Other religions" category, largely as a result of immigration from non-Western countries.

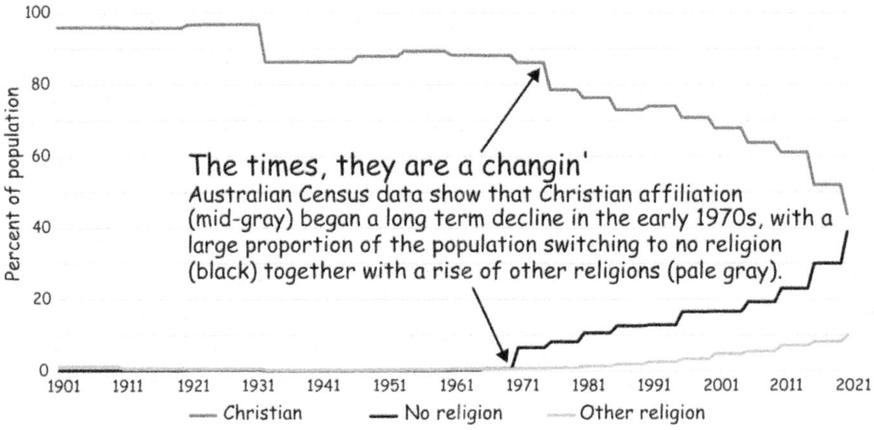

Figure 20.1: Religious affiliation over time (Australia)

But does Christian affiliation, as recorded in census data, paint a reliable picture of the true religious landscape? Anecdotally, some people tick a religious box on their census form, even though they do not practice the religion.

To find out whether the de-Christianisation of the West is as legitimate as Australian census data suggests, we look under the hood of society using another metric.

Changing standards in Hollywood

Since 1930, American motion picture studios had voluntarily adhered to what was known as the Hays Code, which provided guidance on the self-censorship of offensive content. In accepting the code, the industry was tacitly

2 ABS sources: the archived 2008 Year Book (cat. No. 1301.0) for 1901–1966 (web.archive.org/web/20120527023156/http:/www.abs.gov.au/ausstats/abs@.nsf/7d12b0f6763c78caca257061001cc588/636F496B2B943F12CA2573D200109DA9). 2021 Census page, from for 1971–2021 (abs.gov.au/articles/religious-affiliation-australia). Accessed October 5, 2025.

acknowledging, as the preamble of the code stated, that "entertainment and art are important influences in the life of a nation".[3] During the heyday of the Hays Code, Hollywood blockbusters such as *Quo Vadis* (1951), *The Robe* (1953), *The Ten Commandments* (1956) and *Ben Hur* (1959) were characterised by biblical themes, top talent, and outstanding commercial and critical success. But in the 1960s, the system progressively broke down, partly due to the influence of European cinema, which did not have an equivalent to the Hays Code. Finally, in 1968, the Motion Picture Association of America (MPAA) disbanded the code entirely, instituting instead a rating system based on the age of viewers. The MPAA issued a press release in October 1968, announcing that the new rating system would become operational on 1 November that year.[4]

With self-censorship now officially abolished, the motion picture industry rapidly ramped up the production of content which, just a few years before, had been acknowledged as unacceptable. Script writers unapologetically touted the merits of pagan-influenced entertainment. "Greek theater, Shakespeare, and actually everything of quality known to Western persons has a significant undertow of sex or violence or a combination thereof," wrote Lew Hunter in *Screenwriting* (1993).[5] Hollywood had flipped from producing biblically themed movies, which contemporary audiences supported, into content inspired from a range of sources including the ancient Greeks. Again, the Bible was out, replaced by paganism.

The following chart (Figure 20.2) demonstrates the increase in profanity in movies and TV over time. As a proxy, the chart tracks utterances of "Jesus Christ" (language that was proscribed under the Hays code, at least when spoken irreverently) in these media from the 1930s to the 2010s.[6] The data has some limitations: it is not granular, with only a single point per decade; it does not distinguish between reverent and irreverent uses of "Jesus Christ"; and it does not capture the full range of profanity—data for the individual words "Jesus" and "Christ" are not included. But a clear trend is evident over the decades, which affords a back-of-the-envelope view of anti-Christian sentiment in these media.

In movies, utterances of "Jesus Christ" remained low from the 1930s to the 1960s, before increasing noticeably in the 1970s. This is in line with expectations, given the abolition of the Hays Code in late 1968. From the new baseline of the 1970s, annual utterances increased around five-fold into the 2010 decade.

3 The Motion Picture Production Code, 2023.
4 Motion Pictures Association, 2023.
5 Hunter, 1993, p. 23.
6 The Movie Corpus, The TV Corpus, 2023.

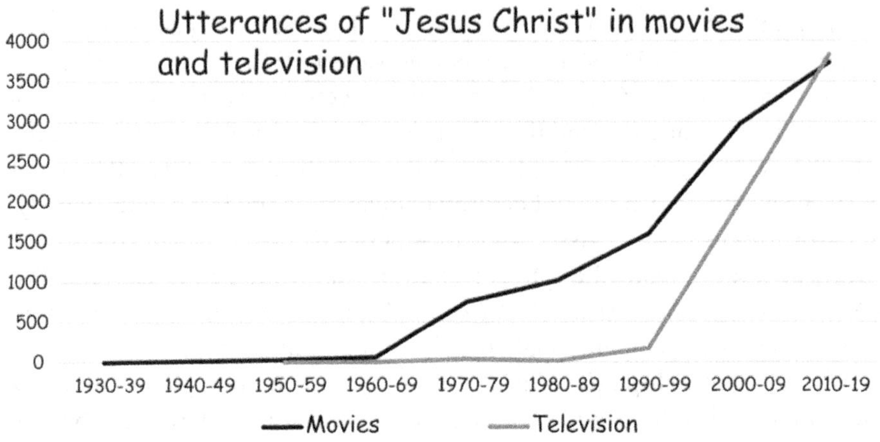

Figure 20.2: A surrogate measure of profanity in movies and television

The data for TV is perhaps more revealing. Television is a far more accessible medium than movies because viewers do not have to purchase a ticket. Accordingly, we would expect TV to be far less profane than movies, and this was indeed the case through the 1990s. But in the 2000s, the profanity of TV shows increased markedly. By that stage, even the most vulnerable members of society, children, could potentially access profane content. Of course, by that same time, the internet had become increasingly popular as an accessible entertainment platform, so in a sense the profanity of TV was, in terms of access for minors, perhaps a moot point.

What this graph shows, however, is the kind of content that the public was willing to consume. The increasing level of profanity over time shows that the de-Christianisation of the West was indeed underway. Language that was deemed unacceptable in American entertainment in the 1950s was completely normalised by the 2000s.

Anno Pagano

Because of the way this revolution in the West has occurred, with a coterie of elites convincing the masses to reject Christ via neopagan origins, we denote the Western post-Christian era *Anno Pagano* (Pagan Year). And, in line with the convention of earlier dating systems, which treat the inaugural year—in this case, 1968—as year 1, then the year of publication of this book, 2025 AD, corresponds to 57 AP.

However, Anno Pagano has in some respects been a stealth revolution. The dating system of the calendar has, officially, continued unchanged. Many of our institutions have the same look and feel as they once did; but there is a new set of taboos. In some instances, law enforcement and the judiciary have been repurposed to guard neopagan dogmas and punish neopagan heresies. Recall that a century ago, British intellectuals like Bertrand Russell could openly defy the official religion of their day[7] without fear of sanction. Those freedoms are fading fast in the neopagan era. Journalist Greg Sheridan has cautioned: "Being overtly Christian today is intensely unfashionable if not downright illegal".[8]

The rise of anti-science

For several decades, Richard Dawkins has been a combative and outspoken advocate of atheism in the Anglosphere—carving out for himself a public persona which is probably equivalent to Thomas Huxley in Darwin's day. Dawkins has long advocated for society to move on from what he considers to be the irrationality of religion and embrace secularism instead. Among his many activities, Dawkins founded the Richard Dawkins Foundation for Reason and Science (RDFRS) in 2006. "The foundation's mission is to realize Richard's vision to remove the influence of religion in science education and public policy, and eliminate the stigma that surrounds atheism and non-belief", reads the summary blurb on the Foundation's homepage.[9] Dawkins' efforts, we should note, are in line with the agenda of the X Club. This group also sought to promote "a purely naturalistic worldview", along with "the infrastructure of science within education and government".[10] The X Club, like Dawkins after it, wanted the "rationality" of science to permeate the entire society and its government.

However, unlike the members of the X Club, Dawkins has lived in an era in which Western Christianity has been eclipsed. It is Dawkins' privilege to witness outcomes which the X Club could only set in train. Intuitively, then, Dawkins and his colleagues should be able to point to a substantial and growing "rationality dividend". That would be the expectation, but, as we shall see, Dawkins, and some of his colleagues from the secular origins camp, have found that the opposite has occurred.

In the mid 1980s, one of Dawkins' collaborators, Lawrence Krauss (whose ideas we discussed in Chapters 2 and 9), was commencing his career at Yale University. Krauss could not help but notice what he considered to be the dubious intellectual developments that were germinating in the humanities departments across campus. These included deconstructionism, which held that it was not possible to derive objective truth from a scholarly work without "deconstructing" the ideological biases of the authors. As far as Krauss and his colleagues were concerned, this type of thinking amounted to a "lack of

---

7 In Russell's case, Christianity (refer Chapter 2).

8 Sheridan, 2025, Chapter 14.

9 RDFRS, n.d.

10 Thyer, 1995.

intellectual standards … . After all," they scoffed, "stars emit frequencies of light that are not only independent of human biases but were emitted before humans even evolved!"[11]

Krauss's attitude toward these developments in the humanities departments strikes us as cavalier. We saw in Chapter 6 that sixteenth century humanism, which started in the humanities, eventually represented a serious threat to science. In other words, there is a historical precedent for the spread of humanistic ideas out of the humanities towards other disciplines which might have seemed at arms length intellectually. Philosopher of science Thomas Kuhn had noted that the early humanists involved themselves with efforts aimed at "stopping science".[12] Nisbet further noted that it was only the rise of Protestantism that effectively countered the anti-progress of the early humanists.[13]

Sure enough, the modern incarnation of the humanists took little time to train its deconstructionist critiques on the scientific enterprise.[14] Richard Dawkins was eventually provoked to respond. In 1998, Dawkins published an essay, "Postmodernism Disrobed", in which he rejected aspects of postmodernist discourse.[15] In this essay, Dawkins introduced readers to feminist philosopher Luce Irigaray's conception of "sexed" science.

> Irigaray argues that $E=mc^2$ is a 'sexed equation'. Why? Because it privileges the speed of light over other speeds that are vitally necessary to us.[16]

Dawkins regarded this line of thinking as a "daffy absurdity". He critiqued a further example of Irigaray's sexed science that "privileges rigid, solid things". Irigaray argued that science which was related symbolically to the male sex organ (because it concerns rigid bodies) receives more attention from scientists than science that that she symbolically related to the female sex organs (like fluid mechanics).[17]

That was more than a quarter century ago. Dawkins' broadside against "absurdities" did not, evidently, have much effect. Indeed, anti-science developments seemed to appear with increasing frequency. In one episode, Dawkins felt the need to personally intervene in a science education dispute that arose in New Zealand. This involved what Dawkins described as "the ludicrous move to incorporate Maori 'ways of knowing' into science curricula

---

11 Krauss, Ed. (2025), Introduction and Overview.

12 Kuhn, 2003, p. 127.

13 Nisbet, 2009, p. 118.

14 A key task of this book has been to deconstruct the secular origins account. Our first chapter, "Why worldviews matter" could have been titled "Understanding the biases behind origins accounts". While we believe that deconstructionism is applicable to the origins domain, it has limited relevance to empirical science. This latter view is consonant with the views of Krauss and Dawkins, quoted above.

15 This essay was reprinted as article 1.7 in Dawkins, 2004, p. 55.

16 This quote is from Sokal and Bricmont, as cited by Dawkins, 2004, p. 58.

17 Ibid.

in New Zealand, and the frankly appalling failure of the Royal Society of New Zealand to stand up for science".[18]

A pertinent question for the secular origins camp to answer is: Why has this materialised now? We saw above that Australia, as a bellwether for the West, has been de-Christianising; and recent censuses in New Zealand have also shown an increasing trend toward "no religion". In New Zealand's 2023 census, this category surpassed 50% for the first time.[19] Under these conditions, shouldn't the secular science community expect to see New Zealand becoming notably more rational than in the past, with its scientific community leading the way?

Krauss, Dawkins and a group of like-minded academics have lifted the lid on what they regard as anti-science developments in contemporary academia. This airing of dirty laundry comes in the form of a book, *The War on Science: Thirty-Nine Renowned Scientists and Scholars Speak Out About Current Threats to Free Speech, Open Inquiry, and the Scientific Process* (2025). The book is edited by Lawrence Krauss, who claims in the introduction that science is being attacked "in at least four major areas: the sociology of science, the infrastructure of science, academic freedom and free speech, and finally, the scientific and scholarly enterprise itself."[20]

Alas, the developments that Dawkins and Krauss find so disturbing are not limited to the academic sphere. Dawkins, for his part, has apparently been irked by the increasing fascination with horoscopes and astrology in contemporary society. A central feature of pagan Greek and Roman life, astrology sank into a lengthy nadir in the West as Christianity rose in the fourth century. Interest in astrology then began resurging in the late nineteenth century[21] to the position of prominence it holds in newspapers and magazines today. "I always thought that by the twenty-first century science and reason would have long since cleaned up, and yet every day of the week we're encouraged to retreat into the fog of the superstitious past," stated Dawkins in his *Enemies of Reason* program.[22]

Where is the rationality dividend? Why haven't science and reason "cleaned up" as Dawkins thought they would? Why do 25–30% of the British and U.S. populations[23] [24] believe in something which, according to Dawkins in the *Enemies of Reason* program, "denies scientific progress"?

Around 160 years have passed since the X Club membership began their covert machinations for what they thought would be the rational, secular state,

---

18 Dawkins, 2021.
19 Stats NZ, 2024.
20 Krauss (ed.), 2025, Introduction and Overview.
21 Garg, 2023.
22 Refer youtube.com/watch?v=dlMiKrwCRQ0 starting at 2:23. Accessed September 27, 2025.
23 Ibid., starting at 0:38. Accessed September 27, 2025.
24 Rotolo, 2025.

but now that it has arrived, we can see that it is not behaving in the way they might have predicted. In Dawkins' own words, we see displays of "daffy absurdity", "ludicrous moves", and "the fog of the superstitious past". Arguably, some of these examples reveal not only bizarre thinking, but the emergence of anti-science. There is a risk that neopagan science may be in the early stage of "petering out". What has happened?

In a nutshell: teach pagan origins, reap a pagan society.

Neopaganism: a big tent

In our view, Neo-Darwinism has served as a Trojan horse for neopaganism. Evolutionary theory itself might *appear* to many people to be rational and grounded in science,[25] but it underpins some of the irrational, inconsistent and polarising elements of neopaganism. We think Dawkins has been blindsided; he has been focused mainly on his particular battle in the realm of origins science and has only more recently noticed some of the ill-effects of the wider culture war. As Dawkins enters the autumn of his life and reflects on his long career, the excesses of the neopagan society around him could serve as a living rebuke for the evolutionary origins he has championed.

Dawkins himself espouses a number of views which may be considered consonant with neopaganism. Most importantly, he has promoted the pagan origins account, which provides the basis for the non-monotheistic worldview that other elements of neopaganism require. We view Dawkins' particular place in the neopagan tent as the *philosophic* corner. Congregating here are the scientific types who regard nature as intrinsically sacred and awe inspiring. Elsewhere, *civic* neopagans are present. These submit to a pagan-consonant morality, and some engage in activism and political lobbying to impose their worldview more broadly. Looking into another recess of the neopagan tent, we discern the *mythical* corner. This is where, in contemporary times, the New-Age types congregate. The crowd is large—where would an online news outlet be these days without its horoscopes?[26]

Despite Dawkins' apparent reservations about views held by some neopagans, the big tent of neopaganism is marked by its general attitude of inclusiveness toward those holding compatible religious perspectives. All these groups, which share a revulsion of Christianity and its transcendent truth claims, are bedfellows for better or worse.

To the extent that Dawkins is aware that evolutionary theory is consonant with neopaganism, he may have assumed his philosophic subtype would be dominant, given that it was Neo-Darwinism that ushered pagan origins into

---

25 Arguably, evolutionary origins is inconsistent and irrational.

26 Ancient Babylon, whose astronomy was based on the zodiac and was closely tied to astrology, is the origin of modern horoscopes.

schools. Decades ago, appeals to "science" carried the day. But that has changed. In contemporary public discourse, it is the civic and mythical neopagans who have proven to be profoundly influential.

Some may view Richard Dawkins and his ilk as having been exposed as out of their depth. Arguably, they have been engaged in a spiritual battle—which they would neither recognise nor agree with—and may struggle to help society find a way forward.

The confounding of the evolutionists represents a major turnaround in the origins debate. Readers will recall that "liberal" Christian Reverend Driver (refer Chapter 6) explained why the creationist-minded Christians of his day, like Philip Henry Gosse, were confounded: the claims of "science" contradicted what they believed was religious truth. But the tables have been turned; it is the evolutionists who are confounded today. Dawkins and his colleagues have been blindsided by the religious and moral perspectives of civic and mythical neopaganism overriding the neutral claims of "science". Whereas the issues that baffled Gosse have been resolved (or are in the process of being resolved) by creationist research, the issues facing Dawkins are of a different order. To continue pushing evolutionary "science", which may be Dawkins' instinctive reaction, will push society further in the direction of neopaganism, and risk further "absurdity".

In our view, this amounts to a Catch-22 scenario that may portend the decline of Dawkins' worldview in an intellectual sense. This state of affairs also serves as another test of a worldview, which complements the first five tests we investigated in Chapter 2.

The sixth criterion: the worldview values objective truth

This criterion examines the extent to which the worldview promotes and values objective truth in a way that enables the scientific enterprise to prosper. The preceding discussion should already give readers sufficient understanding of the failure of the secular worldview to satisfy this criterion. The fact that *The War on Science* has been written by a range of contemporary scientists, including from the origins domain, is evidence enough that the current academic environment, which has materialized on the watch of the secularists, does not pass this test.

An important corollary of this finding is that the legacy of the X Club lies in tatters. Their desire to free society from "theologies which are based on … the Book of Genesis",[27] has not delivered the outcomes they sought. We contend that the intellectual heirs of the X Club—the likes of Dawkins and Krauss—now find themselves defending against the excesses of the neopagan society they have helped usher in. In the Introduction we quoted Dawkins expression of

27 Huxley, 1893, p. 147.

concern that Christianity in the UK might be replaced by another religion. In our view this is precisely what has happened. Christianity has been replaced by the neopagan religious perspective. The publication of *The War on Science* indicates, arguably, that some academics are finding this state of affairs to be "dreadful".

On the other hand, biblical Christianity has been a great enabler of objective truth. In Chapter 6 we reviewed Loren Eiseley's observation that science owed its start "to the Christian conception of the nature of God" leading to the "sheer act of faith that the universe possessed order and could be interpreted by rational minds".[28] We can add that in view of the Christian God as the creator and sustainer of all things, there is only one set of truth, that is, God's truth. The pagan conception that allowed for the sycretization of foreign Gods into the pantheon is completely repudiated in the Old Testament, where any mixing with pagan religions was strictly forbidden. Truth was to be protected and preserved. In the New Testament, one of the most direct encounters between Christian and pagan conceptions of truth occurred when Christ told the Roman governer Pontius Pilate His purpose in coming into the world was "to testify to the truth" (John 18:37). Pilate replied with the cynical question: "What is truth?" (John 18:38).

Ultimately, it was the Christian emphasis on truth, strengthened during the Reformation, that enabled the scientific enterprise to continue progressing despite the machinations of the humanists.

The immediate outlook for neopagan origins

Intellectually, secular origins began fragmenting in the 1960s with Kimuran Neutral Theory and the Big Bang. This trend is now accelerating, with influential voices such as Thomas Nagel, The Third Way, and others openly turning their backs on Neo-Darwinian theory. Such fragmentation does not yet amount to a change in worldview, but rather an intellectual weakening of the status quo.

Neopagan origins will be increasingly on the defensive going forward. As pointed out in the Introduction, Richard Dawkins, in acknowledging developments in contemporary UK society, has admitted to being a cultural Christian. We suspect that many in the UK and the broader West will similarly look at the trajectory of the West and draw a conclusion: the more Christianity is marginalized, the more bizarre and fragmented society appears to become. With the neopagan framework now in place, responsibility for problems in society are going to be sheeted home to the incumbents, and many of these new problems have commenced and are continuing on the neopagan watch. The

28 Eiseley, 1958, p.62.

Epperson decision has not ushered in a more rational era, but arguably, a "war on science".

The outlook for young-earth creationism

Young-earth creationists today are in a much surer position than they were prior to publication of *The Genesis Flood* in 1961. Creationist materials are readily accessible, with the major ministries sharing resources freely online. It is easier than ever to subscribe to the periodicals of a creation ministry, to access a wealth of research, and for individuals to become involved in some way.

Creationism, like any other branch of science, can never provide all the answers. Hebrews 11:3 tells us that a complete origins account will never be decisively proven; some element of reasoned and reasonable faith will always be necessary. But the origins enterprise is primarily about progressing a workable system. Progress is marked by the development of new theories, the falsification of wrong ideas, and achieving a logical consensus around new data. Viewed through these metrics, the strength of the creationist position is unquestionably improving.

At this point in church history, there is a lack of consensus on how to harmonize Genesis and origins science. Theistic evolution and old-age interpretations continue to attract support. In this book, we have endeavoured to outline for the church at large the strengths of the creationist position. This began with our observation that secular origins is hobbled by the lack of a plausible infinity. In origins science, applying the dictum *etsi Deus non daretur*—"as if God does not exist"—forces secularists to invoke impersonal infinities in their origins scheme. A classic example is long ages. This serves as a quasi-infinity which is often marketed by the evocative moniker "deep time". While this kind of infinity is necessary for the secular camp, it is philosophically redundant for all those who believe in the God of the Bible, whose infinite qualities are adequate to explain all aspects of creation. And, if a long-ages interpretation is discarded, then all evolutionary origins schemes, which depend on long ages for their plausibility, also drop away.

We hope that these sorts of discussions, taking place within the church and its institutions, are undertaken in a spirit of humility, and that they enhance Christian unity.

Conclusion

We commenced this chapter by reviewing some controversies that Richard Dawkins and Lawrence Krauss have been personally involved in. However, we suspect that Dawkins and Krauss are emblematic of a large sector of society—everyday atheists. These people are likely to be evolutionists, and many of them will be more at home in the philosophic form of neopaganism than the other types. They may be dismayed at aspects of their society, be it the war on science, or the aggressiveness of the new ideologues. They have come to realise, belatedly (like our quote of Dawkins in the Introduction), that the Christian ethos is at once valuable and vulnerable.

What many of these atheists fail to realise is that they are the facilitators of this state of affairs. Darwinism was not the enabler of a rational era free from superstition and dogma, as the X Club and Richard Dawkins expected. It was the Trojan horse through which neopaganism conquered the West. Atheists are arguably part of the big tent of neopaganism, which, in its civic and mythical forms, comprises significant anti-science components. And, unlike Christianity, which is unifying, neopaganism seeks to be inclusive—to absorb opposing camps and viewpoints as long as they pay lip service to the overarching scheme and are antagonistic to biblical Christianity. These disparate viewpoints can lead to polarisation and deep divides of the sort that Dawkins and Krauss have experienced with their fellow neopagan-consonant academics. In fact, there is a risk that neopagan society itself will degenerate into an uncontrollable rabble.

*

In closing this volume, it is important to acknowledge the critical work undertaken by the creationist movement since the publication of *The Genesis Flood*. These relatively small but powerful ministries have made the case that a straightforward reading of Genesis is compatible with modern science. In our view as young-earth creationists, this amounts to a rebuilding of the intellectual foundations for Christianity in the West. (Indeed, we hope this book empowers young Christian men and women to enter any field of science, including origins science, without feeling pressured to abandon their faith or worldview.)

But in acknowledging these ministries, we must also be mindful that they themselves emerged from a foundation of even earlier Christians who held to a straightforward reading of Genesis when it was scientifically unfashionable to do so. During the period from the collapse of the scriptural geologists (1820s) through the beginnings of the modern creationist movement in the early 1960s, these Christians refused to be swayed by anti-Christian origins science despite the professional and personal costs this stand exacted. The legacy of these Christians of the past, and of the modern creationist ministries, is a developing understanding of the Genesis origins account that the authors hope will prove to be unifying in the church of the future.

References

Huxley, T. H. (1893). *Darwiniana Essays*. New York: D. Appleton and Company.

Dawkins, R. (2004). *A Devil's Chaplain: Selected essays by Richard Dawkins*. London: Weidenfeld & Nicolson.

Dawkins, R. (2021, December 4). Myths Do Not Belong in Science Classes: Letter to the Royal Society of New Zealand. Retrieved from richarddawkins.net/2021/12/myths-do-not-belong-in-science-classes-letter-to-the-royal-society-of-new-zealand/

Eiseley, L. (1958). *Darwin's Century: Evolution and the Men Who Discovered It*. New York: Doubleday

Garg, G. (2023). How Astrology Returned to Favor in the West. Retrieved September 27, 2025, from fairobserver.com/culture/book/birth-of-horoscopes-whats-your-sun-sign/

Hunter, L. (1993). *Screenwriting*. London: Robert Hale Limited.

Krauss, L. (Ed.) (2025) *The War on Science: Thirty-Nine Renowned Scientists and Scholars Speak Out About Current Threats to Free Speech, Open Inquiry, and the Scientific Process*. Krauss, Lawrence M. (ed.). The War on Science. Simon & Schuster, 2025.

Kuhn, T.S. (2003). *The Copernican Revolution: Planetary Astronomy in the Development of Western Thought*. Cambridge, Massachusetts: Harvard University Press.

Motion Pictures Association. (2023, December 23). MPAA Ratings at 50: Digital Archive. Retrieved from motionpictures.org/press/mpaa-ratings-at-50-digital-archive/

Nisbet, R. (2009). *History of the Idea of Progress*. New Brunswick: Transaction Publishers.

RDFRS (Richard Dawkins Foundation for Reason and Science). Teaching the Value of Science and Advancing Secularism. Retrieved August 30, 2025, from richarddawkins.net/

Rotolo, C. (2025). 30% of Americans Consult Astrology, Tarot Cards or Fortune Tellers. Retrieved September 27, 2025, from pewresearch.org/religion/2025/05/21/3-in-10-americans-consult-astrology-tarot-cards-or-fortune-tellers/

Sheridan, G. (2025). *How Christians can Succeed Today: Reclaiming the Genius of the Early Church*. Sydney: Allen & Unwin.

Stats NZ. (2024, October 3). 2023 Census population, dwelling, and housing highlights. Retrieved December 26, 2024, from stats.govt.nz/information-releases/2023-census-population-dwelling-and-housing-highlights/

The Motion Picture Production Code. (2023, December 29). Retrieved from productioncode.dhwritings.com/multipleframes_productioncode.php

The Movie Corpus, The TV Corpus. (2023, December 29). Retrieved from english-corpora.org/

Thyer, B. A. (1995). The X club and the secret ring: Lessons on how behavior analysis can take over psychology. *Behavior Analyst*, 18(1), 23–31.

Appendices

Appendix A
The problems of suffering and evil

The problem of suffering

The upside-down life of the Christian

In the New Testament, Christ leaves immensely challenging instructions for His disciples. Love your enemies. Pray for those who persecute you and spitefully use you. "Turn the other cheek" when dealing with an oppressive political apparatus.[1] There are many more such commands in the Sermon on the Mount (Matthew 5–7), which are in such opposition to what we find natural that they are, in fact, impossible in our own strength. Let's look at a couple of on-the-record examples of ordinary (in the sense of not being well-known) Christians as they follow Christ into the "upside-down" kingdom of true Christianity.

Loving your enemies

Damascus, 2013. The Arab Spring protests that roiled the Middle East a couple of years ago have petered out in most places, but here, in Syria, those portentous events spiralled into a bitter civil war. The main opposition force, the Free Syrian Army (FSA), is advancing on the capital, with the goal of removing President Bashar al-Assad from power. Al-Assad, from the Alawite minority, has become unpopular for a range of reasons, including his alleged deliberate stacking of important posts in the Syrian government with fellow Alawites rather than well-credentialled people from the Sunni majority. As the FSA battles its way into the capital, al-Assad's forces counterattack with all the force they can muster, in a desperate attempt to defend the regime. The Damascenes, already used to military checkpoints, are now exposed to gunfire as the civil war encroaches further into their city.

Just after midnight, a battered car comes to a stop in a northern suburb of the city.[2] The car is pock-marked with bullet holes, having recently driven

---

1 We do not interpret "turn the other cheek" as a denial of the right of Christians to self-defence or to limited expressions of civil disobedience. The Bible approves the principle of self-defence, governments protecting their populations using "the sword", and Christians practicing civil disobedience when they must in order to "obey God rather than man". Typically, the latter relates to preaching the Gospel or distributing the Bible.

2 This narrative is paraphrased from *Killing Christians*, by Tom Doyle, W Publishing Group, an imprint of Thomas Nelson, Nashville, 2015 (pp. 27–28 and 30–31).

through a gunfight between Alawite and al-Qaeda forces. Three men alight and, without uttering a word, enter a darkened house nearby. Inside the dwelling, they maintain complete silence as they descend a stairwell to the basement. Then, opening the basement door, they take in an unlikely sight.

The room is filled with 30 or more people, all engaged in fervent worship. Hands are raised as impassioned voices rise together in song. They are Christians; some from a Sunni background, others Alawite. Outside, these same people groups are killing each other mercilessly, but here, the love of Christ has dissolved such enmities. The atmosphere of love in the room is clearly evident—and the whole scene is entirely incongruent with the civil war raging above.

As the music ebbs away, the Syrian Christians bring out bowls and towels for their next activity: a foot-washing session. They are following the example of Christ, who, in washing the feet of His disciples, demonstrated an attitude of servanthood. As we watch this incredible sight, we ask ourselves: why is this meeting taking place in the dead of night, in the soundproof basement of a darkened house? The answer is provided by a speaker who takes the floor after the foot washing comes to an end. "He has called us out," says the speaker, referring to Jesus Christ. "Out of the world, out of this war, out of our families to stand for Him in Syria!"[3]

We lean in and listen carefully to what the man says next: "We will die for this. The Jesus message is upside down compared to the ethnic and religious hatred that's destroying our country."[4]

In one respect, these words are surprising: why would citizens that are interacting peacefully with each other incur the death penalty? The answer is that in demonstrating such love, the members of the group have effectively identified themselves as Christians (John 13:35) and are viewed as traitorous members of their communities.

In another respect, the scene we have here is truly amazing. For all the UN sponsored peace talks and other high-level negotiations that pervaded the Syrian civil war, no man-originated program or scheme was able to deliver the kind of love between warring communities that the Christian faith delivered.

Notice also how the speaker phrases it: "The Jesus message is upside down." This is an important observation. Christianity teaches us that reality often runs counter to what we would expect. Here in Damascus, enemies are turned to friends. People from opposing groups that we would expect to despise each other have instead become each other's servants.

3 Doyle, 2015, p. 31.

4 Ibid.

Feeding the hungry

Our second example is from North Korean escapee, and more recently, TED Talk speaker[5] Joseph Kim, who recounts his tragic yet heartwarming story in *Under the Same Sky*.[6] Growing up in North Korea in the wake of the famine of the late 1990s, the teenaged Kim, largely bereft of family, was forced to look out for his own survival. He fell in with a succession of knockabout types, who, like Kim, did what they must to live another day.

While spending a summer on a plot of land in the mountains, Kim made a new friend: an ex-convict who had done time for escaping to China. Such an escape was possible because, in wintertime, the Tumen River that served as border between the two countries froze over. To those desperate enough to try their luck in China, the ice-covered Tumen River offered a narrow, if slippery path to the opposite bank—as long as the North Korean guards stationed nearby didn't put a bullet into your back.

The ex-convict regaled Kim with his exploits in China. Kim (who was then known as Kwang Jin), was amazed at the stories of the wealth and food available in China, but one interesting detail stuck in his memory:

> "Kwang Jin, if you ever go to China, the churches will give you money."
>
> ...
>
> "Why would these church people just give you money?"
>
> ...
>
> The ex-convict tried to explain, but I could see even he didn't understand what he was saying. ... What he was describing—people freely giving their hard-earned cash to complete strangers—was plain crazy.[7]

Not long after that, Kwang Jin found himself wandering aimlessly near the Tumen River. He made a spur-of-the-moment decision to cross into China. Taking his life in his hands, he madly dashed across the ice in broad daylight, with his heart thumping, but encountering no trouble. He'd made it.

After days of wandering, a kindly woman urged him to go to a Christian church, telling him that churches were identified by their crosses. He soon found one and knocked on the door. A man came to the door and eyed him dubiously, but handed over a package of food and told him to come back the next day.

> [S]oon after the sun rose, I was back at the church's door. This time, the man gave me twenty yuan, the equivalent of about three dollars.

---

5 Kim, 2013.

6 Kim, 2016.

7 Kim, 2016, pp. 186–187.

So it's true, I thought. Christians are real.[8]

Kim concluded that all Christians were wealthy, so he traveled about, hitting up all the churches in the area for as much money as he could. But one church stood out. It was the only one where he had "found real kindness and understanding."[9] Entering the church for the first time, he had told his story to a middle-aged woman, who turned out to be the pastor's wife. "Listening kindly, she handed me fifty yuan, just like that. I was overjoyed."[10]

But a few days later, in that same church, he learned something unexpected when he overheard church members talking about the pastor's toothache and the fact he could not afford dental care.

> I felt as if something—a sharp sword—had cut through me. The pastor's family was poor! And he was suffering from health problems. Yet his wife had given me fifty yuan as if it were nothing. I became curious about who these people were and what made them so different from the men and women who turned me away.[11]

The difference, of course, is Christ. Christ the Lord, commanding His people to live this "upside down" life of godly love for all our neighbours. Do these neighbours hate and despise us? Love them, says Christ. Do they turn up in the guise of hungry strangers at our door? Love them—but not just with words of comfort. Christ requires us to love in deed and in truth (1 John 3:18), and this is precisely what the Christians did when Kim came knocking. They handed out free cash, food, and accommodation to an unknown teenager simply because Christ requires that sort of love from His disciples.

The inversion of suffering

For those that take on this upside-down life, Christ does something amazing in return. He reciprocates in a way only He can—by ensuring that everything in our lives, whether circumstances, health or relationships—and the suffering that can be associated with them—work in concert for our own good (Romans 8:28).

This is a gamechanger. Even where the circumstances of the Christian's life bring suffering, God has promised to work these circumstances for the good of the Christian. Consequently, Christians have a completely different outlook on suffering to secular humanists. Whereas secular humanists may view suffering in negative or pragmatic terms, Christians, in discharging their responsibility to obey the "upside down" commands of Christ, are promised in return the privilege of having Christ ensure their suffering works for their ultimate good.

---

8 Kim, 2016, p. 223.
9 Ibid., p. 226.
10 Ibid., p. 225.
11 Ibid., pp. 226–227.

This is not to say that suffering is enjoyable. But Christians recognize it as discipline from God, and therefore as serving a positive purpose, as Hebrews 12:5–11 declares.

> And you have forgotten the exhortation that addresses you as sons:
>
> My son, do not take the Lord's discipline lightly
> nor lose heart when you are reproved by him,
> for Lord disciplines the one he loves
> and punishes every son he receives.
>
> Endure suffering as discipline: God is dealing with you as sons. For what son is there that a father does not discipline? But if you are without discipline—which all receive—then you are illegitimate children and not sons. Furthermore, we had human fathers discipline us, and we respected them. Shouldn't we submit even more to the Father of spirits and live? For they disciplined us for a short time based on what seemed good to them, but he does it for our benefit, so that we can share his holiness. No discipline seems enjoyable at the time, but painful. Later on, however, it yields the peaceful fruit of righteousness to those who have been trained by it (Hebrews 12:5–11).

Romans 5 lists the positive effects that suffering provides:

> And not only that, but we also rejoice in our afflictions, because we know that affliction produces endurance, endurance produces proven character, and proven character produces hope. This hope will not disappoint us, because God's love has been poured out in our hearts through the Holy Spirit who was given to us (Romans 5:3–5).

The values of Christianity run opposite to those of the world because the world system runs on pride, whereas Christians are called to live humbly before God. Hence, the Syrian Christians did not fight with each other but served each other in a spirit of humility. Similarly, the Chinese Christians did not turn away the destitute Kim but served him in that same spirit.

We can go one step further. Not only does God respond to the humble by working circumstances in their favor, but he also actively resists the proud. (Pride was the original sin of the devil (Ezekiel 28:17).) One of the key truths in the Bible is expressed in 1 Peter 5:5b—"God resists the proud but gives grace to the humble." The first part of this verse is also represented by the proverb, "Pride comes before destruction" (Proverbs 16:18a).

In Table A.1, we list several types of pride that are found in the Bible, and their result. Table A.2 lists the issues that the humble may face, and how God assists them.

Table A.1: God's judgement on pride

Type of Pride	Reference	God's judgment
Military	2 Kings 18–19	Sennacherib's army lost its leaders and Sennacherib was killed by his sons.
Political	Daniel 5:18–20	Nebuchadnezzar's kingdom was taken from him as he succumbed to temporary madness.
Intellectual	Isaiah 44:25	God will turn wise men backward and make their knowledge foolishness.
Wealth/possessions	Luke 12:16–21	A wealthy man lost his life after thinking he had laid up wealth for many years.
Spiritual—Satan	Ezekiel 28:17	A powerful angel (henceforth known as Satan) became proud and was cast out of heaven.
Spiritual—man	Acts 12:20–23	Herod was struck by an angel and died after accepting the praise of men rather than giving glory to God.
Social	Matthew 23:5–7	The Pharisees and scribes loved the best seats at feasts and at the synagogue, and loved to be greeted in the marketplaces, an attitude which Jesus condemned.
Physical characteristics	Psalm 147:10 1 Samuel 16:7	The Lord takes no pleasure in the legs of a man; man sees the outward appearance, but the Lord sees the heart.
Future plans	James 4:13–16	God condemns arrogant boasting about future plans as evil.

Table A.2: How God assists the humble

Issue faced by the humble	Reference	God's actions
Foolishness	1 Corinthians 1:27	The wise put to shame by the foolish.
Weakness	"	The strong put to shame by the weak.
Lowliness	Luke 1:52	Exalted, even as the mighty are put down from their thrones.
Hunger	Luke 1:53	Filled with good things, even as the rich are sent away empty.
Hard pressed	2 Corinthians 4:8	Hard pressed, but not crushed
Perplexed	"	But not in despair
Persecuted	2 Corinthians 4:9	But not forsaken
Struck down	"	But not destroyed
Dying	2 Corinthians 6:9	And behold we live
Chastened	"	And yet not killed
Sorrowful	2 Corinthians 6:10	Yet always rejoicing
Poor	"	Yet making many rich
Having nothing	"	And yet possessing all things.

God has ordained His creation such that it is stacked against the proud and favourably disposed toward the humble. Within this paradigm, everyone operates with agency and takes moral responsibility for their actions, but somehow, mysteriously, God configures events such that, in the long run, the proud are resisted and the humble given grace. In this paradigm, the web of intellectual futility entrapping Classical thought, which Augustine identified (refer Chapter 6), is merely one instance—the intellectual instance—of a much broader web, or moral order—whose owner and operator is God.

God calls on prideful people to humble themselves under His mighty hand (1 Peter 5:6). The irony for those that don't is that they will potentially be subject to humiliation. We have already seen the sad example of Sennacherib; another notable Bible character who was humiliated was Haman, a high-ranking official in the Persian empire in the fifth century BC, whose downfall is described in the book of Esther. Let's consider a contemporary example of the humiliation of those who flout God's will.

Case study: the fall of eugenics

In Chapter 7, we discussed the eugenics movement that commenced with Francis Galton's investigations in the 1880s. The main presupposition underpinning this movement was that the universe was mechanistic, rather than moral. From there, the eugenicists held that governments could introduce policies aimed at limiting the reproductive rights of certain kinds of people. This, they believed, would bring about an overall lift in the quality of humanity in their nations. As we have already seen, this flies in the face of biblical teaching: a society achieves blessing by turning back to God. The universe is moral, not purely mechanistic.

By the early 1900s, several eugenics societies were publicly advocating these policies, with the movement hitting its stride in the 1920s and 30s. And then, world war broke out. The Nazis, it was later discovered, had been practising eugenics themselves. In fact, it was the Nazis who demonstrated most fully the true horror of what a society run along eugenicist lines was like. Unfettered by traditional moral considerations, the regime was able to target groups it considered "unhygienic". The results were catastrophic. The eugenics movement lost credibility overnight.

As we read documents produced decades ago by the eugenics movement, we do so in the knowledge that all the policies they deemed so urgent would be swept into the dustbin of history in just a few short years—a sobering reversal for the movement's advocates.

Persecution of Christians in God's moral order

Persecutors of Christianity aim to restrict the spread of the faith or destroy it altogether, but these activities often backfire. Indeed, the church often shines brightest under persecution, whereas a more liberal environment can result in complacency.

As a brief case study, we will consider the persecution of Christians in an African context—the Ethiopian communist regime—referred to as the "Derg", meaning "committee of equals"—which ruled from 1974 to 1991.

The Derg was a communist military junta that gained power following a coup d'etat against the Ethiopian emperor, Haile Selassie. The regime moved quickly to leave its ideological imprimatur on Ethiopian society, assisted by thousands of "advisors" from the Soviet bloc. Compulsory education in socialist ideology, along with "People's Courts" that handed out harsh penalties to those suspected of opposing the regime, became the order of the day. And, as with any other society organised along Marxist lines, Christians were soon targeted. Former missionary to Ethiopia Richard McLellan, who lived through the so-called "Red Terror" instituted by the Derg, noted that "Christian leaders were killed and thousands imprisoned, some of them for years, because they refused to raise a clenched fist and yell out the Communist slogans".[12]

However, amid this upheaval, the Christian faith shone. Although the Derg shuttered churches and confiscated Bibles with fanatical intent, Christians continued to meet, albeit in their private homes, and often took the risk of inviting non-Christian friends and neighbours. The church's witness to Christ continued underground.

And the church flourished. McLellan notes that "in this way, the numbers of the believers doubled during the seventeen years of the Derg—and then doubled again!".[13]

Similar themes have been noted in communist China. Following the communist takeover in 1949, and the expulsion of foreign missionaries, the church has grown anyway—by some accounts now numbering approximately 100 million people.

Of course, the persecution brought about by these regimes is real and those that live through it suffer horribly. But the gospel always advances and endures. By the time the Derg was swept away in 1991 with the collapse of its Soviet support base, Ethiopian Christians were more numerous than when the Derg had first gained power.

Restoration of lost years

People that flout God's order will find themselves on the receiving end of God's moral order. However, if those same people repent, the mysterious ways of God are somehow able to restore blessings from the lost years. In a remarkable promise in Joel 2:25, God promises the ancient Israelites that if they turn to Him with all their heart (Joel 2:12), He "will restore to [them] the years that the swarming locust has eaten." Nineteenth century British pastor C. H.

12 McLellan, 2006, p. 71.
13 Ibid., p. 78.

Spurgeon (1834–1892) preached on this text; we quote the opening section of his 1886 sermon below:

> L[ost] years can never be restored literally. Time once past is gone for ever. Let no man make any mistake about this, or trifle with the present moment under any notion that the flying hour will ever wing its way back to him … . It will strike you at once that the locusts did not eat the years: the locusts ate the fruits of the years' labour, the harvests of the fields; so that the meaning of the restoration of the years must be the restoration of those fruits and of those harvests which the locusts consumed. You cannot have back your time; but there is a strange and wonderful way in which God can give back to you the wasted blessings, the unripened fruits of years over which you mourned. The fruits of wasted years may yet be yours. It is a pity that they should have been locust-eaten by your folly and negligence; but if they have been so, be not hopeless concerning them. "All things are possible to him that believeth." There is a power which is beyond all things, and can work great marvels. Who can make the all-devouring locust restore his prey? No man, by wisdom or power, can recover what has been utterly destroyed. God alone can do for you what seems impossible; and here is the promise of his grace: "I will restore to you the years that the locust hath eaten."[14]

To those that follow God, even the fruits from the lost years of their lives can be restored. Indeed, for those who have wasted years behind them, becoming a Christian allows them to claim that promise for themselves—not for their own direct benefit—but instead for the glory of Christ and the extension of His kingdom.

The problem of evil

Secular humanists have long suggested that Christianity is internally incoherent because of "the problem of evil". Atheist George H. Smith explains the problem as follows:

> If God does not know there is evil, he is not omniscient. If God knows there is evil but cannot prevent it, he is not omnipotent. If God knows there is evil and can prevent it but desires not to, he is not omnibenevolent. *If, as the Christian claims, God is all-knowing and all-powerful, we must conclude that God is not all-good.* The existence of evil in the universe excludes this possibility.[15] (Emphasis added)

Hence, for atheists, the notion of the Christian God is incoherent. They view God as a human construct which does not pass the coherence test we outlined in Chapter 2.

14 Spurgeon, 2023.
15 Smith, 2016, chapter 3.

But let's unpack the third sentence (italicized) which is the crux of the argument. It amounts to the proposition that "if God can prevent evil, but chooses not to, he is not good". In the Christian worldview, we can modify this to a more readily answerable question "Since God *has* allowed evil to exist up until the end of earthly history, for what compelling reason has He done so?" The Bible provides an answer in both the Old and New Testaments.

Evil allows goodness to be understood

There are several questions to address in answering this. What is evil? Where does evil come from? What is the relationship between evil and the biblical concept of sin? As Creator, is God morally responsible for sin? What purpose does evil serve in God's creation?

An abbreviated answer to the first question notes that philosophers and theologians typically distinguish at least two types of evil: moral and physical. Physical evil is said to arise from natural disasters such as earthquakes or diseases. Moral evil results from human actions and decisions, such as theft and murder.

The Bible is clear in its answer to the second question. A sinful act by Adam and Eve in the Garden of Eden, at the instigation of the devil, led to evil in the world. In the Genesis account, eating the forbidden fruit was the original sin of mankind, resulting in various curses from God, including death. God made clear to Adam that breaking this commandment would be disastrous. At the level of the individual, the origin of sin is psychological. Christ taught that sinful thoughts are upstream of sinful actions, and indulging sinful thoughts is sinful (Matthew 5:21–22, 27–28 and 15:18–19).

To the third question, the Bible makes it clear that both physical and moral evil stem from the fall of Adam. All of us—except Christ—share this legacy by virtue of our position in corporate man (Adam). Further, Adam's position of authority over the earth and living creatures (Genesis 1:28) meant that God's curse on Adam (Genesis 3:17–19) also applied to the earth and living creatures, and specifically included death for sentient animals. At the level of the individual, sinful acts inevitably result in evil.

Is God morally responsible for mankind's sin? God did not *create* sin. Sin begins in thought and is manifested by action, both of which are behaviors, rather than objects in themselves. The moral agent who harbors sinful thoughts or carries out sinful behaviors is the sinner. Recalling that the creation was very good when God completed it (Genesis 1:31) we deduce that if anything, God created angels and humans with a bias away from sinful behavior. When angels sinned, they did so entirely as individuals (angels have no corporate nature), but when Adam sinned, he did so both individually and as the corporate representative of humanity. God was not morally responsible for the sin of angels nor mankind.

To the final question: what purpose does evil serve in God's creation? Both the Old and New Testaments provide answers to this question, underscoring the importance of the issue.

The example from the Old Testament is given in the book of Job.[16] God allows Job to suffer evil; Job maintains his integrity through terrible trials and suffering, which indicates to both God and Satan that Job's goodness is not based on external circumstances. This is a highly significant point. Early in the book of Job, Satan suggests that a man will only show "goodness" if there is a payoff—a comfortable life, good family circumstances, material wealth and so on. Satan raises an important question. Is goodness only good because it has no evil to contend with? Is goodness just a charade that people maintain to increase their chances of success in life? Is the basis of human goodness nothing more than quid pro quo?

In the book of Job this question is answered. God allows Satan to inflict a sequence of physical evils that destroy Job's family, his prosperity and his health. Neither his wife nor his friends offer any useful comfort or counsel to him. Job did not suffer these physical evils because of personal failings or sin, but because he was a part of corporate humanity.[17] In holding on to his integrity (Job 2:9), Job shows the essential character of goodness. This would not have been possible in the absence of evil. (The foundation of Job's goodness is shown to be an immovable faith in God,[18] Job's faith in his Redeemer and in the bodily resurrection of the dead.[19])

The same question Satan asked about goodness could also be asked about other attributes that God desires man to reflect—love, mercy, kindness, courage, graciousness, holiness, faithfulness, humility and righteousness. Without the existence of evil in history, it would be impossible for God to effectively communicate the most important parts of his character to mankind. We see this as the compelling reason for God to allow evil to exist during earthly history.

Turning to the New Testament book of Romans, Satan's question from the book of Job is again answered, but more directly:

> Or has the potter no right over the clay, to make from the same lump one piece of pottery for honor and another for dishonor? And what if God, willing to display his wrath and to make his power known, endured with much patience objects of wrath prepared for destruction? And what if he

---

16 Many Christian scholars regard the Book of Job as among the oldest of the Old Testament books, further indicating the importance of the question of evil.

17 An important corollary of this is that a person who has suffered severe misfortune cannot be assumed to be wicked. Luke 13:1–5 records Jesus addressing this very concept.

18 Job: "Even if [God] ... kills me, I will hope in him" (Job 13:15).

19 Job: "But I know that my Redeemer lives, and at the end he will stand on the dust. Even after my skin has been destroyed, yet I will see God in my flesh. I will see him myself; my eyes will look at him, and not as a stranger. My heart longs within me" (Job 19:25–27).

> did this to make known the riches of his glory on objects of mercy that he prepared beforehand for glory—on us, the ones he also called, not only from the Jews but also from the Gentiles? (Romans 9:21–24).

Paul teaches clearly that God the Father sovereignly ordained evil to exist during earthly history to make known "the riches of His glory" to "objects of mercy".

An additional response: God respects personal promises

There is another aspect to why evil exists in the world.

Hypothetically, God could have killed Adam and Eve immediately after He convicted them of eating the forbidden fruit. He then could have created new, perfect humans as replacements, solving the problem of evil immediately. However, God did not do this. Why not? Returning to Genesis, we observe God blessing mankind in Genesis 1:28: "God blessed them, and God said to them, 'Be fruitful and multiply, fill the earth, and subdue it. Have dominion over the fish of the sea, over the birds of the sky, and over every living thing that moves on the earth.'"

This blessing was made to Adam and Eve both as representatives of corporate humanity and to them as individuals. God held to this promise even after Adam and Eve sinned; the blessing remained for them and their offspring. God did not destroy the line of Adam and Eve.

Hence, counterintuitively, the presence of sin and suffering in the world speaks to God's trustworthiness in personal relations. God does not love "mankind" in the abstract; God loved Adam and Eve, just as he loves *us* and He loves *you*.

Accountability for evil

A final but important aspect to the problem of evil concerns moral culpability. (This involves predestination,[20] a doctrine about the sovereignty of God the Father over creation which implies that Christians who demonstrate their love of Christ by seeking to be obedient to Him can be assured of their salvation. We espouse a moderate, biblically based stance on predestination,[21] in line with many evangelicals.)

---

20 The New Testament contains references to God's predestination. Refer Romans 8:29–30; Ephesians 1:4–5, 11; 2 Timothy 1:9.

21 Most evangelical Christians accept the doctrine of predestination because it is biblical but do so in a moderate and balanced way. Evangelicals tend to avoid extreme views which overemphasize the doctrine of predestination to the point of undercutting other important biblical doctrines. Hence evangelical Christians also emphasize:

- God desires the salvation of all humans (Ezekiel 18:23, 33:11; 1 Timothy 2:3–4; 2 Peter 3:9).
- The way of salvation is open to all people (John 3:16; Romans 10:9–13; Acts 2:21; 1 Timothy 2:3–6; Titus 2:11; Revelation 22:17).
- Humans have responsibility and agency in making choices (Deuteronomy 30:19–20; Joshua 24:15; Proverbs 16:9; Ezekiel 18:30–32; Matthew 7:13–14; Romans 14:12; Galatians 6:7–8; James 1:14–15; Revelation 3:20).

If God predestined Adam to sin, why are Adam and his descendants held accountable for sin?

The answer to this question is found in the trinitarian nature of God. Note that the atheistic dilemma posed by Smith, quoted earlier in this chapter, references some attributes—but not the trinitarian nature—of the Christian God. Smith's criticism therefore applies most directly to unitarian conceptions of God.

If God were a *unitarian* god and predestined an individual to sin, it would not logically follow that he could also act as judge. A god acting in this way would be unjust.

However, God is a Trinity, with each Person of the Godhead having different roles. The Father predestines, but it is the Son who will judge. Christ has several characteristics as judge: He is human,[22] therefore our peer; as part of corporate mankind, He suffered from evil; His life and death were predestined—the course of His life was laid out in numerous Old Testament prophecies; He was tempted in all the ways we are but was without sin; His resurrection speaks to His sinlessness and thus His qualification to be a judge.

Christ judges, but He does not predestine and does not determine the time of the end of the world (implying Christ does not determine when the judgement is to occur) (Matthew 24:36, Mark 13:32). Hence, all humans can and will be held accountable for their words and actions by a Peer, and judged accordingly at the Last Judgement, even though their personhood and the course of their life is, from the perspective of God the Father, predestined.

The Bible mentions several principles concerning the conduct of the Final Judgement. For example, Matthew 7:1–2 states:

> Do not judge, or you too will be judged. For in the same way you judge others, you will be judged, and with the measure you use, it will be measured to you.

Individuals will be judged according to the standard which they themselves apply to others. This may have both an individual and corporate aspect relating to the nature of mankind.

We leave it up to those who promote strictly unitarian concepts of God to explain how they reconcile the following considerations with God's character:

- If a Final Judgement is not accepted, the justice of God is impugned.
- If predestination is not accepted, the omniscience and power of God over creation is denied.

- Individuals are saved "by grace" (the loving act of God to make available the gift of salvation), "through faith" (individual acceptance of the gift with gratefulness and reciprocated love) (Ephesians 2:8–9). Both grace and faith are necessary to a restored relationship with God.
- The trinitarian notion of relationship is more relevant to predestination than endless debates pitting determinism against free will.

22 Christ is fully human, but also fully Divine.

- If a unitarian God both predestines and judges, how is this just?

Conclusion

Everyone, whether they acknowledge it or not, lives their lives in God's moral order. The proud are resisted by God; their suffering will largely be meaningless, and their thinking will be turned to foolishness. As they resist God, He will turn the tables on them and ensure that *their actions further His own glory*. To use metaphorical language, it is as if every fist shaken against God loops another stitch in the victory tapestry that adorns the walls of heaven.

On the other hand, God gives grace to those who live humbly before Him. The most foolish among the humble will put to shame the wise of the world. Their suffering will be turned to blessing, and even the fruits of their lost years can be restored. God works everything for their ultimate good and promises never to leave them nor forsake them.

For the Christian, the Bible is coherent: it explains the world around us, including suffering and evil. Christians are not immune from suffering and should even expect it. But God promises Christians that our suffering is never meaningless (Romans 8:28).

Appendix B
Radiometric dating

The RATE project

An issue which has troubled Christians and caused some to depart from the faith is the discrepancy between the biblical timeline and the ages of the earth and universe claimed by secular science. In view of the importance of this issue, a group of creationist scholars commenced an eight-year research project called Radioisotopes and the Age of The Earth (the RATE project). The results were published in 2005.[1] The key investigative aim of the RATE group was to understand radiometric dating in detail and assess its reliability and dependability. Major contributors included three physicists, a hard and a soft rock geologist, a geophysicist, an atmospheric scientist, and a scholar of the Hebrew language,[2] all of whom had been awarded PhDs.

The RATE project was funded by Christian individuals and organizations to the tune of $1.25 million USD. These contributions were used to cover direct costs of laboratory analysis ($250,000), and wages and travel expenses. A scientific literature review over three years culminated in a book outlining a five-year plan for future externally peer-reviewed research. However, in the event, lack of funding meant that the RATE project concluded in 2005 with two books having been published. Volume 1 contained the literature review, and Volume 2 contained the results and analysis. In the discussion that follows we will explain the rationale behind radiometric dating, then present some of the RATE group findings.

Inferring from point-in-time observations

Radiometric dating is sometimes explained by analogy to an hourglass. Imagine coming across an hourglass resting on the ground with sand flowing through the narrow point. Suppose at that point in time both the upper and lower chambers each contain about half of the sand. A person might argue that about 30 minutes has elapsed since the hourglass was upended. But without

---

1 Vardiman, Snelling, & Chaffin, 2005.

2 Larry Vardiman (atmospheric scientist), Steven A. Austin (geologist), John Baumgardner (geophysicist), Steven W. Boyd (Hebraist), Eugene F. Chaffin (physicist), Donald B. DeYoung (physicist), D. Russell Humphreys (physicist), Andrew A. Snelling (geologist).

witnessing the placement of the hourglass, the time frame is not certain—it can only be inferred from the evidence, rather than directly measured. This is analogous to the secular humanist reasoning underlying radiometric dating: scientists observe phenomena after the event (whether this be the beginning of the universe, the formation of planet Earth, or some other past phenomenon), and seek to infer an age based on the evidence at hand.

In the hourglass example, the timeframe is uncertain because a variety of scenarios may fit the evidence. Someone might have upended the hourglass one minute before the observation, when sand was already evenly divided between the chambers. Right away this gives us two possible scenarios: the first, where the upending occurred 30 minutes ago, and now this second one in which it occurred one minute ago. In fact, there are many other possible scenarios which could also explain the evidence, as long as the starting distribution of the sand is made to match each specific timeframe.

If the sand had become clumped, the narrow part of the hourglass might have become blocked for a period, before it was bumped by a third party, re-establishing flow. In this explanation, hours may have elapsed since the inversion of the hourglass. The point of this discussion is that a point-in-time observation of an hourglass does not give certainty about the elapsed time since the hourglass was flipped, unless corroborated by eyewitness or other supporting evidence. To assert a definite timeframe of 30 minutes is to rely on several unstated assumptions (presuppositions), some of which may be unprovable. The presuppositions include:

1. The upper chamber was full at the time of inversion.
2. The lower chamber was empty at the time of inversion.
3. The flow through the narrowing was known at the time of observation and can be reliably calculated for all other times.
4. No sand has been removed from the upper chamber since inversion except by passage through the narrowing. No sand has been added to the upper chamber since inversion.
5. No sand has been added to the lower chamber since inversion except through the narrowing. No sand has been removed from the lower chamber since inversion.

Although the list of requirements might appear pedantic, each of these five types of assumption are required for our simple example of the hourglass to work. They are also necessary for radiometric dating—which is acknowledged by experts in the field.

As an extra check, the hourglass should have been validated by comparing it with another "gold-standard" timepiece to confirm that it is not subject to intrinsic inaccuracy or unexpected variation.

How radiometric dating works

Many rocks contain a variety of chemicals, including radioactive isotopes. Radioactive isotopes are unstable forms of chemical elements that release radiation as they break down into more stable elements. (For example, uranium-238 can break down into lead, which is more stable). The word isotope refers to the fact that in the atomic nucleus, isotopes have *the same number of protons,* which gives them identical chemical properties. However, isotopes differ in that they have different numbers of neutrons. For example, isotopes of uranium all have 92 protons in their nucleus. But uranium-238 has 146 neutrons, uranium-235 has 143 neutrons and uranium-234 has 142 neutrons, giving each a distinct atomic mass. Other isotopes exist in nature. For example, lead comes in four different isotopes lead-204, lead-206, lead-207 and lead-208. Lead-204 is unique among the lead isotopes because it does not result from any known spontaneous radioactive transmutation, but it only accounts for about 1.4% of natural lead. Therefore, more than 98% of lead on earth appears to result from various radioactive transmutations. Lead-206, for example, comes from the spontaneous radioactive transmutation of uranium-238; here, uranium-238 eventually transmutes into lead-206, releasing helium nuclei and beta particles in the process—these are alpha and beta radiation, respectively. (In this example, the lead-206 is known as the "daughter isotope" as it derives from uranium, the "parent".) In radiometric dating, the "upper chamber of the hourglass" contains a radioactive isotope of a particular type, for example the uranium-238 isotope. The narrowing of the hourglass represents the spontaneous radioactive transmutation of the upper chamber isotope into the "lower chamber" isotope, which in this case is the lead-206 isotope. The relative amounts of uranium-238 and lead-206 (amounts in the "upper and lower chambers") indicate how much transmutation has occurred. Using the assumptions above, a time can be inferred based on the rate at which the transmutation occurs. Rates of radioactive transmutation are experimentally known to decline exponentially over time; this rate is usefully quantified by what is referred to as the "half-life". For example, the half-life of uranium-238 is almost 4.5 billion years. So, after 4.5 billion years, half the uranium-238 atoms in a sample will have transmuted to lead-206. After another 4.5 billion years, half of the remaining uranium-238 will also transmute to lead-206, leaving a quarter of the original amount after 9 billion years (2 half-lives); three quarters will have transmuted to lead-206. This type of analysis works in reverse: if there is three times as much lead-206 in a sample as uranium-238, and our stated assumptions above hold, then we can infer 9 billion years has elapsed. In a nutshell, this is the rationale behind radiometric dating.

Some rocks are composed of materials which already contain some daughter isotope. This leads to a problem of determining the amount of material in the "lower chamber" (daughter isotope) when the rock is first formed. Without knowing this, the age of a rock cannot be inferred using radiometric dating. Scientists developed the *isochron* technique to try to bypass this problem. This uses isotope ratios in such a way as to avoid having to assume an initial amount of daughter isotope. Indeed, the use of isochrons theoretically enables the initial amount of daughter isotope in the sample to be inferred. We won't go into further detail here, except to note in passing that the fact that the isochron method has been developed indicates the initial amount of daughter isotope in a sample matters (refer item (2) in the list above).

RATE group findings

The RATE group analyzed the scientific literature and confirmed that there is strong evidence that substantial amounts of radioactive transmutation has occurred in geological samples of the earth's crust. There were several lines of evidence, including the presence of transmuted isotopes present with the parent isotope, and also physical evidence such as microscopic radiohalos; fission tracks in rock consistent with radioactivity; and the presence of helium in rock which is a known product of a certain type of transmutation process (alpha decay in which helium nuclei are ejected from the parent isotope). This confirmed the importance of investigating this topic. Substantial isotopic transmutation in geological specimens is proven.

The RATE group concluded, based on several lines of technical evidence, that the use of radiometric dating to assess absolute ages of geological specimens is unreliable from a scientific and biblical perspective. What this means is that while radiometric techniques can be used, for example, to rate specimen A as older than specimen B (relative ages), they cannot be reliably used to estimate the absolute age of specimen A (for example, 12 million years old).

A major finding of the RATE project was that transmutation of radioactive isotopes appeared to have been accelerated during brief periods of earth's history.[3] This result was deduced from the interpretation of scientific evidence in light of a straightforward reading of Genesis. The RATE group concluded that most of the transmutation occurred during the events of the creation week, but a significant amount also appeared to have occurred during the Genesis flood.[4] Their investigations led them to reject the explanation that daughter isotopes were created ready-formed in significant amounts during the creation week—instead, daughter isotopes were formed during the creation week via accelerated transmutation from parent isotopes.[5] There is currently no

3 Vardiman, Snelling, & Chaffin, 2005, p. 735.

4 Ibid., p. 737.

5 Ibid., p. 738.

consensus non-miraculous explanation among creationists for accelerated transmutation during the Flood period.

One piece of evidence in favor of the interpretation of accelerated transmutation concerned the helium content of zircon crystals from a drill core containing Precambrian granodiorite.[6] The single atoms of helium gas readily 'leak' through crystal lattices because they are so small. The quantity of helium in the zircon crystals was surprisingly high, considering the expectation for helium to diffuse out of the crystals over time. While the amount of the helium daughter isotope matched the expectation of 1.5 billion years of decay at today's rates, if the specimen were anywhere near that old, most of the helium should have long since disappeared. When the rates of diffusion of helium were plotted against temperature, the results fit a creation model time of 6000 years much better than was predicted by radiometric analysis of the Uranium-238/Lead-206 content using secular origins science presuppositions (1.5 billion years). This makes sense if the same amount of decay occurred in a much shorter time. If the zircon helium clock result was accepted as real, then the transmutation of the Uranium-238 isotope to Lead-206 must have been accelerated at some period in the past to account for the quantity of helium which remained.

Another piece of evidence which the RATE group advanced was polonium-210 radiohaloes, which are found in granites world-wide. (Radiohaloes are small, spherical discolorations in granite or other rock, formed when a radioisotope within the rock decays.) These require special conditions to form in granite, in view of polonium's short half-life of at most 140 days. The granite must be both crystalline and cool enough for the halos not to be destroyed by heat (<150 degrees Celsius). To account for the halos, a substantial amount of polonium must be produced rapidly from radioactive reactions of uranium. The RATE researchers concluded that the radioactive transmutation must have been accelerated.[7] Once again, this explanation fits the Genesis account.

Snelling investigated fission tracks in several formations in the Grand Canyon. He found that some zircons exhibited large ratios of daughter isotopes which matched the density of the associated fission tracks. Under a creationist paradigm, this strongly suggested that accelerated transmutation occurred after the creation week and probably during the Flood.[8]

Austin demonstrated that both whole-rock and mineral isochron ages conflicted for four different isotope pairs in the same rock. If the isochron technique was correct, each isochron should predict the same age of the rock. This was not so, causing doubt in the utility of radiometric dating to establish

6 Ibid., p. 739.
7 Ibid., pp. 743–746.
8 Ibid., p. 747.

absolute dates reliably.[9] Snelling reported a similar result when he analyzed rocks with a variety of ages from recent to the Precambrian period.[10]

There is some relatively new evidence in scientific literature that half-life can shorten under certain conditions. Rhenium-187 exhibits this behavior when highly ionized. This counters the prevailing scientific assertion that half-lives are constant under all experimental conditions.[11]

The RATE group found that transmutation by alpha decay in geological samples generated older age estimates than transmutation by beta decay, leading them to distrust absolute age determinations based on radiometric dating.[12]

Although the RATE group investigated standard geological "clocks" involving isotopes of uranium, thorium, potassium, rubidium and samarium, they also found interesting results related to carbon-14 dating, which we turn to next.

Limitations and uses of Carbon-14 dating

Carbon-14 is used to estimate the ages of formerly biological materials, such as wood or coal, which contain carbon. The carbon-14 isotope is thought to be generated when high energy cosmic rays hitting the upper atmosphere indirectly cause a transmutation of atmospheric nitrogen to carbon-14. This is a weakly radioactive isotope different from the main (non-radioactive) carbon-12 isotope. Carbon-14 partakes in exactly the same sorts of chemical reactions as carbon-12. It is oxidized to carbon dioxide in the atmosphere which is in turn incorporated into plants during photosynthesis. If the plant is eaten, an animal incorporates some of the carbon-14 into its body. The consequence is that both plant and animal material may contain carbon-14. After death, no further carbon-14 is incorporated into the plant or animal.

Carbon-14 transmutes spontaneously back to nitrogen with a half-life estimated at 5730 years. This relatively short half-life means that radiometric dating with carbon-14 is not useful past about 50,000 years—after this elapsed time there is so little carbon-14 left that it is difficult to measure precisely. A second implication is that where carbon-14 is detected in a sample, the age of the sample must be less than about 100,000 years. The reason for this is readily illustrated by a hypothetical example which starts with a ball of pure carbon-14 with the same mass as the earth (about 6×10^{21} metric tons). After one million years, less than two grams would remain; the rest would have transmuted into nitrogen.

---

9 Ibid., 2005, p. 748.
10 Ibid., p. 749.
11 Ibid., pp. 750–751.
12 Ibid., p. 748.

Unexpected results of Carbon-14 dating

A high precision technique called Accelerator Mass Spectrometry revolutionized carbon-14 dating in the late 1970s. This method accurately detected extremely small quantities of carbon-14 directly without having to wait for the atoms to transmute.[13] This technique was faster and could utilize much smaller samples than previous methods.

This new method resulted in unexpected findings that had not been revealed by the older techniques. By the time the RATE group convened, there were at least 70 independent reports in scientific journals which detected carbon-14 in geological formations, spanning most of the geological record.[14] The geological record, however, is claimed to span billions of years. In other words, this new line of evidence called into question the secular ages. The presence of carbon-14 indicated that these deposits averaged 100,000 years old, sharply contrasting with the conventional scientific narrative of millions or billions of years.[15] The secular interpretation was to dismiss these findings as representing contamination from an unknown source.[16]

A member of the RATE group replicated these findings on 10 samples from the U.S. Department of Energy Coal Repository which were sent for analysis at a well-credentialled independent laboratory. Several diamonds were also analyzed for carbon-14. The analyses proved positive. Recall that in the conventional narrative, diamonds formed in the earth's mantle one or more billion years ago. Secular scientists would argue that any carbon-14 contained in these diamonds must have arisen by neutron bombardment of nitrogen-14 contaminating the diamond. However, the amount of neutron activity detected in mines is orders of magnitude less than that required to create the measured quantum of carbon-14.[17] Once again, this evidence supports recent creation and a young earth—certainly less than the millions or billions of years claimed by secular science.

Calibration of Carbon-14 dating

We mentioned earlier that a geological clock ought to be compared with another "gold-standard" clock for validation. However, this is difficult to achieve with carbon-14 dating because the rate of formation of carbon-14 varies across time and location. Therefore, instead of "validating" the results, external evidence is used to "calibrate" carbon-14 dating.

---

13 Ibid., p. 753.
14 Ibid., p. 753.
15 Ibid., p. 754.
16 Ibid., p. 753.
17 Ibid., p. 756.

With carbon-14, it is not a simple matter to calculate an absolute date for a specimen. Solar flares can affect the quantity of cosmic rays hitting the upper atmosphere, and therefore the amount of carbon-14 produced. Similarly, the strength of the earth's magnetic field, which repels solar flares, is also known to impact carbon-14 dating. This is because changes in the earth's magnetic field over time affect the rate of carbon-14 production in the upper atmosphere. Volcanic eruptions release large quantities of carbon-12 as carbon dioxide which also affects the crucial carbon-14 to carbon-12 ratio in the atmosphere over time. In addition, the level of carbon-12 in the atmosphere changes over time with human and other activity. For instance, creationists believe that the Genesis flood caused a dramatic decrease in the amount of carbon in the biosphere (most of which is carbon-12) due to sequestration (trapping) of biomass in geological formations. As carbon-14 continued to be added to the biosphere via transmutation of nitrogen in the upper atmosphere, the carbon-14/carbon-12 ratio of atmospheric carbon dioxide rises. Then, when organisms die, their carbon-14/carbon-12 ratio reflects that of the atmosphere at the time of their death. Hence, after the Flood, increased carbon-14 would be present in dead organisms. If this effect were ignored, carbon dating of pre-flood biological materials (including coal and fossils) would result in spuriously long ages.

Another assumption used in carbon-14 dating is that the half-life of carbon-14 has remained the same as today across the timespan of measurement, which the RATE group contested.[18]

Even the effects of historical nuclear bomb tests must be accounted for in carbon-14 dating. Thus, scientists use calibration curves to adjust dates generated by carbon-14 analysis to correct for known effects. In today's environment, these are based ultimately on the scientific presuppositions of secular scientists. One long-standing assumption which has been challenged recently is that carbon-14 mixed rapidly within each hemisphere, allowing use of standardized curves for the northern and southern hemispheres. However, high precision data has suggested that there have been regional variations in carbon-14 levels which may impact various historical controversies.[19]

Because of potential confounders (a scientific term) affecting carbon-14 dating, calibration with independent "clocks" is required. One method has been to use tree rings to date ancient trees and then calculate a carbon-14 date for a specimen of that tree. Creationists have raised a concern that it is not possible to reliably calculate an age for a tree using tree rings because the number of tree rings can exceed one per year under certain environmental conditions.[20] Layers of sediment deposited annually in lakes, called "varves",

18 Ibid., p. 754.
19 Manning, Griggs, Lorentzen, & Lange, 2018.
20 Carter, 2022.

have also been used for calibration, but secular presuppositions about the carbon-14 content and ratios in the distant past conflict with the biblical presuppositions of creationists.[21] Historical artefacts such as items from the tomb of the Egyptian Pharoah Tutankhamun have been analyzed, but once again this intersects with the disputed Egyptian chronologies of the secular humanists.[22] And comparing carbon-14 dates with radiometric dates based on other isotopes merely establishes relative inferred ages, not absolute dates.

Creationist researchers have suggested that carbon-14 dates of 35,000–45,000 years "should be recalibrated to the biblical date for the Flood". The flood greatly upset the carbon-12/14 balance on which the dating method is based. Refer to Chapter 4 of *The Creation Answers Book* for further details.[23]

Comparisons of various techniques

Attempts to compare other radiometric dating techniques against a gold standard typically involve comparison against alternate radiometric techniques and geological correlation. The latter is typically interpreted according to secular scientific presuppositions. Creationist writer Tas Walker has stated:

> … evolutionary geologists […] will not accept a radiometric date unless they think it is correct—i.e. it matches what they already believe on other grounds … .[24]

If the radiometric date is too old, the scientist has a range of handy explanations:[25]

> The rock might have contained crystals formed long before the rock solidified, and the crystals caused the older date.
>
> Some very old material might have contaminated the lava as it flowed through the earth.
>
> The lava itself might have contained old material.

If the radiometric date is too young, another set of convenient explanations are invoked:[26]

> Perhaps the rock chemicals had been disturbed by groundwater or weathering.
>
> A localized heating event may have disturbed the chemicals without leaving a clear trace.

21 Ibid.
22 See *The Exodus Problem and its Ramifications*, Volumes 1 and 2 (Courville, 1971).
23 Batten, Catchpoole, Sarfati, & Wieland, 2018.
24 Walker, 2021.
25 Ibid.
26 Ibid.

Walker went on to quote a government publication from the Queensland Department of Resource Industries (1990):[27]

> [...] the relative ages [of the radiometric dating results] must always be consistent with the geological evidence. ... if a contradiction occurs, then the cause of the error needs to be established or the radiometric results are unacceptable.

If a radiometric date fits the secular paradigm, it is touted as strong or irrefutable evidence favoring the point in question. If, however, a problematic radiometric date cannot be interpreted away, then it is "unacceptable"—it is quietly swept under the carpet (from a creationist perspective). If geological correlation is used to validate radiometric dating, and radiometric dates are said to validate the ages implied by the geological record, then circular reasoning is being employed.

Conclusions and areas for future research

In concluding their work, the RATE group advised that their results needed further confirmation and elaboration. The group identified promising areas of research, such as radioisotopes in meteorites, which were not addressed by the RATE project. The concept of accelerated radioactive decay has not yet been fully accepted within creationist circles, let alone by the wider scientific community.[28] Further research is required to verify and expand this important topic from a biblical perspective.

From the point of view of creationists, there are several important takeaways from the RATE project:

The interpretation of radioisotope dating is heavily dependent on presuppositions. (These are more of a philosophical or religious nature, rather than scientific.)

Secular presuppositions will lead to very long timeframes, whereas biblical presuppositions will lead to shorter timeframes consistent with the Bible. Carbon-14 dating that comes back as 50,000 years old is compatible with 5,000 years due to the varying amounts of carbon isotopes in the environment around the time of the Flood. The presence of carbon-14 in oil, coal and diamonds could be interpreted as consistent with ages of around 5,000 years.

In Part 3, we introduced Coulson's SFP concept. The main point of this concept was that the amount of process is the key issue, and inferring absolute dates from the amount of process could be unreliable. This potentially applies directly to radiometric dating involving the creation week.

---

27 Ibid.

28 Vardiman, Snelling, & Chaffin, 2005, p. 768.

Christians need to be aware that all the long ages reported by radiometric dating in government, educational, academic or entertainment sectors will have been informed by the presuppositions of secular science, which deny the historicity of the Bible.

Appendix C
Abiogenesis calculations

Justification of probabilistic approach

Before we discuss probability calculations for abiogenesis or evolution, we need to justify our approach. Two influential schools of thought apply to the interpretation of probabilities—the objective and subjective approaches. The objective approach is synonymous with the so-called *frequentist* interpretation; the subjective approach commonly employs a *Bayesian*[1] methodology. These approaches are not necessarily mutually exclusive; an informed mathematician or statistician would typically regard both as legitimate interpretations.

The approach we take in the following calculations is subjectivist; it allows us to calculate a reasonable probability of outcome for a single event.

Famous aeronautical scientist and mathematician Richard von Mises (1883–1953) argued for the objective (frequentist) position in *Probability, Statistics and Truth*, first published in German in 1928. Von Mises rejected the subjective interpretation of probability, likely because of his background in the hard sciences.

We can illustrate the frequentist position with an example. Consider the statement "the probability of a coin landing heads-up is 0.5". To a frequentist, this means that if the coin is tossed repeatedly over time, as the number of tosses increases, the proportion of heads to the total number of throws is expected to get closer and closer to 0.5. (The exact fraction could vary a little due to random effects, especially at the outset when the number of throws is small, but as the number of tosses increases, the result will gradually converge closer to 0.5). An alternative but equivalent frequentist interpretation is where many identical coins are tossed simultaneously, the proportion of coins landing heads-up should be very close to 0.5. Once this empirical evidence is obtained, if the fraction of heads is very close to 0.5 it would be reasonable to conclude that the coin is probably a "fair coin"—one where the odds of heads is the same as the odds of tails.

Thus, in the frequentist interpretation probabilities apply to situations involving many identical trials (preferably virtually unlimited in number),

1 Named after the English statistician and Presbyterian minister, Thomas Bayes (1701–1761).

whether simultaneous or sequential.[2] The frequentist interpretation is commonly used in the "hard sciences"—fields such as physics or thermodynamics. As a frequentist, von Mises argued that *no other interpretation* of probability made sense. Frequentists like von Mises consider their concept of probability to be highly objective.

However, the frequentist paradigm does not mean that *interpretation* of data is not subjective. Consider three people given data corresponding to 100 tosses of a coin, where the coin was (unfortunately) badly damaged after the throws were completed. The data showed 47 heads occurred in 100 completed throws. Our first frequentist might conclude that the coin was probably fair, and random variation and a relatively small sample size of only 100 throws resulted in 47 heads. Our second might conclude that the coin is weakly biassed away from heads—and towards tails—and this resulted in the slight shortfall of heads below the expected 50. A third frequentist might be concerned that information about the surface onto which the coins were thrown was lacking—if thrown on sand, then on three or four occasions the coin may have landed on its edge, and thus the coin could have been biassed to heads, but the head shortfall reflected edge landings, which did not count as either heads or tails. Thus, in this example, the limited amount of data (the coin was damaged so further tests would not be conclusive, and the surface onto which the coin was thrown cannot be reliably assessed) leads to three different subjective interpretations. Consequently, although the frequentist position may be regarded by proponents as highly objective, at the data interpretation stage, subjectivity comes into play.

The frequentist position rose to prominence through much of the twentieth century for two main reasons: it was applicable to the hard sciences and its calculations were much easier to process than the Bayesian approach. (Later, major advances in computing power and algorithms, specifically the Markov chain Monte Carlo algorithm, enabled the Bayesian approach to be more widely used).

Let's move to an entirely different type of scenario where probabilities are used. The weather forecast informs me that tomorrow in my city, the chance of rain is 25%. What does this mean? There is only one tomorrow in my city, so a strict frequentist might complain that probability cannot be meaningfully applied to a single event. However, in medical science, and soft sciences such as economics, probabilities and statistics are often applied to individual events or persons. John Maynard Keynes provided a justification for this in his book *A Treatise on Probability* (1921), in which he argued that a probability can be viewed *as a degree of rational belief.* To a frequentist, probabilities do not apply to a single coin toss. But using a subjective interpretation of probability, a

2 Von Mises, 1981, p. 11.

probability of heads of 0.5 on a coin toss means that we believe that the coin is a fair coin, and that the chance of a head is as high as the chance of a tail. Subjective interpretations of probability can have a strong mathematical basis, and the Bayesian approach has become prominent in this role. Bayesian thinkers accept up-front that they often deal with a subjective degree of belief.

The classic example illustrating Bayesian probabilistic reasoning is the case of a diagnostic test for a disease applied to an individual within a population. Imagine a population in which 1 in 100,000 people are afflicted by a rare disease. Further, there is an imperfect diagnostic test that is positive in 80% of cases where the disease is present, and negative in 90% of cases when the disease is absent. Before the test, the probability that a randomly selected individual has the disease is 0.001%, in accordance with the population prevalence of 1 in 100,000. If the test is positive, the chance that the person genuinely has the disease is not 80%, but 0.008%, which is negligible. Because the disease is so rare in that population, most positive tests are so-called "false positives"—the person does not have the disease, but the test erroneously gives a positive result. This probabilistic reasoning helps us to interpret the positive test result and could assist in determining what to do next. This illustrates how we can adjust our subjective assessment of the likelihood of an event using mathematical modeling. The use of Bayesian probability assessments is accepted in diverse fields.

In our abiogenesis calculations, because we are modeling the chances of a singular event, we must use the subjective interpretation of probability. This also aligns with origins science as a "soft science", not a "hard science", like physics and chemistry. We do not, however, need to use a Bayesian approach.

Probability of abiogenesis

Introduction

In the following analysis, a significant number of probabilities are calculated. We can give an example of the type of probability calculation that is most relevant to this Appendix. Imagine planning to roll a die (plural: dice) ten times and determining the chances of four showing up *at least once*. The calculation is $1-(1-1/6)^{10}=0.84$. This is a specific example of a general probability formula: if a die has x faces and therefore the probability of a particular face ending up on top in one die roll is $1/x$ then the probability of getting this face on top *at least once* in N die rolls is $1-(1-1/x)^N$. This type of calculation is unwieldy when dealing with a die with a very large numbers of faces, or equivalently, events with a probability much less than $1/6$.

For example, imagine a die with 10,000 faces numbered 1 through 10,000 (instead of a 6-sided die). Here the chance of getting a four in one roll is $1/10000 = 0.0001$. However the probability of getting a four *at least once* in 10 rolls is $1-(1-1/10000)^{10}=0.0009996$. It turns out that in this latter case, because the chance of getting a four is very small (0.0001) a simpler formula can be used to approximate the probability of getting at least one four in ten rolls, namely $10 \times 1/10000 = 0.001$. Comparing the results of the two formulas, it is apparent that they are almost identical ($0.001 \approx 0.000996$).

This can be stated for the general case: if the probability of an event occurring in one die roll is y, the chance of the event occurring at least once in N rolls is approximated by Ny. As y gets smaller, the approximation becomes extremely accurate. This relationship can be proven mathematically.

In this Appendix, we use the probabilistic definition of impossibility given in Chapter 10, that is, a probability less than 10^{-45} is regarded as impossible in the real world.

Calculations applied to a "simple" life-form

Our aim is to calculate the probability that ten proteins and their respective DNA segments assemble randomly *at least once* somewhere on earth during a period of half a billion years. That was the minimum requirement for life which we leniently proposed in Chapter 10.

The probability of obtaining a particular amino acid from a primordial sludge containing equal amounts of twenty different amino acids is one twentieth. The probability of assembling a single ten-amino-acid sequence, or a protein containing ten specific amino acids in specific locations is $\left(\frac{1}{20}\right)^{10} = 9.77 \times 10^{-14}$.

If we allow one hundred possible variants of this protein as potentially having some function (with respect to the ten critical amino acids), we adjust this probability upwards to $9.77 \times 10^{-14} \times 100 = 9.77 \times 10^{-12}$. This tilts the table to favor abiogenesis, because we are allowing more than one specific arrangement of critical amino acids for a given protein with its specific function.

To estimate the probability of assembling ten proteins, each containing a unique ten amino acid sequence (and allowing 100 variants of each), the calculation is $(9.77 \times 10^{-12})^{10} = 7.9 \times 10^{-111}$, a stupendously low probability, certainly less likely than an impossible impossibility.

Next, because the central dogma of molecular biology tells us that DNA cannot be created from these necessary proteins, but DNA is required if a cell is to be able to replicate, it is necessary to calculate the probability of randomly assembling DNA sequences which could create one of these proteins, or similar proteins (like the ninety-nine other possible protein variants we are allowing). Each amino acid is represented in DNA by a "codon"—a sequence of three molecular "alphabetical symbols". The DNA "alphabet" itself comprises four symbols: A, C, G and T. A stop codon is found at the end of a protein coding DNA sequence, like a full stop in a sentence. Therefore, to encode the ten amino acids of interest, thirty DNA symbols are required (i.e. ten codons). Because the protein also needs a stop codon (also having three symbols), the total number of symbols required is thirty-three. Another complication is the fact that amino acids have up to six codon synonyms (in the cases of arginine, leucine and serine) which need to be factored into the calculation, as shown in Table C.1.

Table C.1: Codon synonyms for amino acids

Amino Acid	Synonyms	Number of synonyms
Alanine	GCT, GCC, GCA, GCG	4
Arginine	CGT, CGC, CGA, CGG, AGA, AGG	6
Asparagine	AAT, AAC	2
Aspartate	GAT, GAC	2
Cysteine	TGT, TGC	2
Glutamate	GAA, GAG	2
Glutamine	CAA, CAG	2
Glycine	GGT, GGC, GGA, GGG	4
Histidine	CAT, CAC	2
Isoleucine	ATT, ATC, ATA	3
Leucine	TTA, TTG, CTT, CTC, CTA, CTG	6
Lysine	AAA, AAG	2
Methionine	ATG	1
Phenylalanine	TTT, TTC	2
Proline	CCT, CCC, CCA, CCG	4
Serine	TCT, TCC, TCA, TCG, AGT, AGC	6
Threonine	ACT, ACC, ACA, ACG	4
Tryptophan	TGG	1
Tyrosine	TAT, TAC	2
Valine	GTT, GTC, GTA, GTG	4
Stop codon	TAA, TAG, TGA	3

To simplify, the number of synonyms for all codons can be arbitrarily set to six. This tilts the table to favor abiogenesis because the codons for two amino acids have no synonyms, and several have only two synonyms. Most have less than six, including the stop codon, which has three. The probability of forming one appropriate DNA sequence for one protein is less than $(number\_of\_synonyms)^{number\_of\_aminoacids} \times$ $(number\_of\_stop\_codon\_synonyms) \times$ $(probability\_of\_selecting\_1\_DNA\_symbol)^{number\_of\_DNA\_symbols}$ which works out to $6^{10} \times 3 \times \left(\frac{1}{4}\right)^{33} = 2.46 \times 10^{-12}$. This is multiplied by one hundred to allow other possible variants giving 2.46×10^{-10}.

To form ten DNA sequences corresponding to ten proteins, the calculation is $(2.46 \times 10^{-10})^{10} = 8.1 \times 10^{-97}$.

Combining the probabilities for assembling the ten protein and DNA sequences gives $7.9 \times 10^{-111} \times 8.1 \times 10^{-97} = 6.4 \times 10^{-207}$.

The calculated probability represents the chance of assembling ten amino acids into one of one hundred acceptable sequences, while simultaneously assembling ten DNA nucleotide sequences of thirty-three elements (each which code for one of the one hundred acceptable amino acid sequences), terminated by a stop codon. This final value, 6.4×10^{-207}, can be considered the probability of a single roll of the abiogenesis protein and DNA die. Even before we consider how many rolls of that die might be allowed under interpretations of contemporary science, the sheer minuteness of the number looks problematic because it is vastly less than an impossible impossibility. This has been conceded by scientists such as Sir Fred Hoyle, Chandra Wickramasinghe and others.

Calculating how many rolls of the abiogenesis dice are possible

To estimate the number of times the abiogenesis dice can be rolled, we need to estimate three factors. First, we need an estimate of the number of abiogenesis dice we have, second, the amount of time available for dice rolling, and third, the average time taken for a die roll.

In relation to time, the window for abiogenesis is conservatively estimated to be half a billion years. This estimate considers fossil evidence that suggests single-celled organisms had emerged in the earliest phase of the putative 4.5 billion years of earth's existence. This estimate excludes some time subtracted from the start of the abiogenesis window to allow the earth's temperature to drop to a level compatible with life.

To estimate the number of abiogenesis dice available for rolling, imagine the earth covered by a sludge containing a concentrated mixture of amino acids and DNA nucleotides. (Evolutionists usually use the term "soup" instead of sludge, because it sounds more palatable.[3]) For simplicity, and to favor abiogenesis, we will assume that all the accessible carbon on earth is contained within amino acids and DNA nucleotides in the soup. Our first task, therefore, is to estimate how much amino acid and DNA nucleotide material can be fashioned from the available carbon.

It has been estimated that there are 43,500 billion tonnes of surface carbon on earth, amounting to 0.2% of earth's total carbon.[4] This works out to $43500 \times 10^9 \times 10^6$ grams of carbon. We divide this by 12.01 (the molecular weight of carbon-12, which accounts for 99% of all carbon found naturally) to ascertain the number of moles of carbon and then multiply this figure by

3 We credit Sir Fred Hoyle for this witticism.

4 Amos, 2019.

Avogadro's number, 6.02×10^{23}, to estimate the number of accessible carbon atoms on earth. This number works out to 2.2×10^{42}. So, how much amino acid and DNA material could this amount of carbon conceivably give rise to?

The simplest amino acid, glycine, contains 2 carbon atoms, but the average number of carbon atoms in the twenty types of amino acid found in living organisms is slightly greater than five. To help the case for abiogenesis, we will conservatively assume that all amino acids have only two carbon atoms.

In respect to DNA nucleotides, A (Adenine) has ten carbon atoms, C (Cytosine) has nine carbon atoms, G (Guanine) has ten carbon atoms and T (Thymine) has ten carbon atoms. To tilt the table towards abiogenesis it can be assumed all DNA nucleotides contain nine carbon atoms. For simplicity and to tilt the table in favor of abiogenesis, the carbon atoms in the deoxyribose molecules can be ignored, and the calculation can be limited to one strand of the DNA instead of two (DNA is generally double-stranded).

An initial assumption was that each protein contains one hundred amino acids, with ten of them required, and each amino acid contains at least two carbon atoms. Each roll of the abiogenesis protein die, therefore, contains a minimum of two thousand carbon atoms.

In relation to DNA, to function as a template for a protein containing one hundred amino acids, three DNA nucleotides are required per amino acid (for a codon), plus three for the stop codon, which is three hundred and three DNA nucleotides per protein. To code for ten proteins, three thousand and thirty DNA nucleotides, each of which has at least nine carbon atoms, are required. This equates to at least 27,270 carbon atoms. A single abiogenesis die requires carbon for the proteins and carbon for the DNA (we ignore the requirements of the ribose backbone) which is at least 29,270 carbon atoms (that is, an extra 2000 carbon atoms for the proteins).

With this information, the maximum number of dice available for abiogenesis on earth can be estimated, along with the maximum length of time, in seconds, available for die rolling. We will calculate these shortly, but first, we must estimate the final piece of information needed—how quickly a die roll can be repeated. Once again, we will tilt the table to favor abiogenesis. Enzymes catalyze chemical reactions to speed them up by a factor of thousands to billions. Thus, if the minimum time for assembling one hundred amino acids enzymatically is calculated, this represents an upper limit for how many times per second a protein can be assembled. The bacterium *E.coli* can assemble up to seventeen amino acids per second under rich nutrient conditions[5] which would mean a 100 amino-acid protein could be assembled in under six seconds.

5 Zhu & Dai, 2019, p. 7593.

For DNA, the synthesis rate of DNA polymerase III holoenzyme in *E.*coli is 750 nucleotides per second.[6] This would mean that three hundred DNA nucleotides would be copied in about 0.4 seconds. Since this time is less than that of the enzymatic protein assembly time this latter number can be used for both, which simplifies the calculations and strongly favors abiogenesis. Here, we conservatively assume that random assembly of DNA is as rapid as the enzymatic process. To allow the dice to be thrown as rapidly as possible, it can be assumed that as soon as a DNA or protein molecule is assembled, it disintegrates back to its constituent monomers, whether nucleotide or amino acid. With these considerations in mind, we proceed to our calculations.

The number of dice available is the total accessible carbon atoms divided by the number of carbon atoms in the protein/DNA die, which is $2.2 \times 10^{42} \div 29270 = 7.5 \times 10^{37}$. The total number of times each die can be rolled is half a billion years, in seconds, divided by the rate of 0.4 per second. This is $0.5 \times 10^{9} \times 365.2425 \times 24 \times 60 \times 60 \div 0.4 = 3.9 \times 10^{16}$. The total number of dice rolls altogether is $7.5 \times 10^{37} \times 3.9 \times 10^{16} = 3 \times 10^{54}$. Therefore, the probability of abiogenesis occurring on earth, under very favorable assumptions, is at most $3 \times 10^{54} \times 6.4 \times 10^{-207} = 1.9 \times 10^{-152}$. Recalling that the cutoff of probabilistic impossibility is 10^{-45}, it is clear that abiogenesis cannot have occurred on earth. It is less probable than three impossible events (events whose probability is 10^{-45}) occurring in succession.

If the assumptions are changed to permit abiogenesis to occur on any of 10^{25} planets, and the average amount of planetary carbon is increased tenfold, and the available time extended to 13.8 billion years of time, instead of 0.5 billion years, then the probability is raised to $1.9 \times 10^{-152} \times 10^{25} \times 10 \times 13.8 \div 0.5 = 5.3 \times 10^{-125}$ which is still impossibly impossible. Even if the table were tilted even further by invoking some unknown property of the universe to increase the probability of abiogenesis by a factor of a trillion, the probability is not shifted materially. The chance would then be $5.3 \times 10^{-125} \times 10^{12} = 5.3 \times 10^{-113}$ which is still impossibly impossible. What if we—in a more outlandish flight of fancy—assume there are a trillion trillion trillion universes, like our own, as part of a broader multiverse? The probability is now raised to $5.3 \times 10^{-113} \times 10^{12} \times 10^{12} \times 10^{12} = 5.3 \times 10^{-77}$. It still remains impossible for abiogenesis to have occurred in any of those universes.

But the proximity problem now includes a time span of 13.8 billion years instead of half a billion years, and a spatial extent of a trillion trillion trillion universes instead of just one. Thus, even though we have increased the odds of the required proteins and DNA sequences forming by expanding the temporal and spatial windows, it results in a situation where assembling the components into the tiny space required, at just the right moment in time to create a living

---

6 Uri, n.d.

cell, is impossible. No matter how we cut it, abiogenesis is less likely than an impossible event (using the probabilistic definition of impossibility given in Chapter 10).

The only way to increase the odds to a plausible level is to invoke infinity—but infinity, as we have seen, is not an attribute of the visible universe. It is an attribute of God.

Appendix D
The Intelligent Design Movement

Background

"The theory of intelligent design [ID] holds that certain features of the universe and of living things are best explained by an intelligent cause, not an undirected process such as natural selection".[1] This brief summation of the central concept of ID appears, at first glance, to align with aspects of biblical creation. It should therefore come as no surprise that the secular science community is strongly opposed to ID, dismissing it as pseudoscience. Nevertheless, the movement has grown since its beginning in the 1980s, with some IDers becoming notable voices in the contemporary origins debate. Prominent ID writers include Michael Behe (*Darwin's Black Box* (1996)) and Stephen C. Meyer (*Darwin's Doubt* (2013) and *Return of the God Hypothesis* (2019)). Thomas Nagel, a Harvard philosopher, has written supportively of ID concepts, even though he is a declared atheist.

ID is spearheaded by a public policy thinktank, the Discovery Institute, based in Seattle.

ID and creationism compared

According to Douglas Axe, a scientist who has been associated with the ID movement, ID differs fundamentally from creationism in several important respects. ID is committed to secular science and its presuppositions, and seeks to show how these principles, combined with scientific evidence, lead us to "attribute life to a purposeful inventor".[2] The epithet "Intelligent Design" is used not "to smuggle God into science"; IDers view the task of identifying the Intelligent Designer as "beyond the essential principles of science".[3] ID is not prepared to invoke the God of the Bible, but will accept, and indeed promotes, an unknown intelligence as the agent behind aspects of the world around us.

ID is not distinctively Christian, even though it may include Christians (and even some biblical creationists) among its supporters. The "Discovery Institute is a secular think tank ... [representing] a variety of religious traditions,

---

1 Discovery Institute: FAQs, n.d.

2 Axe, 2016, Chapter 4.

3 Ibid.

including mainline Protestant, Roman Catholic, Eastern Orthodox, Jewish, and agnostic".[4] ID appeals to this broad range of religious traditions by pointedly declining to identify the intelligent agent involved in the design. As a result, "the scientific theory of intelligent design ... has no commitment to defending Genesis, the Bible or any other sacred text".

Like Darwinian evolution, ID points to ancient pagan thinkers as the earliest proponents of its central ideas. In the "Greco-Roman tradition, Plato and Cicero both espoused early versions of intelligent design".[5] This is different to creationism, which sources important presuppositions from the Bible.

In relation to education policy, the "Discovery Institute is not a creationist organization, and it does not favor including either creationism or the Bible in biology textbooks or science classes".[6] The Institute also states that "[f]ar from reducing the coverage of evolution, Discovery Institute seeks to increase the coverage of evolution in textbooks. It believes that evolution should be fully and completely presented to students, and they should learn more about evolutionary theory, including its unresolved issues".[7] We agree that the mechanisms of "evolution" could be presented as causal model science, but there remain two issues for creationists. First, the use of the term "evolution" favors the secular origins view, as explained in Chapter 8. Second, creationists do not accept all the presuppositions of secular scientists and ID proponents. As creationists, we do not believe that "evolution" should be presented to students as a holistic origins account; this is a worldview and forms part of a religious perspective. However, if evolution were presented 'warts and all', so that students came away with a balanced understanding of the theory and its problems, then a strong case could be made for teaching evolution in science classes.

The Discovery Institute promotes the view that the Bible has no place in biology textbooks or science classes, statements indistinguishable from those of secular humanists, who do not want origins taught from a Christian perspective. Notably, ID as communicated by the Discovery Institute is compatible with Deism, polytheism, and even agnosticism, which are non-Christian, or even anti-Christian belief systems.

Although the ID movement embraces almost all the presuppositions of secular science, except perhaps the possibility of intelligent design and cosmic purpose, its critiques of Neo-Darwinism are indirectly helpful to creationists. As creationist writers, we regard their work to be of high quality overall, and have benefited from ID literature in writing this book. However, in the final analysis, ID is not an origins theory we can support. Our chief concerns are that

---

4 Discovery Institute: FAQs, n.d.

5 Ibid.

6 Ibid.

7 Ibid.

the ID movement is not Christian and has embraced almost all the presuppositions of secular origins science.

The view of the secular science establishment

The secular scientific establishment considers that "people of faith who challenge Darwinism are *really* pushing religion", even if the challenge has a scientific basis,[8] and the challenge is in the realm of causal model empirical science (rather than origins science). This fits with the commonly held but incorrect belief that there is a war between religion and science; "all enlightened people" should defend science against this type of religious attack.[9] The scientific establishment only allows critiques of the secular consensus from those that demonstrate a commitment to secularism, atheism or materialism. This indicates that the person carrying out the critique "is a friend of the greater cause".[10] ID, with its inference of a designing intelligence, does not fit this mold.

ID infers the possibility of cosmic purpose and intelligent design of life from scientific evidence. The scientific establishment presupposes there is no cosmic purpose, and the diversity of life on earth can be explained without intelligence. These positions appear to be reverse sides of the same coin, but they are not. To the question whether there is an intelligent designer, secular science *presupposes* its answer *before* looking at the evidence, whereas ID *infers* its answer *after* considering the evidence.

Conclusion

Because ID is non-Christian and, in some instances, may be based on unbiblical presuppositions, we caution Christians to be wary of it, even though ID critiques of Neo-Darwinism are indirectly helpful to creationists. Most creationists and ID proponents do not want to be misidentified with each other. Whereas some creationists wish to distinguish themselves from IDers as a group, secular scholars show a tendency to lump both together.[11]

In the debate between ID and secular origins science, the truth of Romans 1 is on full display. ID proponents infer an intelligence behind the universe,[12] which is in line with Romans 1:20a: "For his invisible attributes, that is, his eternal power and divine nature, have been clearly seen since the creation of the world, being understood through what he has made." However, secular

---

8 Axe, 2016, chapter 4.

9 Ibid., chapter 4.

10 Ibid., chapter 12.

11 To provide one example, Eugenie Scott and Glenn Branch used the term "intelligent design creationism" in an article published in 2009 (Scott & Branch, 2009, p. 91), psychologically linking the intelligent design movement to the creationist movement even though they are two distinct, and in several respects conflicting viewpoints.

12 ID infers intelligent design on the basis of neopagan presuppositions, excluding the a priori commitment to atheism.

scientists try to suppress the arguments of the ID movement, in line with Romans 1:18b: "who by their unrighteousness suppress the truth ...". Yes, there is a God who designed creation. ID proponents admit to this, even if they hold short of identifying the God of the Bible. But even this noncommittal approach is enough for some secular scientists to vigorously oppose them.

Appendix E
Catastrophic Plate Tectonics

Catastrophic Plate Tectonics (CPT) is the most advanced attempt to date at explaining the mechanisms of the global flood of Genesis in a way compatible with contemporary scientific knowledge. (Other theories which have been advanced include the precipitation of a water vapor canopy by Morris and Whitcomb, the hydroplate theory of Walter Brown, and the impact theory (or sub-theory) of Michael Oard.)

The name *Catastrophic Plate Tectonics* indicates that this class of explanations derives from plate tectonics theory, with the word *catastrophic* indicating that the plate tectonic movements envisioned in CPT are billions of times faster and more intense than occur in conventional plate tectonics.

CPT views the Flood as being caused by a sequence of geological events. One of the key mechanisms of CPT is runaway oceanic plate subduction, a process where denser oceanic crust subducts (slides) under continental crust into the underlying mantle layer of the earth. By the time of the Flood an unstable situation existed where denser oceanic crust was hypothesized to be "floating" on top of less dense mantle, related to the normal processes of differentiation which affected the pre-flood earth. Hence a large amount of gravitational potential energy was available to be converted into kinetic energy and heat as the ocean plate subducted.

The movement of oceanic plate into the mantle heated the mantle by friction, lowering its viscosity. A positive feedback process, known as "runaway subduction," accelerated oceanic crust deeper into the mantle with increased heating and further reductions in mantle viscosity. Crustal velocity was estimated at meters per second. Runaway subduction only stopped when the entire ocean floor had disappeared into the mantle.

Austin (et al.), a group of creationist scientists who promoted CPT explanations for the global flood, provided a narrative of what might have occurred, based on their expertise in a range of relevant scientific disciplines. They presented this complex scenario in articles and videos; the account which we present here draws on this material.

CPT reworks the widely accepted theory of plate tectonics to position the global flood as a major determinant of current geography. It provides a holistic explanation for a range of phenomena:

1. The rapid formation and deposition of large amounts of sediments, leading to the formation of much of the geological column (the majority of which is sedimentary rock).
2. Large amounts of sea-floor sediments on land along with marine fossils.
3. Large amounts of carbonates in the geological column.
4. The processes by which large numbers of fossils were rapidly formed.
5. Formation of structures in the earth's surface such as ocean trenches and mountain ranges.
6. The separation of the continents into precursors of the current arrangement.

Runaway subduction

The oceanic subduction was hypothesized to occur along the margins of a Pangea-like supercontinent that existed prior to the Flood. This leads to the supercontinent pulling apart in a manner roughly consistent with patterns visible in the rocks of today's ocean floor.[1]

Computer simulations are a useful method to model historical geological events. Creationist John Baumgardner was "one of the world leaders in numerical models of mantle convection" and the person who first outlined the CPT proposal in 1986.[2] Baumgardner worked as a scientist at the Los Alamos National Laboratory of the U.S. Department of Energy. He had created sophisticated modeling software, TERRA, written in the computer language Fortran. This software came to be used by geophysicists around the world. Another programmer, Jamie Painter, added a graphics module to represent TERRA's results. A secular scientist commented: "As far as the code goes, Baumgardner is a world-class scientist".[3]

The software ran on a supercomputer and used finite element simulation based on known laws of nature to run plate tectonic processes as a long sequence of small steps. At each step, the mantle of the earth was modelled as millions of small 3-dimensional cells, and equations relating to conservation of mass, energy and momentum were solved numerically. In this way, scientifically plausible movements of tectonic plates could be investigated. To make the calculations computable (given the hardware limits of the time) Baumgardner had to use a number of approximations. When he input values consistent with secular science into the model, the simulation yielded results consistent with secular science. When he used values consistent with those expected in the Genesis flood, the results showed accelerated processes. These caused a flood and rearranged the continents into positions approximating

1 Baumgardner, 1994, p.2.
2 Burr, 1997, p. 58.
3 Ibid.

those of today—and this over a time period compatible with a straight-forward reading of Genesis.

The basis of the positive feedback loop leading to runaway subduction was scientific evidence indicating mantle rock (mainly olivine) can weaken dramatically under certain conditions, even without reaching its melting point. As oceanic plate subducts into the mantle, the mantle rock is heated, which leads to weakening and faster sinking. Baumgardner also modeled this process in two dimensions to confirm its plausibility. He later showed that this plastic deformation could occur due to mechanisms besides heating.

Baumgardner noted that seismic studies have revealed that the deep mantle near the outer core shows evidence of variation that is consistent with the presence of relatively cold subducted plate.

Creationist speculation on what might have initiated the geological rupture has been wide-ranging, including impacts or near misses of comets, asteroids or meteorites, but no firm conclusions have been drawn.[4] Horstemeyer and Baumgardner suggested that an external event is not required to trigger runaway subduction and performed numerical simulations on computers to confirm this possibility.[5] In this case, the future catastrophe could have been "baked into the rocks" at an earlier time, perhaps at the time of the curse.

Flood onset—"the fountains of the deep"

As subduction proceeded, areas of the ocean floor tore apart (known as rifting). This allowed molten magma to well up, releasing gases and boiling adjacent ocean water. This generated geyser activity along the lengths of the rifts (potentially thousands of kilometers long) spewing water vapor and gas into the atmosphere—which is hypothesized to correspond to the biblical "fountains of the deep" (Genesis 7:11). This process may have lasted at least 40 days and potentially up to 150 days, resulting in intense rain as the water vapor cooled and precipitated over the surface of the earth.[6] The flood was underway.

Rising floodwaters

The upwelling magma, being hot and less dense, also caused the mid-ocean ridge system to form.[7] These are underwater chains of mountains found in all major oceans. The raising of the ocean floor displaced water from the oceans onto the continents, causing complete inundation.

The oceanic plate subduction was accompanied by frequent, powerful volcanic activity and high intensity earthquakes. The tectonic plate movement

4 Austin, et al., 1994, pp. 611–612.
5 Horstemeyer & Baumgardner, 2003.
6 Austin, et al., 1994, pp. 612–613.
7 Austin, et al., 1994, p. 613.

caused thrust faulting—or rising mountain ranges—in the crust around the continents. Examples are the Himalayas, the Andes, and the Rocky Mountains.

As the oceanic plate subducted at a rate of several meters per second beneath the continental plate margin, the latter was hypothesized to be grabbed and released in a staccato fashion. Calculations showed that this could potentially depress the edge of the continental margin by a kilometer or more, before release and elastic spring-back occurred, sending tsunamis into the continental interiors. Baumgardner estimated that thousands or tens of thousands of tsunamis may have been triggered while the oceanic plates were subducting.

During the period when the waters covered the earth completely, some geological evidence from North America suggests that tidal currents (at least partly caused by the gravitational pull of the moon) moved roughly east to west across continents.[8] In the case of North America, the geology suggests that there were at least five water-flow surges across that continent during the world-wide flood.[9]

Geographical formations

The widespread subduction of the oceanic crust pulled continental crust attached to regions of non-subducting perimeter laterally. Continental collisions subsequently thrust up mountains such as the Appalachians, Himalayans, Caspian and Alps.[10] These were accompanied by earthquakes. The continental margin also thickened during runaway subduction. This may also have contributed to the uplift of mountain ranges as continental plate is "buoyant" in the mantle.

Sediments and their distribution

Geological activity led to the formation of new sediments, while redistributing existing ocean-floor sediments. Other sediments resulted from dissolved minerals precipitating into carbonate particles in the oceans. This precipitation related to magma degassing, and rising ocean temperatures. Another source of sedimentary carbonates was pulverized and resedimented pre-flood shell debris.[11]

The subducting ocean floor transported sediments like a conveyor belt towards the continental edge, which was warped downwards, while tsunami wave action and tidal resonance moved sediments onto the continental surface

8 Wise, 2017.
9 Ibid.
10 Austin, et al., 1994, p. 612.
11 Ibid., p. 613.

in turbulent flows.[12] Baumgardner performed calculations to investigate the generation of enormous quantities of sediments laid down in the flood and suggested that cavitation erosion of bedrock by high-speed water flows was an important mechanism.

Effects on the Earth's magnetic field

The subduction of the oceanic plate occurred over such a wide front that it established mantle-wide flow. Subducting oceanic plate may have descended as deeply as the outer core of the earth. (Creationists have interpreted data from some contemporary seismological studies to indicate relatively cool areas near Earth's outer core.[13]) These cool areas are theorized to have affected the flow of the molten outer core, rapidly changing the earth's magnetic field. Modelling suggests that repeated reversals of the earth's magnetic field may have occurred during this period with periodicity measured in weeks.[14] There is evidence of multiple magnetic reversals in some geological specimens from the ocean floor.

Another mechanism proposed by Baumgardner to explain rapid magnetic reversals was that the massive changes in the mantle caused the earth to enter a period of eccentric spin while the core remained unmoved. This would have a similar effect to a flipping magnetic field.

Ocean floor replacement and the end of the flood

After 150 days, most of the pre-existing oceanic floor was replaced by new material and the rapid plate movement stopped. As the new, hot oceanic floor cooled, its density increased, pushing it lower into the mantle and increasing the depth of the oceans.[15] Water drained off the continents causing sheet erosion phenomena.[16] This was followed by channel erosion as water levels subsided below the level of various geographical features.

After the flood period ended, tectonic instability gradually settled over a period of several centuries.[17] The warmth of the oceans post-flood caused a more temperate climate on earth, especially on the edges of continents. This assisted the geographical dispersion of animals and humans,[18] and led to higher rainfall in the interior of continents from the extra evaporation of the warmer oceans. In high latitudes, the extra snowfall over centuries would have led to the buildup of massive ice sheets. The intense volcanism left much ash and aerosol material in the upper atmosphere, ensuring the summers were cool

12 Ibid., pp. 613–614.
13 Austin, et al., 1994, p. 612.
14 Ibid., p. 612.
15 Ibid., p. 614.
16 Ibid., p. 614.
17 Ibid., p. 615.
18 Ibid., p. 615.

enough to prevent the ice melting back.[19] This post-flood Ice Age would have lasted for several centuries.

Early in the Ice Age, year-round temperatures were more even—summers were cooler and winters warmer. A land bridge then connecting England and France allowed migration of warm weather animals such as hippopotamuses, along with cold climate animals, to places like England, Wales and northwest Europe during the early Ice Age.[20] The mild summers and winters allowed cold and warm weather animals to coexist. However, towards the end of the Ice Age, seasonal variation was more extreme, and the hippopotamuses died off.[21]

Climate change

Five centuries after the flood the polar ice caps had formed and the glaciers in cool areas had advanced significantly. Rapid climate change may have caused the extinction of a range of macro-fauna.[22] The dinosaurs possibly could not cope well with rapid climate change, contributing to their extinction.

The problem of heat

CPT suggests that enormous amounts of heat may have been released from a number of processes acting during the flood year. Heat is released by the processes causing geographical remodeling and also by radioactive decay, which the RATE group believed had been accelerated during the flood year. This heat either remained in the mantle or was conducted to the core or to the surface of the earth. Calculations carried out by Baumgardner show that the amount of heat involved is very high. How this heat was contained or dissipated without causing loss of life on the ark due to hyperthermia is yet unknown. There is presently no consensus among creationists as to how this problem might be resolved in relation to CPT.[23] However, the application of an SFP-like process to the flood period would provide a solution to this problem.

---

19 This cooling resulted from ash blocking sunlight and aerosols reflecting a portion of sunlight back to space. In a more contemporary example, global temperatures are thought to have dropped by an average of around 1.2 degrees Celsius for three to five years after the single eruption of Krakatoa in 1883.

20 Hippopotamus bones have been found with reindeer, musk oxen and woolly mammoths in Ice Age debris. Hippopotamuses were once geographically widespread. (Oard, 2024).

21 Oard, 2024.

22 Austin, et al., 1994, p. 615.

23 Worraker, 2018.

Appendix F
Evidence for a young earth

A straightforward interpretation of Genesis as history, combined with other biblical and historical data, yields a timescale in which the earth was created about 6,000–7,000 years ago and suffered a catastrophic deluge around 4,500 years ago. Clearly, these figures are orders of magnitude different from ages given by secular science. As creationists, however, we expect to see evidence that corroborates this biblically derived timeline. In this Appendix, we will briefly describe some of this evidence. We recommend that interested readers refer to creationist websites (such as creation.com or answersingenesis.com) for more comprehensive lists of this kind of evidence.

In approaching this topic, readers would do well to consider Richard Feynman's advice to individuals (repeated from Chapter 4):

> [Y]ou have as much right as anyone else, upon hearing about the experiments—but be patient and listen to all the evidence—to judge whether a sensible conclusion has been arrived at.

As we approach the following evidence, rather than following secular scientific orthodoxy, which is wedded to long age interpretations, we follow a creationist interpretation, which employs the biblical timescales mentioned above. Nevertheless, this evidence contradicts the secular scientific narrative.

Soft dinosaur tissue: Creationists believe that dinosaurs were carried aboard the ark and therefore lived on the earth in the early part of the post-flood period. The discoveries of soft tissues of dinosaurs—blood vessels, cells and connective tissue—with intact proteins (and even DNA) since the 1990s has intrigued creationists and baffled secular scientists. A creationist report noted that "One description of a portion of the tissue was that it is 'flexible and resilient and when stretched returns to its original shape'".[1] Secular science tells us that the dinosaurs died out 65 million years ago, but it is incomprehensible that these tissues could have survived as intact as they did for that length of time. Nor are these findings limited to a single case or species—soft tissues from T. Rex, hadrosaurids (duckbilled dinosaurs), and *Triceratops* have been reported in secular scientific literature.[2] [3]

---

1 Wieland & Menton, 2005.

2 Catchpoole, 2013.

3 Tay, 2016.

Revived bacteria: Ocean-bed drill core samples taken from a location some 2,300 kilometers northeast of New Zealand contained dormant bacteria which researchers were able to revive. This was an unexpected result; the seabed sediments which the bacteria inhabited were dated by secular science at just over 100 million years old, contained almost no nutrients, and were located under layers of hard rock known to be impenetrable to microbes. The revived bacteria were not "zombies"; they performed "diverse metabolic activities" and reproduced. The astounding fact for the researchers was that no bacteria should have been able to survive under these conditions for the length of time given. Another report suggested that microorganisms detected in salt could still be alive despite the salt being dated to 830-million-years-old.[4] For the creationist, this result is far less surprising: the sediments were likely laid down at the time of the great deluge, meaning that the bacteria had only been encased in their sludgy environs for around 4,500 years.[5]

Tightly bent strata: There are many examples of thick, tightly bent strata (layers) in rock formations. Rocks are not usually known for their flexibility, but these formations show little evidence of cracking around the radii of the bends (which would be expected when deforming brittle rocks), indicating that the layers were still pliable when bent. This accords well with the global flood scenario, in which quickly deposited sedimentary layers were still pliable when subjected to forces associated with the tectonic or seismic events of the time. The layers were therefore "plastic" (able to bend without breaking) when deformed. These bent rocks, which are present in a variety of locations around the world, are very difficult to explain using mechanisms spread over philosophically uniformitarian deep time.[6] [7] Figure F.1[8] shows an example of tightly bent strata exposed in a cliff.

---

4 Cox, 2022, June 7.

5 Catchpoole, 2021.

6 Allen, 2002.

7 Cox 2022, March 16.

8 Source: Wikimedia Commons: Penhalt Cliff Foxhole Point Millook Haven (unmodified) by Chris Gunns, creativecommons.org/licenses/by-sa/2.0/

Figure F.1: Bent rock layers must have been pliable when they were formed

Polystrate fossils: Petrified tree trunks that cross multiple sedimentary strata (layers), known as polystrate fossils, afford strong evidence that sedimentary materials were laid down in relatively short times and in catastrophic conditions. This aligns with the biblical flood narrative. In Joggins, Canada, for example, "we see beds of exposed strata, equivalent to three times the depth of Grand Canyon, that contain copious polystrate lycopod trunks".[9] These trunks must have been buried rapidly; a slow deposition rate of sediments would leave the top of the trunk exposed to the elements, where it would rot away in a relatively short period of time. The creationist article goes on to explain that "[t]his undermines the general assumption that such huge swathes of strata were laid down gradually over millions of years. If this much sediment can be deposited rapidly, then where is the need to assume any deep time at all?" Interestingly, the term "polystrate" appears to be largely confined to creationist groups; secular scientists may use terms such as "upright fossils".

Coal deposits are localized; coal can form rapidly: There is abundant evidence for catastrophic deposition of coal seams related to volcanic activity. In the Mount St. Helens eruption of 1980, large numbers of debranched trees and stumps were uprooted and concentrated by ash and mud flows, in some cases forming floating rafts in nearby lakes, which sunk as they became waterlogged. Within a few years, some of the "debris appear[ed] to have already coalified".[10] Some clay constituents have been found experimentally to catalyze the formation of coal at temperatures below 200°C. For example, a report from the Argonne National Laboratories reported that brown coal could form in two to eight months when lignin (a major component of timber), acidic clay and water were placed in a sealed container and heated to 150°C.[11] Snelling and Mackay suggested that "it is entirely feasible that all of today's coal seams were formed by the global yearlong Noah's flood catastrophe and its aftermath."[12] The Latrobe Valley coal deposits of Australia have several features indicating catastrophic deposition which are inconsistent with the peat swamp theory.[13]

These interpretations are at odds with a major secular theory of coalification which ties in with geological uniformitarianism, namely, the gradual transformation of peat swamps.

Petrification of wood can happen rapidly: Petrification of wood (fossilization) is known to occur under natural conditions in time frames of less than a one hundred years. In the laboratory, under the right conditions, rapid petrification by silicification (involving a solution containing silicon and aluminum compounds similar to volcanic mineral water) at normal temperatures and pressures is well established.[14] A team of Japanese scientists demonstrated rapid petrification of wood in 2004, as reported in a creationist article, which we quote: "The scientists, led by Hisatada Akahane, studied a small lake cradled in the explosion crater of the Tateyama Volcano in central

9 Price, 2020.
10 Snelling & Mackay, 1984.
11 Walker, 2001.
12 Snelling & Mackay, 1984.
13 Walker, 2001.
14 Snelling, 1995.

Japan. A mineral-rich solution gushes from the bottom and fills the 15m pond with steaming acidic water. It cascades over the edge as a waterfall. The scientists found that the naturally fallen wood in the overflow was hard and heavy because it was petrified with a mineral called silica. Yet the wood was less than 36 years old. As an experiment, they fastened pieces of fresh wood in the lake with wire. After seven years the wood had turned into stone, again petrified with silica".[15]

Plateaus and underfit rivers point to a recent, global flood: Plateaus, also known as tablelands, are a very common geological feature. Between 70 and 75% of the earth's landmass is covered in sedimentary rock—which fits the global flood narrative—of which 40% comprises plateaus. These can easily cover tens of thousands of square kilometers; in Africa, 60% of the entire continent is a plateau. Plateaus are formed when floodwater flows off, leaving a flat surface.[16]

Floodwaters flowing across a landmass, toward the ocean, have carved out very large gorges or deep valleys in certain locations, far larger than a contemporary river requires. These waterways have been termed "underfit", meaning that the water flows we observe today do not explain the size of the valley. Perhaps the classic example of an underfit river is the Colorado River, which flows through the Grand Canyon (see Figure F.2). The Colorado reaches a maximum width of around 150 meters, which is vastly underfit in the Grand Canyon, which is almost 30 kilometers (eighteen miles) wide at its widest. Under uniformitarian assumptions, these formations are difficult to explain. Much more likely is that water flows were much higher in the past—for example, as the floodwaters receded from the earth—and are now vastly reduced. Prior high flows are compatible with a historical global flood.[17]

15 Walker, 2007.
16 Neller, 2024.
17 Austin, 1983.

Figure F.2: Underfit rivers can be explained by a prior global flood

Well-known rock formations said to be millions of years old behave as if they are geologically active and likely much younger: In the U.S., the Arches National Park in Utah is famous for its 2000 or more stone arches, one of which is shown in Figure F.3.[18]

18 Source: Wikimedia Commons: Arches National Park, Utah (unmodified) by bobistraveling, creativecommons.org/licenses/by/2.0/deed.en

Figure F.3: Arches National Park, Utah

In the 45 years since 1970, at least 43 arches have collapsed,[19] a rate of almost one per year on average. Applying the principle of uniformitarianism simplistically, 2000 arches might be anticipated to collapse within several millennia, a far cry from the purported 65-million-year history advertised by the National Park Service.[20]

Similar phenomena are observed off the coastline of the Australian state of Victoria. In this locale, a variety of impressive rock features, which the Encyclopedia Britannica describes as having formed over the last 10–20 million years,[21] have collapsed in recent memory. The "London Bridge" collapsed in 1990, one of the nearby "Twelve Apostles" collapsed in 2005, and the "Island Archway" collapsed in 2009. Again, with this observed rate of collapse, why are these rock formations still around for contemporary observers to witness?

Examples of rapid canyon formation have been observed in recent history: Providence Canyon in the U.S., known as "Georgia's Little Grand Canyon" has formed over the last 170 years or so from what was once "rolling hills covered with pine forest". Providence Canyon features several canyons which are up to 50 meters deep, 180 meters wide and 400 meters long.[22]

19 Catchpoole, 2015.
20 nps.gov, 2024.
21 Britannica, n.d.
22 Gibson, 2000.

Elsewhere, Burlingame Canyon near Walla Walla, Washington, was observed to form in six days in 1926. It is up to 35 meters deep and wide, and 450 meters long, but started as a ditch no more than three meters deep and less than two meters wide.[23] In March 1982, a 43-meter-deep canyon developed from a mudflow in one day in the Mount St Helens area.[24] The formation of these canyons in a relatively short space of time shows that, under the right conditions, canyons can form in timescales measured in days or decades, not millions of years.

Flat gaps point to short time frames: Flat gaps occur where two layers of rock, one directly over the other, and believed by secular science to be separated by millions of years, display no significant signs of erosion at their interface. Typically, millions of years of erosion would be expected to result in an irregular, weathered surface, over which the subsequent layer should fit. In a flat gap scenario, there is no evidence of such erosion.

In Figure F.4,[25] the Coconino sandstone is observed overlying the Hermit shale, which according to geologists are separated by about six million years. Yet the upper surface of the soft Hermit shale shows no signs of erosion, being smooth and flat where it abuts the overlying sandstone. This suggests that the sandstone was laid upon the shale before enough time elapsed for erosion to have taken place.[26 27]

---

23 Morris, 2002.
24 Morris & Austin, 2003, p. 75.
25 Source: Wikimedia Commons: Coconino Sandstone rockfall (unmodified) by brewbooks, creativecommons.org/licenses/by-sa/2.0/deed.en
26 Sarfati, 2009.
27 Roth, 2009.

Figure F.4: Flat gap—Coconino sandstone overlying Hermit shale

The enormous lateral extent of some sedimentary layers is compatible with global flood: The Coconino sandstone (a small portion of which is shown in Figure F.4) is estimated to have an average thickness of nearly 100 meters and with equivalent rock layers to the east covers an area of over half a million square kilometers. This is twice the land area of Victoria (a state in Australia) or Colorado (a state in the U.S.). This works out to a volume of sand of about 41,000 cubic kilometers. Secular geologists have theorized that this layer was deposited over eons from dry desert sands. However, geologists agree that the layer above and below were deposited by or in water. Scientist Leonard Brand has examined animal tracks produced in submerged sand and suggested they resembled tracks in the Coconino sandstone, which in turn implies that the Coconino sand may have been submerged when these fossil tracks were preserved. A secular geologist, Glen Visher, has challenged the idea that the Coconino sandstone formed from dry desert sands. He noted that various formations found in the Coconino sandstone are like those which form under water. Both lines of evidence suggest submersion of the Coconino sandstone during its formation, and the enormous geographic extent of this sandstone layer is compatible with formation during a global flood.[28]

---

28 Snelling & Austin, 1992.

Conglomerate is another type of sedimentary rock. It is composed of sand, pebbles, cobbles (larger rocks) and sometimes boulders cemented together. It requires moving water to provide the energy to pick up, transport and deposit these particles and stones. The Shinarump Conglomerate in the U.S. covers about a quarter of a million square kilometers.[29] Today conglomerates are formed in association with rivers, but no rivers today can explain such large and extensive formations. A global flood, on the other hand, could.

Water gaps are well explained by a global flood: Water gaps are geological features, found globally, where a river or stream cuts through a ridge, mountain, or other elevated landform creating a steep-sided valley or gorge. The famous Delaware river water gap is shown in figure F.5.[30]

Figure F.5: Delaware river water gap

Many of these gaps formed during the Genesis flood when the overlying flood waters eroded the ridge or mountain in a top-down fashion which continued as the flood waters receded. They are more difficult to explain from a philosophical uniformitarian perspective because the expectation would be that water would flow around the mountain or ridge, rather than through it.[31]

29 Wieland, 2014.

30 Image is in the public domain.

31 Oard, 2007.

Carbon-14 in coal and diamonds does not correlate with deep time: We earlier saw in Appendix B that carbon-14 has been detected in coal and diamonds. It has also been detected in oil and fossilized wood, indicating an age less than 100,000 years, vastly less than the millions or billions of years proposed by uniformitarian science for these substances.[32] As explained in Appendix B, ages of 50,000 years deduced from Carbon-14 dating under secular presuppositions would more likely correspond to 5,000 years under biblical presuppositions. This is because carbon dating is affected by the trapping of carbon-12 during the global flood.

32 Batten, 2019.

References

Allen, D. (2002). Warped earth. *Creation magazine*, 25(1), 40–43.

Amos, J. (2019, October 2). Scientists estimate Earth's total carbon store. Retrieved November 17, 2024, from bbc.com/news/science-environment-49899039

Austin, S. A. (1983, April 1). Did Landscapes Evolve? Retrieved July 8, 2024, from icr.org/article/did-landscapes-evolve

Austin, S. A., Baumgardner, J. R., Humphries, D. R., Snelling, A. A., Vardiman, L., & Wise, K. P. (1994).

Catastrophic Plate Tectonics: A Global Flood Model of Earth History. *Proceedings of the Third International Conference on Creationism* (pp. 609–621). Pittsburgh: Creation Science Fellowship, Inc.

Axe, D. (2016). *Undeniable: How Biology Confirms Our Intuition That Life Is Designed*. London: HarperOne.

Batten, D. (2019, September 18). Age of the earth: 101 evidences for a young age of the earth and the universe. Retrieved July 8, 2024, from creation.com/age-of-the-earth

Batten, D., Catchpoole, D., Sarfati, J., Wieland, C. (2018). *The Creation Answers Book. Creation Book* Publishers. (Chapter 4). Retrieved from creation.com/cab4

Baumgardner, J. (1994). Computer Modeling of the Large-Scale Tectonics Associated with the Genesis Flood.

Presented at the Third International Conference on Creationism, Pittsburgh, PA, July 18–23, 1994.

Britannica. (n.d.). Twelve Apostles. Retrieved January 13, 2025, from britannica.com/topic/Twelve-Apostles-rock-formation-Victoria

Burr, C. (1997, June 16).The Geophysics of God: A scientist embraces plate tectonics—and Noah's flood. *US News & World Report.*

Carter, R. W. (2022, April 12). How carbon dating works. Retrieved November 17, 2024, from creation.com/how-carbon-dating-works

Catchpoole, D. (2013, December 2). Double Decade Dinosaur disquiet. Retrieved from creation.com/double-decade-dinosaur-disquiet

Catchpoole, D. (2015). A dangerous view. *Creation*, 37(2), 12–15. Retrieved from creation.com/a-dangerous-view

Catchpoole, D. (2021, July). 'Dinosaur Age' bacteria revived from deep sea bed. *Creation* 43(3), 38–39.

Courville, D. (1971). *The Exodus Problem and its Ramifications*. Vols 1 & 2. Loma Linda: Challenge Books.

Cox, G. (2022, March 16). Fantastic folds—Curvy rock layers undermine millions of years. Retrieved from creation.com/folded-rocks

Cox, G. (2022, June 7). "830-million-year-old" microorganisms. Retrieved February 15, 2025, from creation.com/830-million-year-old-microorganisms Discovery Institute. (n.d.). Frequently Asked Questions. Retrieved November 1, 2023, from discovery.org/id/faqs/

Doyle, T. (2015). *Killing Christians*. Nashville: W Publishing Group, and imprint of Thomas Nelson.

Gibson, R. (2000, September). Canyon creation: Faster than most people would think possible, beauty was born from devastation. *Creation*, 22(4), 46–48. Retrieved from creation.com/canyon-creation

Horstemeyer, M. Baumgardner, J. (2003). What Initiated the Flood Cataclysm? *Proceedings of the Fifth International Conference on Creationism*, R.L. Ivey Jr. (Ed.) pp. 155–163.

Kim, J. (2013, June 19). The family I lost in North Korea. And the family I gained. Retrieved from ted.com/talks/joseph_kim_the_family_i_lost_in_north_korea_and_the_family_i_gained?language=en

Kim, J. (2016). *Under the same sky*. Boston: Mariner Books.

Manning, S. W., Griggs, C., Lorentzen, B., & Lange, T. E. (2018, May 29). Fluctuating radiocarbon offsets observed in the southern Levant and implications for archaeological chronology debates. *PNAS*, 115(24), 6141–6146.

McLellan, D. (2006). *Warriors of Ethiopia*. Eastwood, NSW: Dick McLellan.

Morris, J. (2002, September). A canyon is six days. Creation, 24(4), 54–55. Retrieved from creation.com/a-canyon-in-six-days

Morris, J., & Austin, S. A. (2003). *Footprints in the Ash—The Explosive Story of Mount St. Helens*. Green Forest: Master Books. nps.gov. (2024, February 27). Retrieved July 8, 2024, from nps.gov/arch/learn/nature/geologicformations.htm

Oard, M. (2007, June). Do rivers erode through mountains? Water gaps are strong evidence for the Genesis Flood.

Creation, 29(3), 18–23. Retrieved from creation.com/do-rivers-erode-through-mountains

Oard, M. (2024). The hippos of England: A mystery solved by the biblical Flood and its Ice Age aftermath. *Creation*, 46(4), 24–27.

Neller, R. (2024, April 11). Flood Expert Gives 5 Evidences For a Global Flood. Retrieved from youtube.com/watch?v=k5_S8bunhfg

Price, P. (2020, April 16). How the Joggins polystrate fossils falsify long ages. Retrieved from creation.com/joggins-polystrate-fossils

Roth, A. A. (2009, August). 'Flat gaps' in sedimentary rock layers challenge long geological ages. *Journal of Creation*, 23(2), 76–81.

Sarfati, J. (2009, March). 'Millions of years' are missing—Jonathan Sarfati interviews biologist and geologist Ariel Roth. *Creation* 31(2), 46–49.

Scott, E. C., & Branch, G. (2009). Don't Call it "Darwinism". Evolution: Education and Outreach, 2, 90–94. doi: 10.1007/s12052-008-0111-2

Smith, G. H. (2016). *Atheism: The Case Against God*. Amherst: Prometheus Books.

Snelling, A. (1995). 'Instant' petrified wood. *Creation*, 17(4), 38–40.

Snelling A. A. & Austin, S. A. (1992, December). Startling evidence for Noah's flood—footprints and sand 'dunes' in a Grand Canyon sandstone. *Creation* 15(1), 46–50.

Snelling, A., & Mackay, J. (1984, April). Coal, volcanism and Noah's Flood. *Journal of Creation*, 1(1), 11–29. Retrieved from creation.com/coal-volcanism-and-noahs-flood

Spurgeon, C. H. (2023, December 16). Truth Stranger than fiction. Retrieved from spurgeon.org/resource-library/sermons/truth-stranger-than-fiction/#flipbook/

Tay, J. (2016, November 19). Triceratops soft tissue. Retrieved from creation.com/triceratops-soft-tissue

Uri, M. (n.d.). DNA synthesis rate of DNA polymerase III holoenzyme. Retrieved November 17, 2024, from bionumbers.hms.harvard.edu/bionumber.aspx?id=104120

Vardiman, L., Snelling, A. & Chaffin, E. (eds.). (2005). *Radioisotopes and the age of the earth: Results of a young-earth creationist research initiative.* Chino Valley, Arizona: Creation Research Society, Institute for Creation Research.

Von Mises, R. (1981). *Probability, Statistics and Truth*. New York: Dover Publications, Inc.

Walker, T. (2001, March). Coal: memorial to the Flood. *Creation*, 23(2), 22–27.

Walker, T. (2007, July 23). Wood petrified in spring: creationist's rapid claims recognised. Retrieved from creation.com/wood-petrified-in-spring

Walker, T. (2021, June 23). The way it really is: little-known facts about radiometric dating. Retrieved November 17, 2024, from creation.com/the-way-it-really-is-little-known-facts-about-radiometric-dating

Wieland, C. (2014). A hard time finding God. Retrieved February 15, 2025, from creation.com/hard-time-finding-god

Wieland, C., & Menton, D. (2005, 2005). Answering Objections to creationist 'dinosaur soft tissue' age arguments. TJ, 19(3). Retrieved January 13, 2025, from creation.com/images/pdfs/tj/j19_3/j19_3_54-59.pdf

Wise, K. (2017, October 11). In-depth Explanation of What Caused Noah's Flood—Dr. Kurt Wise. Retrieved March 18, 2024, from youtube.com/watch?v=n2ANUKSF2BE

Worraker, W. (2018). Heat Problems Associated with Genesis Flood Models—Part 1: Introduction and Thermal Boundary Conditions. *Answers Research Journal*, 11, 171–191.

Zhu, M., & Dai, X. (2019, August 22). Maintenance of translational elongation rate underlies the survival of Eschericia coli during oxidative stress. *Nucleic Acids Research*, 47(14), 7592–7604.

Index

Made in the USA
Monee, IL
04 November 2026

67385957R10262